# Bentham Briefs in Biomedicine and Pharmacotherapy

## (Volume 3)

# Anthraquinones: Bioactive Multifaceted Therapeutic Agents

Edited by

## Pardeep Kaur
*Department of Botany, Khalsa College*
*Amritsar, Punjab, India*

## Ajay Kumar
*University Centre for Research and Development (UCRD)*
*Chandigarh University, Mohali, Punjab, India*

## Robin
*Agilent Technologies India Pvt. Ltd., Chandigarh, India*

## Tarunpreet Singh Thind
*DST-CURIE Research & Teaching Laboratory (DCRTL)*
*Goverment College for Girls*
*Ludhiana, Punjab, India*

&

## Kamaljit Kaur
*Department of Biotechnology, Khalsa College*
*Amritsar, Punjab, India*

# Bentham Briefs in Biomedicine and Pharmacotherapy

*(Volume 3)*

*Anthraquinones: Bioactive Multifaceted Therapeutic Agents*

Editors: Pardeep Kaur, Ajay Kumar, Robin, Tarunpreet Singh Thind and Kamaljit Kaur

ISSN (Online): 2810-997X

ISSN (Print):  2810-9988

ISBN (Online): 978-981-5313-98-7

ISBN (Print): 978-981-5313-99-4

ISBN (Paperback): 978-981-5322-00-2

©2025, Bentham Books imprint.

Published by Bentham Science Publishers Pte. Ltd. Singapore. All Rights Reserved.

First published in 2025.

need for a court order if at any point you breach any terms of this License Agreement. In no event will any delay or failure by Bentham Science Publishers in enforcing your compliance with this License Agreement constitute a waiver of any of its rights.

3. You acknowledge that you have read this License Agreement, and agree to be bound by its terms and conditions. To the extent that any other terms and conditions presented on any website of Bentham Science Publishers conflict with, or are inconsistent with, the terms and conditions set out in this License Agreement, you acknowledge that the terms and conditions set out in this License Agreement shall prevail.

**Bentham Science Publishers Pte. Ltd.**
80 Robinson Road #02-00
Singapore 068898
Singapore
Email: subscriptions@benthamscience.net

**BENTHAM SCIENCE**

# CONTENTS

# FOREWORD

The book series 'Bentham Briefs in Biomedicine and Pharmacotherapy' is dedicated to a comprehensive understanding of pharmacology and its role in treating different diseases. The third volume, 'Anthraquinones: Bioactive Multifaceted Therapeutic Agents' explores the realms of anthraquinones. These pharmacologically active molecules exhibit a diverse range of structural variations and different applications. Few enzymes in their biosynthetic pathways can bring out huge diversity as they can work on a large number of similar substrates having minor modifications at different positions. This book explores various aspects of anthraquinones, including their chemical structures, biosynthetic pathways, therapeutic potential, and potential applications in cancer treatment and nanotechnology. Anthraquinones are already in use for chemotherapy of various cancers. However, creating nanoparticles of these molecules brings out a new dimension in more effective delivery of these molecules, aimed at better effectiveness and reduced toxicity. Each chapter meticulously unravels the layers of complexity surrounding these compounds, offering a panoramic view of their current scientific and practical relevance. These chapters also illuminate the future pathways they may traverse, especially in the realm of nanotechnology-enhanced therapies. I am confident that this handbook will inspire readers to actively explore, critically analyze, and collaborate in order to further advance the understanding and utilization of anthraquinones in various fields.

**Arun Kumar Sharma**
Department of Plant Molecular Biology
University of Delhi, South Campus
New Delhi – 110021
India

# PREFACE

The exploration of anthraquinones, a class of naturally occurring aromatic compounds, represents a confluence of tradition and innovation, bridging millennia-old therapeutic practices with cutting-edge scientific research. In this third volume of the Bentham Briefs in Biomedicine and Pharmacotherapy series, titled "Anthraquinones: Bioactive Multifaceted Therapeutic Agents," we embark on a comprehensive examination of these compounds, renowned for their dynamic roles in the natural world and their therapeutic potential in medicine.

Anthraquinones, characterized by their distinctive aromatic structure, have been utilized since ancient times, most notably in traditional medicine and as natural dyes. Today, they are the subject of intensive scientific inquiry, particularly in the realm of pharmacology, where their diverse bioactivities including antimicrobial, anticancer, and anti-inflammatory effects offer promising avenues for new drug development. This volume aims to encapsulate the multifaceted nature of anthraquinones, from their chemical and biosynthetic properties to their therapeutic applications and emerging roles in nanotechnology-enhanced drug delivery.

The chapters presented herein are crafted by leading experts in the field, each delving into various aspects of anthraquinone research. The content ranges from detailed analyses of chemical structures and biosynthesis pathways to comprehensive reviews of the therapeutic uses and potential of anthraquinones, particularly in combating challenging diseases like cancer. Furthermore, the incorporation of nanotechnology in anthraquinone applications heralds a new era of precision medicine, where the delivery and efficacy of these compounds are significantly enhanced. We believe that this compilation not only serves as a repository of current knowledge but also as a catalyst for future research, inspiring continued exploration and innovation in the use of anthraquinones.

**Pardeep Kaur**
Department of Botany, Khalsa College
Amritsar, Punjab, India

**Ajay Kumar**
University Centre for Research and Development (UCRD)
Chandigarh University, Mohali, Punjab, India

**Robin**
Agilent Technologies India Pvt. Ltd., Chandigarh, India

**Tarunpreet Singh Thind**
DST-CURIE Research & Teaching Laboratory (DCRTL)
Goverment College for Girls
Ludhiana, Punjab, India

&

**Kamaljit Kaur**
Department of Biotechnology, Khalsa College
Amritsar, Punjab, India

# List of Contributors

| | |
|---|---|
| **Amrit Kaur** | Department of Chemistry, Guru Nanak Dev University, Amritsar, Punjab-143005, India |
| **Ajay Kumar** | University Centre for Research and Development (UCRD), Chandigarh University, Mohali-140413, Punjab, India |
| **Gulshan Kumar** | Department of Chemistry, Banasthali University, Banasthali Newai-304022, Rajasthan, India |
| **Ganavi D.** | Department of Chemistry, Sri Dharmasthala Manjunatheshwara College (Autonomous), Ujire, Karnataka-574240, India |
| **Maushmi S. Kumar** | Somaiya Institute for Research and Consultancy, Somaiya Vidyavihar University, Vidyavihar (East), Mumbai – 400077, India |
| **Manoj Kumar** | Department of Microbiology, Guru Nanak Dev University, Amritsar-143005, Punjab, India |
| **Pardeep Kaur** | Department of Botany, Khalsa College, Amritsar, Punjab-143002, India |
| **Pooja Sharma** | Department of Chemistry, Guru Nanak Dev University, Amritsar, Punjab-143005, India |
| **Prashasthi V. Rai** | Department of PG Studies and Research in Chemistry, Sri Dharmasthala Manjunatheshwara College (Autonomous), Ujire, Karnataka-574240, India |
| **Pratima P. Pandey** | Somaiya Institute for Research and Consultancy, Somaiya Vidyavihar University, Vidyavihar (East), Mumbai – 400077, India |
| **Robin** | Agilent Technologies, India Pvt. Ltd., Chandigarh, India |
| **Roopa Nayak** | Department of Radiation Biology and Toxicology, Manipal School of Life Sciences, Manipal Academy of Higher Education, Manipal, Karnataka-576104, India |
| **Shilpa** | Department of Chemistry, Maharishi Markandeshwar Engineering College, Maharishi Markandeshwar (Deemed to be University), Mullana, Haryana-133207, India |
| **Simrandeep Kaur** | Department of Botanical and Environmental Sciences, Guru Nanak Dev University, Amritsar, Punjab-143005, India |
| **Sandhya Choudhary** | DST-CURIE Research & Teaching Laboratory (DCRTL), Government College for Girls, Ludhiana, Punjab-141001, India |
| **Sukhvinder Dhiman** | Institute of Nano Science and Technology, Mohali, Punjab-140306, India |
| **Sapna Devi** | Department of Microbiology, DAV University, Jalandhar-144012, Punjab, India |
| **Sonika** | Department of Chemistry, Maharishi Markandeshwar Engineering College, Maharishi Markandeshwar (Deemed to be University), Mullana, Haryana-133207, India |

**Tarunpreet Singh Thind**    DST-CURIE Research & Teaching Laboratory (DCRTL), Government College for Girls, Ludhiana, Punjab-141001, India

**Vidushi Gupta**    DST-CURIE Research & Teaching Laboratory (DCRTL), Government College for Girls, Ludhiana, Punjab-141001, India

**Vasantha Kumar**    Department of PG Studies and Research in Chemistry, Sri Dharmasthala Manjunatheshwara College (Autonomous), Ujire, Karnataka-574240, India

**Vijesh A. M.**    PG Department of Chemistry, Payyanur College, Payyanur, Kannur University, Kerala-670327, India

# CHAPTER 1

# Anthraquinones: Integrated Perspectives on Analytical Methodologies and Functional Applications

**Robin[1], Sandhya Choudhary[2], Vidushi Gupta[2], Pardeep Kaur[3]** and **Tarunpreet Singh Thind[2,\*]**

[1]*Agilent Technologies India Pvt. Ltd., Chandigarh, India*

[2]*DST-CURIE Research & Teaching Laboratory (DCRTL), Government College for Girls, Ludhiana, Punjab-141001, India*

[3]*Department of Botany, Khalsa College, Amritsar, Punjab-143002, India*

**Abstract:** Anthraquinones are aromatic organic compounds essential in both nature and industry, known for their diverse applications. Anthraquinones with a chemical formula of $C_{14}H_8O_2$ are commonly found in plants like *Aloe vera* and rhubarb, fungi, lichens (a symbiosis of fungi and algae), bacteria like *Streptomyces* species, and certain animals like crinoids and sponges. It has various biological functions such as antimicrobial properties and anticancer activities. Anthraquinones are synthesized through natural processes like the polyketide and shikimate pathways within plants and extracted using methods such as ultrasound-assisted extraction and super critical fluid extraction for their isolation and purification. In the medical field, anthraquinones play a crucial role in the development of drugs like anthracyclines for cancer and metformin for diabetes treatment, showcasing their therapeutic potential. Industrially, anthraquinones find application as natural dyes in textiles, imparting vibrant colors, and as additives in papermaking to enhance the strength and durability of paper products, highlighting their versatility in diverse industrial sectors. The utilization of analytical techniques such as ultraviolet-visible (UV-Vis) spectroscopy is essential for determining the absorption spectra of anthraquinones, while high-performance liquid chromatography is crucial for separating and quantifying individual compounds, emphasizing their indispensable roles in accurate anthraquinone analysis. The compound's significance extends beyond its bioactivities, playing a vital role in various industrial applications, which underscores the ongoing research and interest in exploiting its properties for innovative solutions in healthcare, manufacturing, and environmental sustainability.

**Keywords:** Anthraquinones, Biological activities, Biosynthesis, Chemical synthesis, Gas chromatography, High-performance liquid chromatography,

---

\* **Corresponding author Tarunpreet Singh Thind:** DST-CURIE Research & Teaching Laboratory (DCRTL), Government College for Girls, Ludhiana, Punjab-141001, India; E-mail: tarunthind@gmail.com

Industrial applications,  Mass spectrometry, Natural pigments, Polyketide pathway, Quinones, Shikimate pathway, Super critical fluid extraction, Ultrasound-assisted extraction.

## INTRODUCTION

The group of quinones and their derivatives including benzoquinones and naphthoquinones, which constitute the large number of natural pigments are called anthraquinones. These are the aromatic organic compounds with the chemical formula $C_{14}H_8O_2$, where keto groups are located on the central ring (Fig. **1**). The compounds belonging to this class are abundantly produced from natural sources like plant parts such as roots, rhizomes, flowers, and fruits, while others are present in lichens, fungi, and animals [1].

Anthraquinones are of tremendous use in biological properties such as inhibiting bacterial and fungal growth and in industrial applications by acting as a natural dye and are used in bleaching pulp for papermaking. These are phenolic compounds widely present as a skeleton of 9,10-anthraquinone. The name anthraquinones was given by Carl Graebe and Libermann in the year 1868. The synthesis of these phenolic compounds includes the oxidation of anthracene in the presence of oxidant chromium (VI). These compounds are studied in plants belonging to the *Rubiaceae* family, such as *Morinda*, *Rubia*, and *Gallium* species. These quinones are derived from anthracenes and possess a broad spectrum of bioactivities such as anticancer, cathartic, anti-inflammatory, and diuretic, and also play a potential role in autoimmune diabetes [1–3]. Anthraquinones show their potential applications in various industries, such as in medicine, with their uses as drugs or as an anticancer agent.

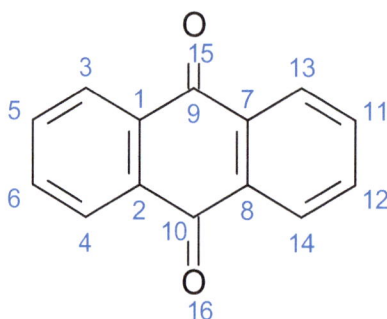

**Fig. (1).** Structure of anthraquinone.

## BIOSYNTHESIS OF ANTHRAQUINONES: ENZYMATIC MECHANISMS AND GENETIC INSIGHTS

Anthraquinone synthesis includes 2 pathways *viz.* the polyketide pathway and the shikimate pathway.

**Polyketide Pathway:** This pathway is common in bacteria and fungi and is carried out in the presence of enzymes polyketide synthases that result in the formation of an intermediate during the anthraquinone synthesis (Fig. **2**) [4].

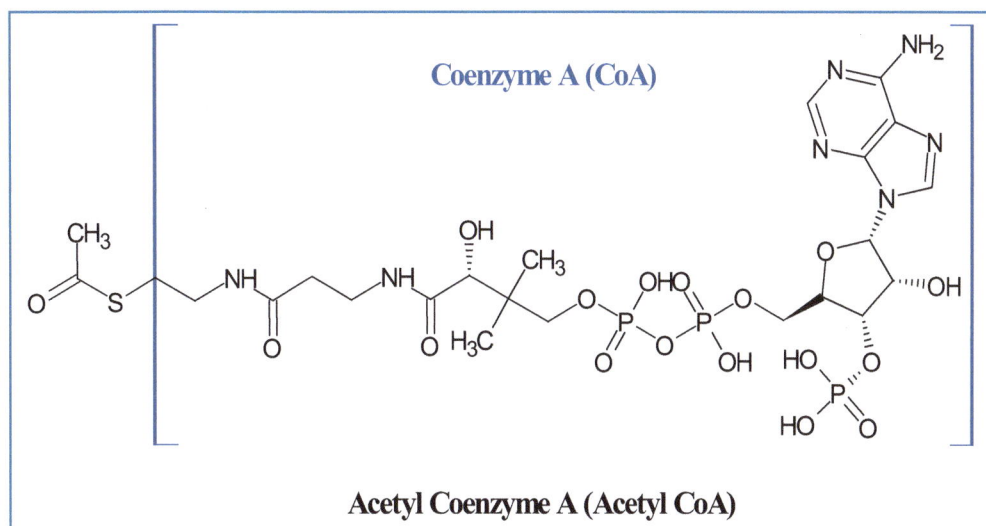

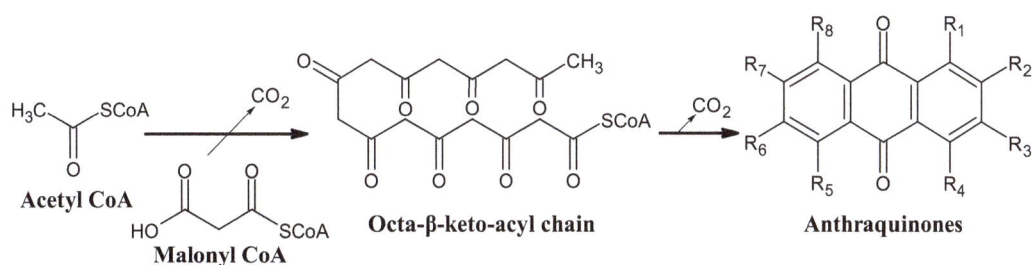

**Fig. (2).** The polyketide pathway.

**Shikimate Pathway:** α-ketoglutaric acid and shikimic acid result in the formation of o-succinoylbenzoic acid, which is further added to mevalonic acid and results in the formation of 1,2 dihydroxylated anthraquinones (Fig. **3**) [1].

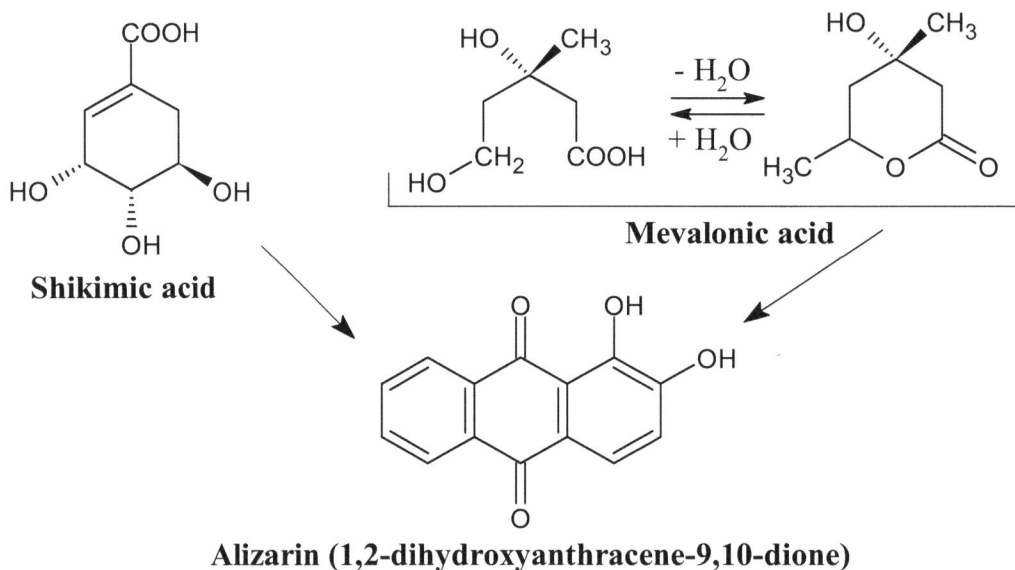

**Alizarin (1,2-dihydroxyanthracene-9,10-dione)**

**Fig. (3).** The shikimate pathway.

## Enzymatic Mechanisms and Genetic Insights

Anthraquinones represent a significant class of naturally occurring organic compounds, primarily known for their vivid coloration and diverse medicinal properties. Recent scientific investigations have shed light on the complex biosynthetic pathways and genetic underpinnings responsible for anthraquinone production in various organisms.

In the study conducted by Liu *et al.*, the focus is on a specific prenyltransferase, RcDT1, in *Rubia cordifolia*, illustrating its indispensable role in the biosynthesis of alizarin-type anthraquinones. This research enhances our understanding of the enzymatic steps that dictate the structural assembly of anthraquinones, providing potential for metabolic engineering to augment the production of these valuable compounds [5]. Further elucidating the biochemical intricacies, Schmalhofer *et al.* explore the role of polyketide trimming in shaping dihydroxynaphthalene-melanin and anthraquinone pigments. Their findings reveal the critical enzymatic activities necessary for constructing the tricyclic aromatic rings, pivotal for the chemical backbone of anthraquinones. This study not only advances our biochemical knowledge but also opens new avenues for synthetic biology applications in pigment production [6]. From a fungal perspective, the work by Xu *et al.* uncovers a fungal cytochrome P450 enzyme with an unusual two-step mechanism responsible for constructing a bicyclononane skeleton, which links anthraquinone and xanthone moieties. This discovery provides insights into the diversity of enzymatic strategies employed by nature in the biosynthesis of complex organic molecules, highlighting the intricate evolutionary adaptations of fungal metabolites [7]. On a genomic scale, He *et al.* present a comprehensive analysis of the near-complete genome of *Reynoutria multiflora*, revealing the genetic basis of stilbenes and anthraquinones biosynthesis. Their findings emphasize the genetic and metabolic diversity across different species, contributing significantly to our understanding of the molecular genetics underlying secondary metabolite biosynthesis [8]. Also, the comparative transcriptomic analysis by Zhao *et al.* in *Rheum palmatum* identifies key R2R3-MYB genes involved in the anthraquinone biosynthetic pathway. This study not only sheds light on the gene expression profiles related to anthraquinone production but also emphasizes the potential for genetic and enzymatic manipulation to enhance the yield of these pharmacologically important compounds [9].

These recent studies collectively advance our comprehension of the biosynthesis of anthraquinones, unveiling the intricate biological and chemical processes that govern their production. Such insights are crucial for the development of biotechnological approaches to optimize the synthesis and extraction of anthraquinones, with potential applications in pharmaceuticals, cosmetics, and dyes.

## NATURALLY OCCURRING ANTHRAQUINONES (NOAQ)

### Plant-Based

The naturally occurring anthraquinones are isolated from lichens, fungi, and other higher medicinal plants like the ones belonging to the family *Polygonaceae*, *Rhamnaceae*, *Asphodelaceae*, *Rubiaceae*, *Fabaceae*, *Scrophulariaceae*, *Liliaceae*, *Verbenaceae,* and *Valerianaceae*. Anthraquinones are aromatic compounds and are a group of secondary metabolites with 9,10 – dioxoanthracene core [10]. The most common naturally occurring anthraquinones are:

1. Aloe-emodin: Found in *Aloe vera* and other *Aloe* species, aloe-emodin has laxative properties and shows potential in anticancer research [11].

2. Emodin: This compound is present in the roots and bark of numerous plants, such as *Rheum palmatum* (Chinese rhubarb) and *Polygonum cuspidatum*. Emodin has a variety of pharmacological effects, including anti-inflammatory, antibacterial, and anticancer activities [12].

3. Chrysophanol: Chrysophanol occurs in plants like *Rheum* species and *Cassia* species, known for its antifungal and anti-inflammatory properties [13,14].

4. Physcion: Found in species such as rhubarb and *Polygonum*, physcion is recognized for its antimicrobial and antifungal effects [15,16].

5. Rhein: This anthraquinone, derived from rhubarb roots, is noted for its therapeutic potential, including laxative, antiarthritic, and anticancer activities [17].

6. Dantron (Chrysazin): Predominantly used for its laxative effects, dantron is found in *Senna* leaves and Chinese rhubarb [18,19].

These anthraquinones are distinguished by their significant biological activities and are extensively studied for their potential therapeutic benefits, especially in traditional and modern medicine. In contrast to the beneficial effects of anthraquinones, they may lead to the potential damage of cells due to the close similarity in structure between anthraquinones (AQs) and toxic analogue, anthracene [20]. Anthraquinones are commonly found in the form of natural pigments. They are particularly abundant in the family *Rubiaceae*, which includes the madder plant *(Rubia tinctorum)*. The roots of madder are used to extract alizarin, a red dye that historically played a significant role in the textile industry. Plant-based anthraquinones like alizarin and senna (from *Cassia species*) are not only important for their dyeing properties but also for their medicinal uses, serving as laxatives and treating various ailments [21,22].

## Bacterial Based Anthraquinones

Bacteria, especially actinobacteria like *Streptomyces* and *Micromonospora*, are prolific producers of structurally unique hydroxy anthraquinones through polyketide pathways, mainly utilizing the polyketide synthase-II system [23]. Notable among these is aloesaponarin-II, isolated from *Streptomyces coelicolor*, *Streptomyces* sp. GW32/698, GW24/1694, and marine *Streptomyces* sp. M097, along with chrysophanol, another well-known compound from *Streptomyces* sp. GW32/698. Others are 1-O-methyl chrysophanol, derived from *Amycolatopsis thermoflava* SFMA-103, and 9-hydroxyl aloesaponarin-II from *Streptomyces* sp. M097 and *Streptomyces lividans* K4-114. Some other compounds include 1-hydroxy-6-methoxy-8-methyl-anthraquinone from *Streptomyces* sp. GW24/1694, 3,8-dihydroxy-1-methyl-anthraquinone-2-carboxylic acid from *Streptomyces* sp. GW32/698, and 1,5,7-trihydroxy-3-hydroxymethyl anthraquinone and 6,61-bis(1,5,7-trihydroxy-3-hydroxymethylanthraquinone), both from *Streptomyces* sp. ERI-26. Additionally, 2,3-dihydroxy-9,10-anthraquinone was found in

*Streptomyces galbus* ERINLG-127, while 2-hydroxy-9,10-anthraquinone was isolated from *Streptomyces olivochromogenes* ERINLG-261. Further noteworthy discoveries include 1,8-dihydroxy-2-ethyl-3-methyl-anthraquinone from *Streptomyces* sp. FX-58, and saliniquinones A-F, a group of fused anthraquinone-c-pyrones, from marine *Salinispora arenicola*. 5-hydroxy ericamycin, from *Actinoplanes* sp. strain 4731, and galvaquinones A, B, and C, isolated from *Streptomyces* sp. inoverrucosus and *Verrucosispora* SN26-14.1, also exhibit notable biological activity. Other potent compounds include uncialamycin from *Streptomyces uncialis*, tiancimycin A from *Streptomyces* sp. CB03234, yangpumicin A from *Micromonospora yangpuensis* DSM 45577, and dynemicin from *Micromonospora chersina* sp. nov. These bacterial anthraquinones are often studied for their antimicrobial properties and anti-cancer treatments [23]. Despite the numerous reviews available on anthraquinones from plant and fungal sources, there is a noticeable gap in the literature concerning bacterial anthraquinones.

## Animal and Fungal Based Anthraquinones

Animal-origin anthraquinones are relatively rare compared to their plant-derived counterparts. Anthraquinone-like compounds have been detected in some marine organisms, and they can also be bioaccumulated through food chains. Studies that touch upon animal-origin anthraquinones often investigate marine sponges or other organisms that might acquire these compounds from symbiotic bacteria or algae [24]. These sponges host a wide range of symbiotic bacteria, which are often the true producers of various bioactive secondary metabolites, including anthraquinone derivatives. Compound lunatin (1,3,8-trihydroxy-6-methoxyanthraquinone) was isolated from the fungus *Curvularia lunata* associated with the marine sponge *Niphates olemda* from Indonesia. The sponge-associated microbial symbionts play a crucial role in producing these compounds, contributing to the rich chemical diversity found in marine environments. Urdamycinone E and urdamycinone G are produced by the actinomycete *Streptomyces fradiae*, found in association with marine environments. Dehydroxyaquayamycin was isolated from the fungus *Penicillium oxalicum* associated with marine sponges. Stachybotrins D-F is produced by the marine-derived fungus *Stachybotrys chartarum* isolated from marine sediments. Stachybocins E-F, stachybosides A-B, and stachybotrin G are found in *Stachybotrys chartarum* Additionally, gymnastatins A-E, gymnastatins F-H, and gymnasterones A-B are isolated from the fungus *Gymnascella dankaliensis* associated with a marine sponge *Haliclona* sp. Certain sea urchins derived fungus and marine species synthesize anthraquinone compounds [25,26]. These molecules are thought to play a role in the defense mechanisms of these organisms, offering protection against predators and microbial infections [25,27]. However, further research is needed to fully understand the prevalence and specific mechanisms of anthraquinone production in animals.

A variety of bioactive anthraquinones are produced by different fungal species which include dothistromin, bisdeoxydothistromin, bisdeoxydehydrodothistromin, 6-deoxyversicolorin C, averufin, nidurufin, and averythrin, all produced by *Dothistroma pini*. Averythrin is also found in *Aspergillus versicolor*. Other anthraquinones, such as macrosporin and 6-methylxanthopurpurin 3-methyl ether, are synthesized by several *Alternaria* species, including *Alternaria bataticola*, *A. porri*, *A. solani*, and *A. cucumerina*. Altersolanols A and J, along with nectriapyrone, are produced by *Diaporthe angelicae* (anamorph *Phomopsis foeniculi*), a pathogen affecting fennel. *Stemphyfium botryosum*, a pathogen of lettuce, produces stemphylin and dactylariol, while *Cryphonectria parasitica* generates several anthraquinones, including rugulosin, skirin, chrysophanol, and emodin. Emodin is also produced by other fungi such as *Aspergillus fumigatus*, *Hormonema dematioides*, *Aspergillus glaucus*, *Pyrenophora tritici-repentis*, and *Gliocladium* sp. Additionally, *Drechslera* species produce catenarin, helminthosporin, and cynodontin. Physcion is another common anthraquinone produced by *Aspergillus fumigatus*, *Microsporum* sp., and *A. glaucus*. Questin and isorhodoptilometrin are synthesized by *Aspergillus* sp. YL-6, while isorhodoptilometrin is also found in *Gliocladium* sp., and chrysophanol is produced by *Phoma foevata*. These anthraquinones exhibit a wide range of biological activities, including phytotoxicity, antibacterial, antifungal, antiviral, and cytotoxic properties, making them significant for both agricultural and medicinal applications [28]

**Insect Based Anthraquinones**

Anthraquinones of insect origin, primarily from the Hemiptera order include kermesic acid, produced by *Kermococcus ilicius* on oak (*Quercus coccifera*), and carminic acid, derived from *Dactylopius coccus*, a bug that feeds on *Opuntia* cactus and is widely used in food coloring. Additionally, lac insects such as *Kerria lacca* produce a variety of anthraquinones, including laccaic acids A, B, C, D, E, and F, which give these insects their characteristic crimson coloration. These pigments are synthesized *via* the polyketide pathway [29,30]. Insects, such as the lac insect (*Kerria lacca*), produce anthraquinone derivatives as part of their shellac secretions [31]. The red pigment carminic acid, derived from the scale insect *Dactylopius coccus*, is another example of an insect-based anthraquinone. These compounds are used in the food and cosmetic industries as natural dyes [32].

Despite the growing interest in insect-based anthraquinones, several gaps in research limit the full realization of their potential. One of the major gaps is the limited identification of insect species that produce these compounds. While a few species, such as those in the *Pyrrhocoridae* family, have been identified, the vast majority of insect species remain unexplored. There is a need for comprehensive surveys of different ecosystems, particularly in biodiversity hotspots, to discover additional sources of anthraquinones. Expanding this knowledge will provide a broader base for future research and applications. Moreover, the overall yield of anthraquinones from insects is often low, making it challenging to scale for industrial applications without optimizing farming practices and production systems.

## TOXICITY OF NATURAL ANTHRAQUINONES

Natural anthraquinones, while valuable for their diverse applications in dyes, medicines, and cosmetics, have raised concerns regarding their toxicity. These compounds, found in various natural sources like plants, fungi, and insects, exhibit a range of biological activities that can be both beneficial and harmful depending on their concentration, exposure time, and the specific anthraquinone compound.

### Toxicity Mechanisms

The toxicity of natural anthraquinones is primarily attributed to their ability to generate reactive oxygen species (ROS) and their interaction with cellular macromolecules. For instance, anthraquinone-like emodin, obtained from the rhubarb plant (*Rheum palmatum*), has been studied for its potential to cause oxidative stress and DNA damage in cells. Emodin, specifically, can intercalate into DNA, affecting DNA replication and transcription, and leading to cytotoxic effects [20,33]. Aloe-emodin, found in *Aloe vera*, is another anthraquinone derivative that has been studied for its genotoxic and carcinogenic potential. Research has shown that aloe-emodin can cause DNA damage and exhibit mutagenic effects in mammalian cells, raising concerns about the long-term use of *Aloe vera* extracts containing this compound [20]. *Senna*, a medicinal plant containing anthraquinone glycosides, is used as a laxative. However, excessive consumption of *Senna* has been linked to hepatotoxicity and damage to the gastrointestinal tract, illustrating the need for cautious use of anthraquinone-containing herbs [20]. While natural anthraquinones have important industrial and medicinal applications, their potential toxicity necessitates careful evaluation and regulation. Understanding the specific mechanisms of toxicity and conducting comprehensive toxicological studies is crucial to ensure the safe use of natural anthraquinones in various products.

## CHEMICAL SYNTHESIS OF ANTHRAQUINONES

The chemical synthesis of anthraquinones is an important area of study in organic chemistry, given their extensive use in dyes, drugs, and other industrial applications. These compounds are synthesized through various chemical reactions, with the aim of producing anthraquinone derivatives with specific properties.

### Basic Synthesis Methods

One of the classic methods for synthesizing anthraquinones is the Friedel-Crafts acylation of benzene or its derivatives with phthalic anhydride (2-benzofuran-1,3-dione) in the presence of a Lewis acid catalyst like aluminum chloride ($AlCl_3$). The general reaction process is shown in Fig. (4). This reaction typically produces 9,10-anthraquinone (anthracene-9,10-dione), which serves as a precursor for further chemical modifications [34].

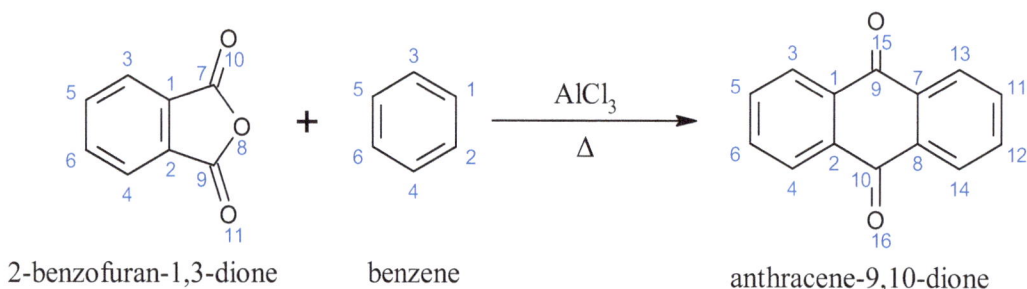

2-benzofuran-1,3-dione          benzene                    anthracene-9,10-dione

**Fig. (4).** Friedel–Crafts acylation.

### Synthetic Pathways

#### *Vat Dye Synthesis*

A well-known application of anthraquinone synthesis is in the production of vat dyes. For example, anthraquinone-2-sulfonic acid, a key intermediate in the synthesis of anthraquinone dyes, is obtained by sulfonation of anthraquinone. These dyes exhibit excellent fastness properties on textile fibers [35].

#### *Medicinal Chemistry*

In medicinal chemistry, the synthesis of anthraquinone derivatives like doxorubicin, a potent anticancer drug, involves complex multi-step processes that attach various functional groups to the anthraquinone core, enhancing its biological activity and solubility [36].

## Advanced Synthetic Techniques

Recent advancements in synthetic chemistry have led to the development of more efficient and environmentally friendly methods for producing anthraquinones. For instance, catalytic oxidation processes have been utilized to synthesize anthraquinones directly from aromatic precursors, reducing the environmental impact of traditional methods [37]. This approach involves the reaction of benzene with phthalic anhydride (PHA) under optimized conditions to produce anthraquinone with high efficiency and minimal environmental impact. The Ni-modified Hβ zeolite acts as a solid catalyst, enhancing both the conversion rate and the selectivity of the reaction due to the synergistic effects of Lewis and Brønsted acid sites. The catalyst's high stability and reusability further contribute to the green nature of the process. This method not only reduces the cost and environmental footprint but also simplifies the production process, making it an attractive alternative to conventional synthesis routes [37].

## BINDING INTERACTIONS

### Anthraquinones with Bovine β-Lactoglobulin

Anthraquinones, due to their diverse biological and chemical properties, are of significant interest in the study of protein-ligand interactions. The binding interactions between anthraquinones and bovine β-lactoglobulin (β-LG) are particularly noteworthy, given the role of β-LG as a major whey protein with the ability to bind various small molecules, thereby affecting their bioavailability and activity [38].

### *Binding Mechanism and Specificity*

The binding of anthraquinones to β-LG typically occurs at specific sites, namely the calyx-shaped cavity in the β-LG molecule. This cavity allows β-LG to accommodate a variety of ligands, including fatty acids and various hydrophobic compounds. Anthraquinones, with their planar structures and hydrophobic nature, fit into this cavity, leading to complex formation. This interaction is primarily driven by hydrophobic forces, although hydrogen bonding and van der Waals interactions also contribute [38,39].

## *Examples of Interactions*

### *Aloe-emodin and β-LG*

Studies have shown that aloe-emodin, an anthraquinone derivative found in *Aloe vera*, exhibits strong binding affinity to β-LG. This interaction has been explored using techniques like fluorescence quenching and molecular docking to elucidate the binding mechanism and the structural basis of the interaction [39].

### *Rhein and β-LG*

Rhein, another anthraquinone compound present in rhubarb and other plants, is known to bind to β-LG. Research has focused on understanding how this binding affects the structural conformation of β-LG and the implications for rhein's bioavailability and pharmacological activity [38].

### *Relevance to Drug Delivery and Nutraceuticals*

The study of anthraquinone-β-LG interactions is crucial for the development of drug delivery systems and nutraceutical formulations. By binding to β-LG, anthraquinones can potentially have altered absorption and metabolism in the body, influencing their therapeutic efficacy and safety [38,39].

## RECENT ADVANCES IN THE EXTRACTION OF ANTHRAQUINONES

The process involves solvents to extract anthraquinones from plant materials or synthetic compounds. The extracted anthraquinones are then purified and analyzed using various techniques, such as chromatography and spectroscopy. The processes are particularly essential for isolating and studying beneficial anthraquinones with rapid and energy-efficient extraction procedures in high amounts, which have a wide range of industrial and medicinal applications [40,41].

### Ultrasound-Assisted Extraction (UAE Method)

Ultrasound-Assisted Extraction (UAE) is a widely acknowledged method for the efficient extraction of bioactive compounds, including anthraquinones, from various plant matrices. The technique employs ultrasonic waves to induce cavitation in the solvent, which facilitates the disruption of cell walls and enhances the mass transfer of analytes from the sample matrix to the solvent (Fig. **5**). This process is highly favored for its rapid extraction times, reduced solvent consumption, and the ability to preserve the structural integrity of sensitive compounds [42].

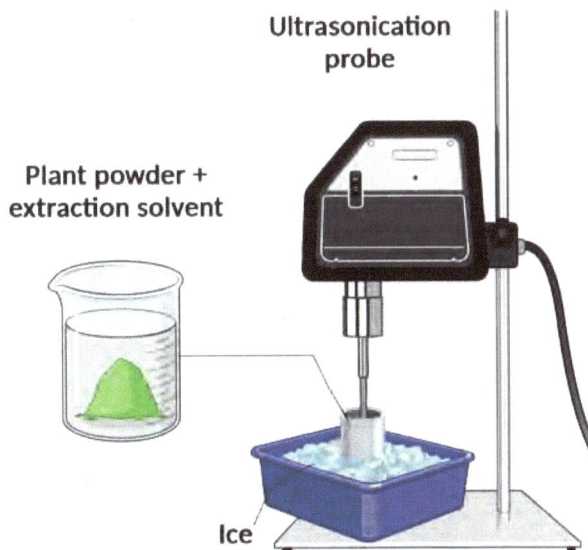

**Fig. (5).** Ultrasound-assisted extraction (UAE).

The traditional methods of extraction, such as Soxhlet and maceration, often involve prolonged extraction times and higher solvent consumption, which can lead to the degradation of these thermosensitive compounds [42,43]. In contrast, the UAE offers a gentler and more efficient alternative, ensuring higher yields of intact anthraquinones. The optimization of UAE parameters such as solvent type, concentration, temperature, and ultrasonic frequency is crucial for maximizing the extraction efficiency of anthraquinones [44]. Studies indicate that solvents like ethanol and methanol, often in combination with water, are effective in extracting these compounds due to their ability to penetrate cell matrices and solubilize a wide range of phytochemicals [43]. Its ability to reduce processing times, minimize solvent usage, and maintain compound integrity makes it a preferred method in the pharmaceutical and bioactive compound extraction industries [45,46]. Thus, UAE is a powerful and eco-friendly technique for extracting valuable compounds like anthraquinones [44,47,48].

## Super Critical Fluid Extraction (SCFE)

Super Critical Fluid Extraction (SCFE) is a highly efficient and innovative technology used for the isolation and purification of compounds, such as anthraquinone, from various matrices. SCFE utilizes supercritical fluids, which are substances at conditions above their critical temperature and pressure, where they exhibit unique properties between those of a gas and a liquid [49,50]. Carbon dioxide ($CO_2$) is the most commonly used supercritical fluid due to its moderate critical temperature and pressure, non-toxicity, and inertness [51].

The extraction of anthraquinone using SCFE is significant in the pharmaceutical and dye industries because of its efficiency and environmental friendliness [52,53]. The SCFE process offers a cleaner alternative to traditional solvent extraction methods, reducing the use of harmful organic solvents and enhancing the purity of the extracted anthraquinone. In the SCFE process, the supercritical $CO_2$ penetrates the matrix and dissolves the target compound, in this case, anthraquinone [54]. The solubility of anthraquinone in supercritical $CO_2$ can be optimized by adjusting the temperature and pressure, allowing for selective extraction and improved yield (Fig. **6**).

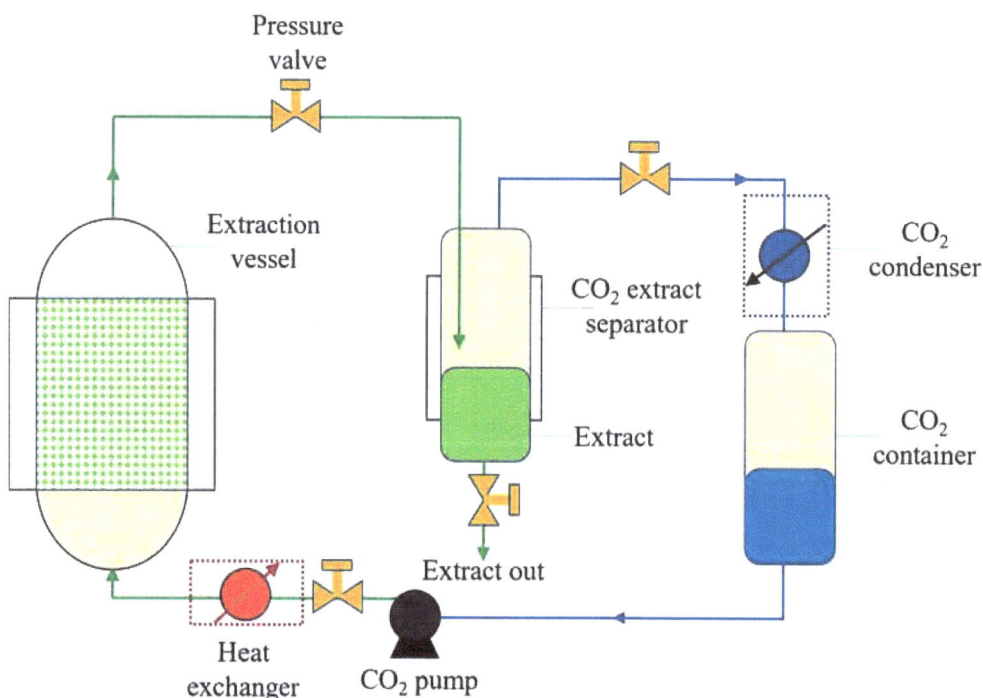

**Fig. (6).** Supercritical fluid extraction (SCFE).

After extraction, a reduction in pressure causes $CO_2$ to revert to its gaseous state, leaving behind the pure anthraquinone without residual solvents. Research has demonstrated that SCFE provides a higher purity of anthraquinone by minimizing thermal degradation, preserving its chemical integrity with a lower environmental impact compared to conventional methods [54,55]. Thus, SCFE stands out as a superior technology for extracting anthraquinone, offering significant advantages in terms of efficiency, selectivity, and environmental sustainability. As the demand for cleaner and more sustainable industrial processes increases, the role of SCFE in the extraction of valuable compounds like anthraquinone is likely to expand further.

## Ionic Liquid/Salt Based Aqueous Two-Phase Extraction System (ATPS)

The extraction of anthraquinones using ionic liquid/salt aqueous two-phase systems (ATPS) has garnered significant attention in the field of separation science due to its efficiency and environmental sustainability [56]. These systems utilize the unique properties of ionic liquids combined with salts to create two immiscible liquid phases, facilitating the selective extraction and purification of target compounds such as anthraquinones, which are prominent for their pharmacological activities (Fig. **7**) [57].

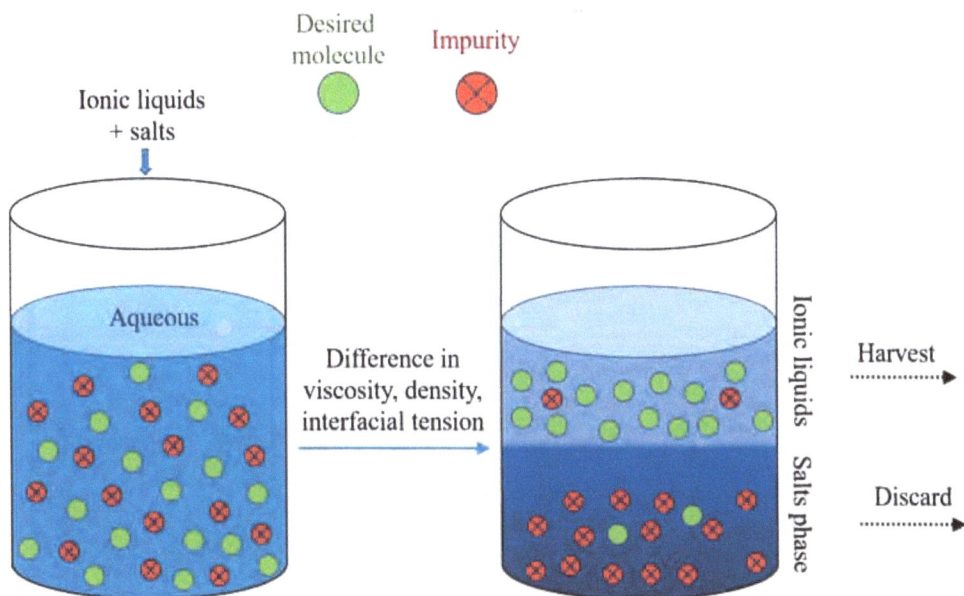

**Fig. (7).** Ionic liquid-based salt-induced liquid-liquid extraction.

Tan *et al.* demonstrated the viability of these systems in isolating and purifying anthraquinones from aloe, exploring the impacts of various system parameters on the partitioning behavior and extraction efficiency. Their work underscores the potential of ionic liquids and salts in creating ATPSs capable of selectively and effectively isolating these bioactive compounds [57]. Similarly, Deng *et al.* furthered this research by employing hexafluoroisopropanol-based deep eutectic solvents (DESs) in combination with salts to form ATPSs for extracting anthraquinones from *Rhei Radix* et *Rhizoma* samples. This study highlighted the novel application of HFIP-DES-based ATPSs, marking a significant advancement in the extraction methodologies for complex plant matrices [58]. Wang *et al.* extended the application of ionic liquid-based systems to the extraction of polyphenols and anthraquinones from *Polygonum cuspidatum*, comparing their efficacy against traditional extraction methods. Their research provided valuable insights into the optimization of extraction processes, emphasizing the advantages of ionic liquid-based salt-induced liquid-liquid extraction techniques [59]. Further exploration by Sun and colleagues into ultrasound-assisted ionic liquid solid–liquid extraction coupled with aqueous two-phase extraction presented a novel approach for extracting naphthoquinone pigments in *Arnebia euchroma*. This method leveraged the unique properties of ionic liquids in an ATPS, optimizing extraction efficiency and selectivity [60]. The review by Buarque, *et al.* encapsulates the broad utility of ATPSs based on ionic liquids and deep eutectic solvents, emphasizing their role in the selective recovery of non-protein bioactive compounds. This comprehensive overview not only synthesizes current findings but also sets the stage for future research directions in the extraction of bioactive compounds using environmentally benign solvents [61]. These collective efforts in the scientific community reflect a growing trend towards the use of ionic liquid/salt ATPSs in the efficient and sustainable extraction of valuable compounds like anthraquinones, highlighting their significant role in advancing green chemistry and separation technology.

## Deep Eutectic Solvents (DES) Extraction

Deep eutectic solvents (DES) are increasingly recognized as efficient and eco-friendly extraction media for bioactive compounds such as anthraquinones. These solvents, formed by combining a hydrogen bond donor (HBD) and a hydrogen bond acceptor (HBA), offer a green alternative to traditional organic solvents like methanol and acetone due to their low toxicity, biodegradability, and tunable properties (Fig. **8**) [62]. DES systems have demonstrated significantly higher extraction yields compared to conventional solvents. For instance, anthraquinones extracted from *Rheum palmatum* (Chinese rhubarb) using a DES system (choline chloride and lactic acid) yielded 12.3 mg/g, while ethanol only produced 9.7 mg/g [63]. Additionally, DES-based extraction methods retain up to 90% of the antioxidant capacity of anthraquinones, compared to 65% when ethanol is used, due to the stabilizing hydrogen bonding network in DES [64].

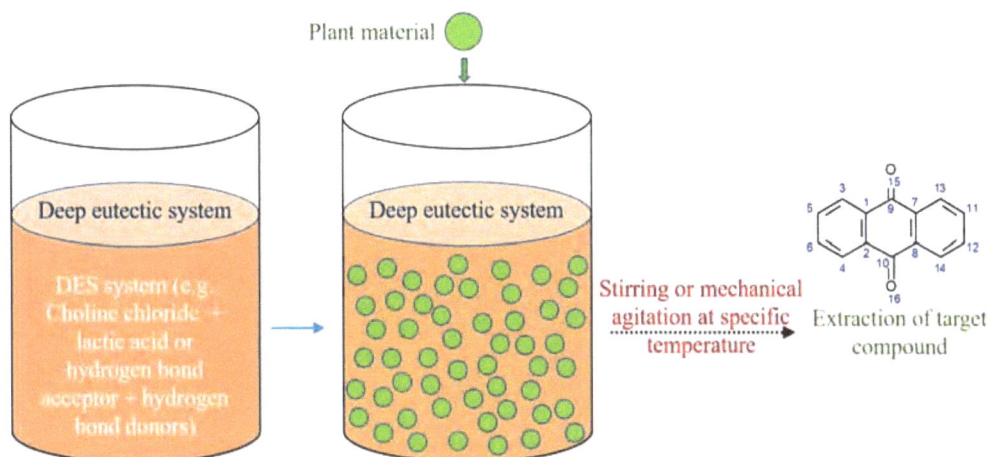

**Fig. (8).** Deep eutectic solvents (DES) extraction.

The sustainability and reusability of DES systems further enhance their industrial applicability. In a study where a choline chloride-lactic acid DES was reused five times, it maintained over 80% of its original extraction efficiency [65]. DES flexibility also allows customization for different target compounds, improving the extraction process for both hydrophilic and hydrophobic bioactive compounds [66].

Moreover, the application of non-conventional techniques such as ultrasound-assisted extraction (UAE) and microwave-assisted extraction (MAE) in conjunction with DES has been shown to further improve extraction efficiency and reduce environmental impact, supporting the principles of Green Analytical Chemistry [65]. Therefore, DESs offer a greener, more efficient alternative for the extraction of anthraquinones, providing higher yields, better bioactivity preservation, and sustainability advantages [66, 67].

## Pressurized Hot Water Extraction (PHWE)

Pressurized hot water extraction (PHWE) represents a significant advancement in the extraction of bioactive compounds such as anthraquinones, noted for their therapeutic properties including antibacterial, antiviral, and anticancer activities. The innovative approach of using hot water under high pressure as a solvent capitalizes on the enhanced solubility and diffusivity of water at elevated temperatures, leading to more efficient extraction processes (Fig. **9**) [68].

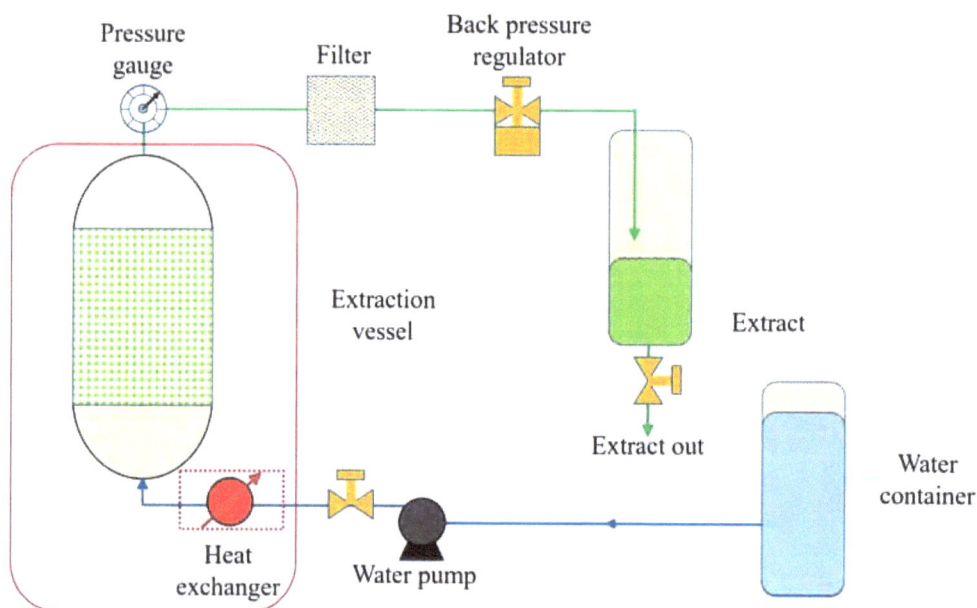

**Fig. (9).** Pressurized hot water extraction (PHWE).

A study by Shotipruk *et al.* on the roots of *Morinda citrifolia*, commonly known as Noni, showcases the effectiveness of PHWE in isolating anthraquinones [69]. Their research revealed that operational parameters like water temperature and flow rate dramatically influence the extraction yield. Specifically, at 220 °C, alongside an optimized flow rate, the method achieved maximum extraction efficiency, underscoring the critical role of temperature control in PHWE [69]. Further research by Vázquez *et al.* extended the application of PHWE to *Heterophyllaea pustulata* Hook f., a plant native to South America [70]. This study provided a detailed analysis of how temperature, pressure, and water flow rate affect the extraction yield of anthraquinones. By tailoring these conditions, PHWE was proven to be a powerful technique for extracting a wide range of anthraquinones with significant biological activities [70].

These studies collectively demonstrate the potential of PHWE as a green and sustainable extraction method. The technique's ability to operate without organic solvents not only reduces the environmental footprint but also aligns with the principles of green chemistry. PHWE, therefore, stands out as a promising approach for the efficient and eco-friendly extraction of anthraquinones, contributing to the advancement of natural product research and development.

# ANALYTICAL METHODS FOR ANTHRAQUINONE DETECTION AND QUANTIFICATION

## Spectroscopic Techniques

### *UV-Visible Spectroscopy*

UV-Visible Spectroscopy (UV-Vis) is a widely employed technique in the analysis of anthraquinones, a group of compounds known for their applications in dyes and biological activities [71]. UV-Vis spectroscopy measures the absorption of ultraviolet and visible light, revealing key insights into their electronic structure. It provides quick and non-destructive analysis but lacks specificity in complex mixtures.

### *Principles and Applications*

UV-Vis spectroscopy relies on the excitation of electrons from bonding ($\pi$) and non-bonding (n) orbitals to anti-bonding ($\pi^*$) orbitals. In anthraquinones, these transitions are primarily $\pi \rightarrow \pi^*$ and n $\rightarrow \pi^*$, due to their conjugated aromatic systems. The absorbance is affected by factors such as the structure of the molecule, the substituents on the anthraquinone ring, the solvent environment, and the pH conditions [72].

### *Quantitative Analysis*

Anthraquinones typically exhibit characteristic absorption bands in the UV-visible spectrum due to their conjugated ring structures and electron transitions between molecular orbitals. These absorption bands are indicative of the anthraquinone's molecular structure and can be used to identify and quantify specific anthraquinone derivatives in various samples. The UV-Vis spectroscopic analysis of anthraquinones, including purpurin, alizarin, carminic acid, and 2-(hydroxymethyl)-9,10-anthraquinone, in dimethylsulfoxide (DMSO) revealed key insights into their protonation and deprotonation equilibria. Purpurin exhibited a main absorption band at 486 nm, with new peaks emerging at 360 nm and 518 nm upon the addition of triethylamine (TEA), indicating a shift to deprotonated forms. Alizarin showed an absorption peak at 432 nm, with new peaks forming at 355 nm and 570 nm as TEA was added, marking the formation of mono- and di-anions. Both compounds displayed isosbestic points, confirming equilibrium between protonated and deprotonated states [71]. Carminic acid, with initial absorption peaks at 280 nm and 504 nm, showed increasing absorbance at 535 nm and 567 nm, corresponding to its di-anionic and tri-anionic forms as TEA was introduced.

Isosbestic points at various wavelengths confirmed the stepwise deprotonation process. In contrast, 2-(hydroxymethyl)-9,10-anthraquinone did not exhibit significant spectral changes upon TEA addition, indicating stable protonation. This study highlights how substituents and molecular structure influence the protonation behavior of anthraquinones in aprotic solvents [71].

The detection of alizarin and purpurin, two anthraquinone pigments in historical artworks, is often performed using UV-visible spectroscopy to understand the composition and degradation processes of the pigments [72]. For 9,10-anthraquinone, the absorption spectrum typically features two main bands: one in the 275–340 nm region and another in the 210–275 nm region, with vibronic structures contributing to the observed peaks at 317 nm and 328 nm. Computational approaches, such as time-dependent density functional theory (TD-DFT), have been employed to model these spectra, with the adiabatic Hessian and Franck-Condon approximations effectively capturing the experimental spectra's vibronic couplings. In the case of alizarin, the presence of two hydroxyl groups results in a p→π* absorption band with a maximum of around 425 nm in acidic conditions. The absorption spectrum of purpurin, which has three hydroxyl groups, shows prominent vibronic structures, with absorption maxima observed at 458, 486, and 520 nm. These differences in spectral properties highlight the impact of molecular substitutions on the UV-Vis absorption behavior of anthraquinones. Theoretical models confirm that adding hydroxyl groups shifts the absorption from the near UV to the visible range, with vibrational couplings playing a crucial role in shaping the spectral progression [72].

Anthraquinones, such as those isolated from *Rubiaceae* plants, exhibit π→π* electronic transitions with absorption bands typically found between 220-350 nm and one notable n→π* absorption band close to 400 nm. The position and intensity of these bands are significantly influenced by the nature and positions of autochromes groups, such as hydroxyl and methoxy groups, which can cause bathochromic shifts (redshifts) or hypsochromic shifts (blue shifts) in the absorption maxima. For instance, the presence of hydroxyl groups tends to extend the conjugation and induce bathochromic shifts, while methylation can lead to hypsochromic shifts [73].

UV-Vis spectra recorded in ethanol showed that anthraquinones with ortho hydroxyl groups, like 1,2-dihydroxyanthraquinone, have maximum absorption bands at 432 nm, while para-substituted isomers exhibit shifts to longer wavelengths, such as 481 nm. Theoretical methods, particularly Time-Dependent Density Functional Theory (TD-DFT) with hybrid functionals like PBE0 and

B3LYP, were used to predict the absorption maxima ($\lambda_{MAX}$), with a good correlation between the calculated and experimental values. Solvent effects, modeled using the polarizable continuum model (PCM), also demonstrated a bathochromic shift when polar solvents were used, further stabilizing the excited state of the anthraquinones. These findings help establish structure-property relationships, crucial for understanding anthraquinone behavior in various applications, including their roles as dyes and bioactive compounds [73].

## Environmental and Pharmaceutical Monitoring

 The UV-Vis spectroscopic analysis of anthraquinones plays a crucial role in both environmental and pharmaceutical monitoring, particularly due to their photophysical properties and their presence in various industrial and biological applications. In environmental monitoring, anthraquinone dyes are common pollutants in wastewater from textile and dyeing industries, with complex aromatic structures that make them resistant to degradation. These dyes are monitored using UV-Vis spectroscopy, as they show distinct absorption peaks in the UV-visible range, typically around 593 nm, which is used to assess their concentration and the efficiency of removal processes like adsorption or biodegradation. Studies have shown that the presence of heavy metals, carbon, and other additives can significantly influence spectral characteristics, allowing for optimization of wastewater treatment methods [74,75].

In pharmaceutical applications, anthraquinones are analyzed using UV-Vis spectroscopy to monitor their interaction with biological molecules, such as DNA, or their potential as therapeutic agents. For instance, UV-Vis spectroscopy is used to study the binding behavior of anthraquinone derivatives with DNA, as these compounds can intercalate into DNA strands, disrupting replication and transcription processes. In cancer research, this property is harnessed to develop drugs, with absorption bands around 466 nm being indicative of complex formation between anthraquinones and DNA. This technique provides critical insights into drug efficacy and the stability of anthraquinone-based therapeutics in biological systems [76,77].

## Fluorescence Spectroscopy

Fluorescence spectroscopy is an optical technique used to study the interaction between light and matter, particularly focusing on how certain substances (fluorophores) emit light after absorbing it. It plays a crucial role in biological, pharmaceutical, and chemical studies due to its sensitivity, selectivity, and ability to provide insight into molecular interactions.

## *Principle and Application*

The process begins with the absorption of light, where the energy excites a molecule from its ground state to a higher energy state. After a short time (typically in the range of nanoseconds), the excited molecule returns to the ground state by emitting light of a longer wavelength than that absorbed. This emission of light is termed fluorescence, and the difference between the excitation and emission wavelengths is called the Stokes shift [78]. The excitation wavelengths for anthraquinone compounds typically range around 280–300 nm, while emission wavelengths are in the 340-450 nm range, depending on the specific anthraquinone and the molecular environment. For example, when binding with BSA, the emission peak of alizarin is around 348 nm, which shifts with changes in concentration and environment, providing valuable data for interaction analysis [79–81].

Fluorescence quenching occurs when the intensity of fluorescence decreases due to the interaction between the fluorophore and another molecule (quencher). This process can be static or dynamic. Static quenching occurs when a non-fluorescent complex forms between the fluorophore and quencher before excitation. Dynamic quenching results from collisions between the fluorophore and quencher during the excited state [79,80]. In anthraquinone studies, static quenching has been identified as the dominant quenching mechanism when anthraquinones interact with proteins such as bovine serum albumin (BSA) or human serum albumin (HSA) [79,81].

Anthraquinones, due to their planar structures and light-absorbing properties, are ideal candidates for fluorescence spectroscopy. They are commonly studied for their interactions with biological molecules like proteins and DNA. In protein binding studies anthraquinones such as alizarin, emodin, and derivatives are known to interact with proteins like bovine serum albumin (BSA) and human serum albumin (HSA). In these studies, fluorescence quenching reveals how anthraquinones bind to proteins, providing insights into binding constants (K) and quenching mechanisms. For instance, alizarin's interaction with BSA shows significant quenching, with a quenching constant ($K_{SV}$) of $1.2447 \times 10^5$ L/mol [79,80].

In DNA binding studies fluorescence spectroscopy also helps in analyzing how anthraquinones interact with DNA. Anthraquinones intercalate between DNA base pairs, and fluorescence quenching studies provide data on the binding constants and the quenching mechanism. For example, Fu *et al.* reported that the binding constant for a new anthraquinone derivative (AORha) with calf thymus DNA (ctDNA) was $1.490 \times 10^5$ M$^{-1}$ [73]. Fluorescence spectroscopy can be employed for cellular bioimaging using anthraquinones. Certain anthraquinone derivatives exhibit bright fluorescence in cells, making them useful for tracking and imaging purposes. For example, fluorescence imaging of HepG2 cells treated with AORha showed bright green fluorescence, demonstrating its potential application in bioimaging [81].

## Infrared (IR) Spectroscopy

Infrared (IR) spectroscopy is a vital analytical tool for the detection and characterization of anthraquinone compounds. This technique is based on the absorption of infrared light by molecules, which causes vibrations in their chemical bonds. Each compound has a unique infrared absorption spectrum, serving as a "molecular fingerprint" that can be used for identification [82].

### Principle and Applications

In IR spectroscopy, the anthraquinone molecule absorbs infrared light, which causes changes in the dipole moment of its bonds, leading to vibrational and rotational transitions. The absorption spectrum reflects these molecular vibrations and provides information about the chemical structure of the compound. The molecule exhibits significant conjugation across its fused ring system, leading to distinct IR-active vibrations. The key vibrations occur at characteristic wavenumbers, including 1671 cm$^{-1}$, corresponding to the stretching vibration of the carbonyl group (C=O), and at 1595 cm$^{-1}$, associated with the C-C bond stretching in the side fragments of the molecule. These IR-active vibrations help in identifying structural details through their frequencies and intensities, allowing detailed spectral interpretations [83].

### Chemical Characterization

IR spectroscopy is extensively used to determine the functional groups and structural features of anthraquinone molecules. The IR spectra of anthraquinones show characteristic absorption bands corresponding to the quinone moiety and other functional groups present in the molecule [82]. Chumbalov *et al.* analyzed the chemical composition of anthraquinones and their nitrogen-containing derivatives, such as chrysophanol, using infrared (IR) spectroscopy [84]. The sharp absorption band at 1676 cm$^{-1}$ is characteristic of the carbonyl group (C=O) in anthraquinone. This peak remains unaffected by the introduction of a methyl group (as in 2-

methylanthraquinone), indicating its stability. For chrysophanol, which contains hydroxyl groups in the 4- and 5-positions, a strong intramolecular hydrogen bond forms between the hydroxyl groups and the carbonyl group. This bond leads to a reduction in the absorption frequency of the carbonyl group to 1620 cm$^{-1}$, while the free carbonyl group remains at 1675 cm$^{-1}$. The bond also decreases the intensity of the IR bands, a common feature of systems with strong $\pi$-electron interactions. Substituting the carbonyl group with a nitrogen-containing group, such as =N-OH (as in chrysophanol dioxime), introduces new absorption bands. The frequency of the C=N bond is observed at 1652 cm$^{-1}$. Similarly, in derivatives like the dihydrazone and diphenylhydrazone of chrysophanol, C=N vibrational frequencies emerge around 1658 cm$^{-1}$. The introduction of polar groups like OH or NH$_2$ into the anthraquinone molecule creates strong intramolecular hydrogen bonds, which impact the IR spectra. However, in nitrogen-containing derivatives of chrysophanol, no reduction in vibrational frequency was observed, in contrast to anthraquinone derivatives. The IR spectra of the phenylhydrazone of anthraquinone show specific bands associated with the benzene ring vibrations, at 886 cm$^{-1}$ and 860 cm$^{-1}$ [84]. Zhan *et al.* used near-infrared (NIR) spectroscopy to assess anthraquinones such as emodin, chrysophanol, rhein, and physcion from *Rhei Radix et Rhizoma*. The NIR spectral data, in the range of 4242-5581 cm$^{-1}$ and 6394-7011 cm$^{-1}$, revealed absorbance bands linked to C-H and O-H stretching vibrations. By utilizing partial least squares regression, Zhan's study demonstrated the capability of NIR spectroscopy for rapid, non-destructive quantification of anthraquinones [85].

## *Quality Control in Industries*

In the dye and pharmaceutical industries, IR spectroscopy is employed to assess the purity and composition of anthraquinone products [82]. It ensures that the synthesized or extracted compounds meet the required standards for industrial applications. Studies on anthraquinone dyes often use IR spectroscopy to confirm their chemical structure and to investigate the interactions with textile fibers. In anthraquinone-based dyes, such as those found in historical textiles, IR spectroscopy can differentiate purpurin from alizarin based on specific absorption peaks: alizarin's carbonyl stretch at ~1670 cm$^{-1}$ and purpurin's at ~1700 cm$^{-1}$. These precise measurements allow for rigorous quality control. [86].

## *Environmental Analysis*

IR spectroscopy aids in the detection of anthraquinone pollutants in environmental samples. The method can identify trace amounts of anthraquinone derivatives in complex matrices like soil, water, or air, facilitating pollution monitoring and control [74,75,87].

## Detection in Biological Systems

IR spectroscopy has been utilized to study the incorporation of anthraquinone derivatives in biological systems, examining their potential therapeutic effects or toxicological impact [82,88,89].

## Nuclear Magnetic Resonance (NMR) Spectroscopy

Nuclear Magnetic Resonance (NMR) spectroscopy is a powerful analytical technique used for the structural elucidation and detection of anthraquinone compounds. NMR is invaluable for elucidating the chemical structure and providing detailed molecular insights. It is, however, an expensive and time-consuming technique.

## Principle and Applications

NMR spectroscopy operates on the principle that certain atomic nuclei are capable of absorbing electromagnetic radiation in a strong magnetic field and re-emitting it at specific frequencies, which are characteristic of their chemical environment.

## Structural Elucidation

NMR spectroscopy is particularly valuable for determining the molecular structure of anthraquinones. By analyzing the chemical shifts, coupling constants, and integration of signals in $^1$H and $^{13}$C NMR spectra, researchers can deduce the framework of anthraquinone derivatives and their substitution patterns [90–93]. Anthraquinones, characterized by a quinone moiety at the 9,10 positions of the anthracene skeleton, present a variety of substitution patterns that influence their chemical shifts in NMR spectra. Studies such as that have shown that the substituents on anthraquinone rings, including hydroxyl, methoxy, and carbonyl groups, can be effectively characterized using $^{13}$C NMR. The assignment of carbon frequencies for anthraquinones is essential for structural determination, particularly for derivatives of 9,10-anthraquinone, the core structure found in many natural products, such as anthracycline antibiotics. In the analysis of various anthraquinone derivatives substituent groups like OH, OMe, and COMe shift the chemical environments of the carbons in the anthraquinone structure. Therefore there is a significant importance of precise coupling constants and long-range interactions for complete structural elucidation [92,93]. The structural elucidation of glycosylated and alkylated anthraquinones is aided by observing shifts in proton NMR spectra upon glycosylation or alkylation. The shifts in the proton signals of anthraquinones due to glycosylation, particularly at the 1-O or 8-O positions, follow predictable patterns that are independent of the substituents on the B-ring of the anthraquinone structure. The trend of shifts observed in the A-ring is useful for pinpointing the location of glycosylation. In cases of 1-O-glycosylation, protons at positions 2 and 4 undergo significant downfield shifts ($\Delta'$H-2 = +0.17 ppm, $\Delta'$H-4 = +0.31 ppm). The predictable behavior of these shifts simplifies the structural identification of glycosylated anthraquinones [91]. Dong *et al.* developed a quantitative $^1$H NMR (q-NMR) method to analyze bioactive anthraquinones in *Radix et Rhizoma Rhei*. This study revealed that $^1$H NMR is a fast and accurate technique for determining the concentrations of free anthraquinones, such as emodin, aloe-emodin, rhein, chrysophanol, and physcion. q-NMR provides distinct signals for these anthraquinones, enabling their structural elucidation and quantification without the need for chromatographic separation. The study established specific $^1$H NMR signals for anthraquinones in acetone-d6. For example, the signal for emodin is observed at δH 6.69 (H-7), aloe-emodin at δH 4.81 (3-CH$_2$OH), and chrysophanol at δH 7.64 (H-4), facilitating their identification in complex mixtures [90].

## *Quality Control and Purity Analysis*

In the pharmaceutical and dye industries, NMR spectroscopy is employed to assess the purity and authenticity of anthraquinone products, ensuring they meet the required specifications for industrial use [2,94,95]. 1D and 2D NMR techniques (such as $^1$H NMR, $^{13}$C NMR, HSQC, COSY, and HMBC) were employed to elucidate the structures of anthraquinone compounds. For instance, the compound 2-hydroxy-1,7,8-trimethoxyanthracene-9,10-dione was identified as a novel anthraquinone using high-resolution NMR data. The $^{13}$C NMR and DEPT spectra assist in identifying carbon environments within anthraquinones, ensuring that no unexpected impurities are present. In the same manner, $^1$H NMR spectra can reveal the presence of other organic compounds that could affect purity, such as residual solvents or byproducts from extraction processes [94]. NMR is also valuable in analyzing complex mixtures of anthraquinones and other compounds extracted from natural sources. For example, studies show that solid-state NMR, combined with high-resolution magic angle spinning (HR-MAS) NMR, can characterize both the crystalline and amorphous forms of cellulose and identify anthraquinone derivatives in solid samples like plant extracts [95]. While NMR provides detailed molecular insights, it is often used in conjunction with other analytical methods like FT-IR, Raman spectroscopy, and mass spectrometry to provide a comprehensive analysis of anthraquinones. These methods, when used alongside NMR, offer additional information on functional groups and degradation products, particularly in complex natural matrices [95]. In synthetic organic chemistry, NMR spectroscopy assists in verifying the structure of newly synthesized anthraquinone derivatives and monitoring the progress of chemical reactions [94,96]. The compound 2-hydroxy-1,7,8-trimethoxyanthracene-9,10-dione was confirmed through detailed $^1$H and $^{13}$C NMR spectra, showing key methoxy and aromatic proton signals that allowed for the precise identification of the molecular structure. The use of both HSQC (for correlating proton and carbon signals) and HMBC (to determine long-range coupling) was crucial for verifying the compound's purity [94]. In another study, a novel anthraquinone derivative was synthesized and characterized using $^1$H and $^{13}$C NMR. The spectra revealed the compound's core structure with characteristic aromatic and functional group shifts. Specifically, the $^1$H NMR spectrum showed multiplets corresponding to aromatic protons and functional groups, while the $^{13}$C NMR provided insight into the carbonyl and aromatic carbon environments. The purity of the synthetic anthraquinone was confirmed by matching the experimental NMR data with theoretical predictions using DFT (Density Functional Theory), ensuring the absence of impurities or unexpected isomers [96].

## Interaction Studies

NMR spectroscopy can also investigate the interactions between anthraquinone molecules and other biological or chemical entities, providing insights into their binding mechanisms and functional activities [97,98]. The formation of charge-transfer (CT) complexes between anthraquinones, 1,4-bis{[2-(methylamino)ethyl]amino}-5,8-dihydroxyanthracene-9,10-dione (AQ4) and 1,4-bis{[2-(methylamino)ethyl]amino}anthracene-9,10-dione (AQ4H), and electron donors (pyrene and hexamethylbenzene) was investigated by El-Gogary [97]. NMR spectroscopy showed significant upfield chemical shifts in the aromatic protons, more so for AQ4H due to its stronger electron-accepting ability facilitated by hydroxyl groups. The NMR data was used to calculate the association constants of these complexes, revealing that AQ4H had a higher affinity for pyrene compared to AQ4, indicating a stronger CT interaction [97]. NMR spectroscopy revealed that (AQ4) and 1,4-bis{[2-(methylamino)ethyl]amino}anthracene-9,10-dione (AQ4H) interact with DNA through intercalation and minor groove binding. The anthraquinone derivatives showed notable chemical shift changes, specifically in aromatic proton regions. This shift, along with band broadening, indicated changes in the electronic environment due to DNA binding. AQ4 and AQ4H were found to bind preferentially at guanine/cytosine-rich sites. The NMR studies also highlighted the structural influence of hydroxyl groups in AQ4, which reduced its electron-accepting properties compared to AQ4H [98].

## Chromatographic Methods

### High-Performance Liquid Chromatography (HPLC)

The most widely used technique for anthraquinone analysis, HPLC is versatile, offering high resolution and sensitivity. It allows for the effective separation and quantification of anthraquinone derivatives, even in complex mixtures.

### *Principle and Application*

HPLC operates on the principle of liquid chromatography where compounds are separated based on their interaction with the stationary and mobile phases. For anthraquinones including rhein, emodin, chrysophanol, and physcion, which are often non-polar and aromatic, reverse-phase HPLC is commonly used, employing a non-polar stationary phase and a polar mobile phase to achieve effective separation [99]. A common stationary phase used for anthraquinone analysis is the C18 reversed-phase column. HPLC is particularly suitable for anthraquinone analysis due to its chromophoric nature, which allows for effective ultraviolet (UV) detection. Anthraquinones exhibit strong absorption in the UV region (254 nm), primarily due to their conjugated aromatic systems, making UV detection a cost-effective and sensitive approach for their quantification [99].

### *Detection in Natural Products*

HPLC is instrumental in analyzing anthraquinone content in plants and herbal medicines, like senna, rhubarb, and aloe, where it is used to ensure the quality and consistency of these products [99,100]. UV detection at 254 nm is frequently used for anthraquinones due to their strong absorbance at this wavelength. For example, in the study involving *Rhamnus petiolaris*, the HPLC-UV method was validated, showing high linearity ($R^2 > 0.999$) and sensitivity with detection limits as low as 0.07–0.11 µg/mL for the various anthraquinones [99]. Other detection methods, such as chemiluminescence, have also been explored for enhanced sensitivity. In one study, anthraquinone-tagged amines were detected using HPLC with UV irradiation and luminol chemiluminescence (CL), which leveraged the ability of anthraquinones to generate reactive oxygen species (ROS) upon UV exposure. This approach showed limits of detection for certain analytes (like tryptamine) as low as 124 nM, highlighting the method's high sensitivity [101].

## Pharmaceutical Analysis

In pharmaceutics, HPLC aids in the purity assessment and content determination of anthraquinone-based drugs, such as laxatives and anticancer medications, ensuring they meet the required standards for therapeutic use. The pharmaceutical analysis of anthraquinones using HPLC is enhanced by various detection methods like UV, fluorescence (FL), chemiluminescence (CL), and mass spectrometry (MS). While HPLC-UV offers simplicity, it lacks sensitivity compared to FL and CL, which are more effective for detecting low concentrations. Fluorescence detection is ideal for naturally fluorescent anthraquinones like doxorubicin, whereas non-fluorescent anthraquinones are often derivatized, as seen in the use of anthraquinone-2-carbonyl chloride to improve sensitivity. Chemiluminescence detection, particularly in HPLC-CL systems, is highly sensitive due to anthraquinones' ability to generate reactive oxygen species (ROS) under UV irradiation, which react with luminol to produce a detectable signal. HPLC-MS, offering both qualitative and quantitative data, is widely used for complex matrices, utilizing ionization techniques like APCI or ESI for sensitive detection [40,101].

## Environmental Monitoring

HPLC is used for detecting and quantifying anthraquinone pollutants in environmental samples, like water and soil, helping in the assessment of industrial pollution and its impact. High-performance liquid chromatography (HPLC) is a key technique for monitoring anthraquinones as environmental pollutants due to their toxic effects. HPLC, often coupled with ultraviolet (UV) or fluorescence detection, enables the separation and quantification of these compounds in complex environmental samples like soil and water. While UV detection is commonly used, fluorescence detection offers greater sensitivity, particularly after the derivatization of anthraquinones into fluorescent forms. This method is crucial for detecting trace amounts of anthraquinones, helping to assess contamination levels and environmental risks [40].

## Gas Chromatography (GC)

Gas Chromatography (GC) is a potent analytical technique utilized for separating and analyzing compounds that can be vaporized without decomposition. While GC is less commonly used for anthraquinone detection due to the high molecular weight and thermal stability of these compounds, it can still be employed under specific conditions, especially when anthraquinones are derivatized into more volatile compounds [102].

## *Principle and Applications*

The principle of gas chromatography (GC) for anthraquinone detection is based on the separation of volatile compounds in a gas phase through a column filled with a stationary phase. To facilitate GC analysis, anthraquinones often need to be converted into more volatile derivatives through a chemical derivatization process. Techniques such as silylation or methylation can be applied, transforming the non-volatile anthraquinones into suitable derivatives for GC analysis [102,103]. It has been documented that the retention times of anthraquinones are influenced by their hydroxylation patterns. Anthraquinones without hydroxyl groups exhibited the shortest retention times, while retention time increased with the addition of hydroxyl groups. Specifically, anthraquinones with intramolecular hydrogen bonding, such as 1-hydroxyanthraquinone, showed shorter retention times compared to other hydroxylated forms like 2-hydroxyanthraquinone. Furthermore, blocking hydroxyl groups *via* methylation or trimethylsilylation led to decreased retention times, confirming the role of hydrogen bonding in the chromatographic behavior of these compounds. The study also demonstrated a correlation between increasing retention time and the presence of functional groups such as $CH_3$, $CH_2OH$, and COOH, as observed in anthraquinones like chrysophanol, aloe-emodin, and rhein [102]. Anthraquinones, which are polycyclic aromatic hydrocarbons, have raised health concerns due to their potential carcinogenic effects, leading the European Union to set a maximum residue limit of 0.02 mg/kg for anthraquinone in tea. The study by Pitoi *et al.* demonstrated that GC-µECD provided reliable sensitivity, detecting anthraquinone concentrations as low as 1 µg/L, with good precision (%RSD = 6.68) and strong linearity (r = 0.9973) within the range of 1-10 µg/L. The method's detection limit was 0.5 µg/L, indicating that it could detect anthraquinone at levels relevant to regulatory standards. The application of this method to spiked tea samples showed successful extraction and detection of anthraquinone at concentrations as low as 0.67 µg/L using low-volume liquid-liquid extraction. Although effective, further optimization is required to eliminate sample interferences and improve detection at lower concentrations [103].

## *Residual Solvent Analysis in Anthraquinone Production*

GC is employed to analyze residual solvents in the production of anthraquinone dyes and intermediates. This is crucial for quality control and to ensure the safety and purity of the final products [104]. Many anthraquinones have low vapor pressures, making them traditionally difficult to analyze using GC, derivatization of hydroxyanthraquinones with trimethylsilyl (TMS) ethers or trifluoroacetyl

(TFAA) esters enabled better chromatographic separation and quantification. The retention times of TFAA derivatives were found to be shorter, reflecting the electron-withdrawing nature of the $CF_3CO-$ group, whereas TMS derivatives had longer retention times due to their electron-donating characteristics. These modifications helped overcome issues of irreversible adsorption and tailing peaks that were prominent with non-derivatized hydroxyanthraquinones. Using GC, various substituted anthraquinones (*e.g.*, chloro-, amino-, and hydroxy- derivatives) were successfully separated and identified. Flame ionization detector (FID) response factors varied, with an increase in substituents generally leading to a decrease in relative response. This study concludes that GC, combined with derivatization techniques, provides valuable insights into the optimization of anthraquinone production processes, offering rapid and reliable quantification of residual solvents and by-products [104].

*Forensic Science*

GC can be utilized to identify anthraquinone derivatives used as markers or additives in illicit substances [105,106]. Anthraquinone analysis by gas chromatography (GC) holds significant forensic applications, particularly in detecting substances such as anthraquinone derivatives used in bank security dye packs. Anthraquinones are stable polycyclic aromatic compounds, making them difficult to oxidize and ideal for their role in forensic science. Specifically, forensic analysis of bank security dye packs involves the detection of compounds like 1-(methylamino)anthraquinone (MAAQ), commonly used in theft-deterrent devices. GC-MS (gas chromatography-mass spectrometry) has proven effective in identifying anthraquinone derivatives, including chlorinated variants that form when criminals attempt to bleach stolen, dye-stained currency. In a forensic context, attempts to remove MAAQ stains using bleach lead to the formation of chlorinated derivatives, which can be detected and characterized by GC-MS. For instance, using Ultra CloroxTM as a chlorine source and iron chloride as a catalyst, mono- and di-chlorinated anthraquinone derivatives can be synthesized and subsequently detected by GC-MS. Three chlorinated compounds, two mono-chlorinated (2-chloro-1-(methylamino)anthraquinone and 4-chloro-1-(methylamino) anthraquinone), and one di-chlorinated (2,4-dichloro-1-(methylamino) anthraquinone), were isolated in this study, demonstrating GC's capability in resolving these isomers, which are critical for forensic identification [105,106].

## Spectrometric Methods

### *X-ray Diffraction (XRD)*

X-ray Diffraction (XRD) is a powerful analytical technique used for characterizing the crystalline nature of materials. While XRD is typically not used for detecting anthraquinones directly in mixtures or solutions, it is invaluable for determining the crystal structure and purity of solid anthraquinone compounds, as well as for studying their polymorphic forms, molecular packing, and interaction with other substances in composite materials.

### *Principle and Applications*

The principle behind XRD for anthraquinone detection, or any crystalline substance, involves the interaction of X-ray beams with the crystal lattice of the material. When an X-ray beam hits a crystalline sample like anthraquinone, it interacts with the electrons in the crystal lattice, producing scattered waves. Due to the orderly arrangement of atoms in the crystal, these scattered waves interfere with each other constructively or destructively, leading to a diffraction pattern. The diffraction pattern obtained, consisting of peaks at various angles and intensities, is characteristic of the crystalline structure of anthraquinone. By analyzing these patterns, one can deduce the crystallographic structure, interatomic distances, and other structural details of the anthraquinone compound [107]. X-ray diffraction (XRD) is a key analytical tool for studying the crystal structure and phase transitions of anthraquinones, especially in electrochemical applications like lithium-ion batteries (LIBs). In the research on anthraquinone derivatives, *operando* XRD was used to monitor real-time structural changes during lithiation and delithiation cycles. The study revealed that anthraquinone and its derivatives undergo significant reversible polymorphic transformations as lithium ions are inserted and removed. Each derivative showed distinct behavior, with changes in the diffraction patterns indicating new crystalline phases. For example, 1,5-dicyanoanthraquinone displayed intermediate phases, while dilithium 1,5-dicarboxylate anthraquinone exhibited stable cycling performance, despite evolving diffraction peaks. These findings underscore the importance of XRD in understanding the solid-state reactivity and crystal packing changes in anthraquinones, which is critical for optimizing their performance as organic electrode materials [107].

## Polymorphism Study

Different polymorphic forms of anthraquinones can exhibit varied properties. XRD helps in identifying these polymorphs, which is crucial for understanding their physical, chemical, and pharmacological characteristics. Studies in the pharmaceutical industry focus on the crystalline forms of anthraquinone derivatives used in drug formulation, where XRD is used to ensure consistency and stability in production [108]. Hu *et al.* investigate the impact of alkyl chain length on the polymorphism and melting point of bifunctional anthraquinone derivatives (A-OCn) using X-ray diffraction (XRD) and scanning tunneling microscopy (STM). XRD analysis revealed that shorter chains (n = 1, 4) crystallize in the monoclinic space group, while intermediate chains (n = 3, 5, 6) form triclinic crystals. The study found that hydrogen bond strength decreases and van der Waals interactions increase with chain length, which affects the melting points, exhibiting an odd-even alternation. For example, A-OC8 has the lowest melting point due to reduced packing density. STM imaging of longer chains (n = 7−18) showed an odd-even alternation in self-assembly, further correlating packing density with thermal properties. This work provides a molecular-level understanding of how chain length influences structure and thermal behavior in organic materials [108].

## Material Science Applications

In the field of material science, XRD is used to study the incorporation of anthraquinone molecules in composite materials, such as in dye-sensitized solar cells or functionalized fabrics, providing information on how these molecules interact with the substrate material. Research on anthraquinone-based pigments for historic art preservation, where XRD analysis helps in identifying the specific forms of anthraquinone present in ancient artifacts [22]. For anthraquinones, particularly morindone (an anthraquinone derivative), XRD can be applied to investigate the crystalline nature of the compounds used as dyes in historical and modern textiles. For example, the structural confirmation of morindone extracted from *Morinda citrifolia* or other plants can be achieved through XRD analysis. Furthermore, in the production of bio-colorants, understanding the crystalline structure of anthraquinones *via* XRD is crucial to optimizing dye properties such as stability, solubility, and interaction with textile fibers [22].

## Mass Spectrometry (MS)

MS is often coupled with chromatographic techniques like LC-MS or GC-MS, providing powerful tools for the identification and quantification of anthraquinones [109–111]. These methods offer high sensitivity, specificity, and the ability to analyze complex mixtures.

### *Principle and Applications*

Mass Spectrometry (MS) is an analytical technique used to measure the mass-to-charge ratio of ions. It is a powerful method for identifying the chemical composition of substances, including anthraquinone compounds, by generating spectra of the masses of the molecules and their fragments.

In MS-based anthraquinone analysis, electron impact ionization (EI) is commonly used, as noted in studies such as the one analyzing rhubarb anthraquinones [110]. EI involves bombarding the analyte with high-energy electrons, resulting in the formation of positively charged ions (molecular ions) and their subsequent fragmentation into smaller ions. The fragmentation patterns are highly reproducible and characteristic of specific anthraquinones, allowing for their identification. Each anthraquinone compound, such as chrysophanol, emodin, and physcion, has a distinct fragmentation pattern under EI-MS, making it possible to identify them based on their unique mass-to-charge (m/z) ratios. For instance, the base peak for chrysophanol is observed at m/z 254, while physcion fragments to m/z 284 [110]. The identification of anthraquinones can be further enhanced by employing selected ion monitoring (SIM) or multiple reaction monitoring (MRM), which increases sensitivity by focusing on specific ion transitions. The quantitative determination of anthraquinones is often performed using MS/MS techniques such as gas chromatography coupled with tandem mass spectrometry (GC–MS/MS). In the quantification process, internal standards such as isotope-labeled anthraquinones (*e.g.*, AQ-d8) are utilized to correct for matrix effects and improve accuracy. One of the challenges in analyzing anthraquinones in complex matrices (*e.g.*, food and plant materials) is the presence of co-extracted matrix components such as caffeine or polyphenols that can interfere with detection. Solid-phase extraction (SPE) and dispersive solid-phase extraction (dSPE) clean-up steps are often employed to reduce these interferences, enhancing the sensitivity and reliability of the analysis [109,112].

## Identification and Characterization

High-performance liquid chromatography coupled with mass spectrometry (HPLC-MS) and gas chromatography-tandem mass spectrometry (GC-MS/MS) are key techniques for their identification and quantification. In rhubarb, HPLC-MS using negative electrospray ionization (ESI) identified key anthraquinones, including rhein, aloe-emodin, chrysophanol, emodin, and physcion, with specific [M-H]- ions and MS/MS fragmentation patterns. This method enabled the creation of a standardized fingerprint essential for quality assurance. Similarly, a GC-MS/MS method developed for tea and coffee detected anthraquinone residues with a detection limit of 5 mg/kg, well below the European Union's maximum residue limit of 20 mg/kg. Improved extraction and cleanup procedures, including the use of ethyl acetate and solid-phase extraction, enhanced the accuracy of anthraquinone detection in complex matrices like caffeine-rich beverages, ensuring robust monitoring and safety assessment in food products [111,112].

## Metabolite Profiling

In biological studies, MS helps in metabolite profiling of anthraquinones, tracing their biotransformation pathways in organisms, which is crucial for understanding their pharmacokinetics and toxicity [113]. A study by Yang *et al*. provides a comprehensive liquid chromatography-mass spectrometry (LC-MS) based metabolomics analysis of flavonoids and anthraquinones in *Fagopyrum tataricum* L. Gaertn. (Tartary buckwheat) seeds, identifying 60 flavonoids and 11 anthraquinones across 40 seed cultivars. Key anthraquinones such as aloe-emodin, emodin, and physcion were identified and quantified, with their concentrations analyzed using multiple reaction monitoring (MRM) scans. The study revealed that anthraquinones were more abundant in black Tartary buckwheat (BTB) seeds compared to yellowish-brown varieties (YTB), especially in the hulls, where the fold change ranged from 0.26 to 0.76. These metabolites contributed significantly to seed color variation but did not play a major role in seed shape differences. The higher anthraquinone content in the hulls highlights their potential bioactive properties, suggesting that the often-discarded hulls may possess valuable antioxidant and antimicrobial benefits [113].

## Environmental Monitoring

MS has been employed to detect anthraquinone derivatives in environmental samples, such as soil and water, assessing the impact of industrial waste and pollution [114,115]. Anthraquinones, particularly 9,10-anthraquinone (AQ), are significant environmental pollutants often found in atmospheric particulate matter (PM), posing risks due to their potential for generating reactive oxygen species (ROS). Gas chromatography-mass spectrometry (GC-MS) has been effectively employed for the detection and quantification of anthraquinones in fine particulate matter (PM$_{2.5}$), as highlighted in the work by Sousa *et al.* [114]. This method demonstrated low limits of detection (LOD) for anthraquinones in non-derivatized and derivatized forms, simplifying sample preparation. In case of non-derivatized AQ, the LOD was as low as 0.29 ng/m$^3$, making it a sensitive approach for environmental monitoring. The study emphasized that AQ can be thermally stable, allowing for its direct determination, which is critical in reducing processing time and improving detection accuracy in complex air samples [114].

On the other hand, wooden-tip electrospray ionization mass spectrometry (ESI-MS) represents a rapid, eco-friendly alternative for detecting anthraquinones in complex matrices, such as air, urine, and serum. As explored by Ling *et al.*, this method leverages *in situ* derivatization with cysteamine to enhance the ionization of quinones, achieving LODs as low as 0.16 ng for AQ [115]. Wooden-tip ESI-MS offers advantages in terms of minimal sample preparation and fast analysis, making it ideal for real-time environmental monitoring. Both methods provide complementary insights into the distribution of anthraquinones, with GC-MS excelling in detailed quantification for air pollution studies and wooden-tip ESI-MS offering a versatile, rapid assessment tool suitable for diverse sample types.

*Drug Development*

In pharmaceutical research, MS aids in the structural elucidation of anthraquinone derivatives used in drug formulations, ensuring their safety and efficacy [116]. For instance, the study on knipholone and knipholone anthrone, phenyl anthraquinone derivatives isolated from *Kniphofia foliosa*, highlights their potential in drug development due to their significant biological activities. Knipholone anthrone (**2**) showed potent antiplasmodial activity against *Plasmodium falciparum* 3D7 strain with an IC$_{50}$ of 0.7 μM, while both knipholone (**1**) and knipholone anthrone (**2**) exhibited cytotoxic effects on several human cell lines, with knipholone (**1**) showing stronger inhibition. Additionally, knipholone anthrone (**2**) demonstrated anti-HIV-1c activity, inhibiting viral replication with an EC$_{50}$ of 4.3 μM, though it also caused cytotoxicity in uninfected peripheral blood mononuclear cells (PBMCs), indicating a need for careful evaluation in therapeutic contexts. However, stability analyses revealed that knipholone anthrone (**2**) is unstable in cell culture media, rapidly oxidizing into knipholone, which could affect the observed biological activity. These findings point to the potential of these anthraquinones in therapeutic applications, particularly as antimalarial and anti-HIV agents, but also underscore the need for stability optimization to ensure consistent bioactivity [116].

## Capillary Electrophoresis (CE)

CE is a potent method for the separation of anthraquinone compounds based on their charge and size. It is known for its high efficiency, low solvent consumption, and fast analysis time.

### *Principle and Applications*

Capillary Electrophoresis (CE) is an analytical technique that separates ions based on their electrophoretic mobility with the use of an applied voltage. It is particularly useful for the analysis of anthraquinone compounds such as emodin, rhein, and aloe-emodin due to its high resolution, efficiency, and ability to handle complex mixtures. The separation efficiency depends on the buffer's pH and ionic strength, allowing for optimal migration and differentiation of the anthraquinones in the capillary [2].

### *Pharmaceutical Analysis*

CE is employed to analyze anthraquinone derivatives in pharmaceutical products, ensuring their purity and concentration. This is crucial for maintaining drug quality and therapeutic efficacy [117]. CE techniques are used to study the metabolism and distribution of anthraquinone derivatives in biological systems, facilitating pharmacokinetic and toxicological research [2,117]. Gong *et al.* describe an optimized method for the analysis of anthraquinones in Rhubarb using capillary zone electrophoresis (CZE). Anthraquinones, such as emodin, aloe-emodin, rhein, physcion, and chrysophanol, are the primary bioactive compounds in Rhubarb and play significant roles in its pharmacological effects, including laxative, antibacterial, and antispasmodic properties [117]. The separation of anthraquinones was achieved using a 50 mM borate buffer (pH 8.2) with 25% isopropyl alcohol and 25% acetonitrile as modifiers. The CZE method offered high resolution and reproducibility with a separation voltage of 25 kV at a temperature of 20°C. The running time was 50 minutes, with detection performed at 230 nm. The five anthraquinones showed good linearity within the measured concentration ranges. Limits of detection (LOD) for the compounds were below 0.5 mg/mL, indicating high sensitivity [117].

## *Environmental Monitoring*

CE is used to detect anthraquinone pollutants in water and soil, monitor the environmental impact of industrial discharge, and help in pollution control efforts. CE has been utilized to identify and quantify anthraquinone dyes in textile and food products, providing insights into their composition and concentrations [118,119]. The study by Ahmadi *et al.* (2014) shows that CE can be used to detect purpurin in complex mixtures, but the separation efficiency is highly sensitive to pH conditions. Using a 20 mM borate buffer at pH 9.24 combined with 20 mM sodium dodecyl sulfate (SDS), CE achieves good resolution for anthraquinones, though peak reproducibility for purpurin can be challenging due to its photochemical properties. The study found that purpurin undergoes significant photodegradation when exposed to light under alkaline conditions, with a first-order rate constant for photofading at pH 9.24 being $4.5 \times 10^{-3}$ $s^{-1}$. This rapid fading affects the reproducibility of peak heights in CE separations, suggesting the need for dark conditions or light shielding during analysis [118]. The review of nano-capillary electrophoresis extends this knowledge by detailing the use of microchips and detectors like ultraviolet-visible spectroscopy and laser-induced fluorescence for the analysis of environmental pollutants [119]. The technique allows for the separation of pollutants, including polycyclic aromatic hydrocarbons and dyes like anthraquinones, at low detection limits (nanogram to picogram levels). This capability makes CE highly effective for monitoring trace environmental contaminants, though the challenges posed by photodegradation require careful handling [119].

## *Natural Products Research*

CE is also applied in the separation and identification of anthraquinone compounds in natural product extracts, aiding in the discovery and characterization of new bioactive compounds [120]. Anthraquinones, such as chrysophanol and emodin, alongside bianthraquinones like cassiamin A and B, are compounds of significant interest due to their potential biological activities, including antitumor properties. In the experimental setup, the CE system utilized a Beckman P/ACE System 5000 apparatus equipped with a UV detector at 254 nm. The separation was carried out using a buffer solution composed of 0.1 M borate and 0.05 M hydroxypropyl-gamma-cyclodextrin (HP-g-CD) with 10% acetonitrile at pH 9, under an applied voltage of 20 kV [120]. The results demonstrated that the CE method was successful in separating both anthraquinones and bianthraquinones from the plant extracts. The separation was influenced by various factors, including buffer concentration and the presence of sodium dodecyl sulfate (SDS). It was found that adding HP-g-CD improved the resolution of the compounds, allowing for better separation of the four key compounds: chrysophanol, emodin, cassiamin A, and cassiamin B. Notably, cassiamin B showed promise as a potential anti-tumor-promoting agent. The method provided highly reproducible results, with relative standard deviations (RSDs) for migration times ranging from 0.08% to 0.16%, and recoveries for cassiamin A and B between 99.2% and 103.3% [120]. The ability to analyze such compounds efficiently is crucial for further investigation of their biological activities, including potential applications in cancer prevention and treatment.

## Emerging Techniques

### *Nanomaterials*

Recent advancements include the use of nanomaterials for enhanced detection and quantification of anthraquinones. Nanomaterials have revolutionized the field of analytical chemistry, offering enhanced detection capabilities for various compounds, including anthraquinones. Their unique properties such as high surface area, reactivity, and the ability to modify surface chemistry make them ideal for sensitive and selective detection applications.

### *Nanoparticle-Enhanced Spectroscopy*

Gold and silver nanoparticles can enhance spectroscopic signals through surface-enhanced Raman spectroscopy (SERS), providing a powerful method for detecting trace amounts of anthraquinone compounds. The nanoparticles' surface plasmons enhance the Raman scattering of anthraquinones, leading to highly sensitive detection [121]. A novel silver colloid substrate prepared *via* a solvothermal method results in silver nanoparticles that significantly improve sensitivity and reproducibility. Among the substrates tested, Ag-C2 (0.367 $AgNO_3$ to PVP molar ratio with 2 h reaction) proved to be the most effective, with high sensitivity and reproducibility (RSD of 5.37%). This nanoparticle-enhanced SERS method successfully detected anthraquinones such as alizarin, purpurin, carminic acid, and laccaic acid, enabling precise identification at much lower concentrations than conventional techniques [121]. The application of Ag-C2 allowed for non-destructive analysis of dyed silk fibers and ancient textiles, distinguishing different species of madder plants based on SERS spectra. For instance, purpurin was identified in textiles from Famen Temple (7th-9th centuries AD), while both alizarin and purpurin were found in red silk from the 3rd-5th centuries AD [121]. This method demonstrated high reproducibility and sensitivity, offering advantages for cultural heritage research where sample preservation is essential. The study highlights SERS as a powerful tool for both historical artifact analysis and broader natural product research, providing a non-invasive, sensitive approach to identifying natural dyes.

## *Nanocomposite-Based Sensors*

Nanocomposites, which combine nanoparticles with other materials like polymers or carbon nanotubes, can be engineered to have specific affinities for anthraquinone molecules. These materials are used in sensors to detect anthraquinone derivatives in environmental samples or food products [122]. The nanocomposites exhibit excellent detection performance for anthraquinone. For example, a graphene oxide-based composite functionalized with gold nanoparticles (AuNPs) demonstrated enhanced electron transfer, allowing for more efficient anthraquinone detection. The sensor achieved a detection limit as low as 0.1 μM, with rapid response times of less than 10 seconds, showcasing the effectiveness of such nanocomposites [122].

The integration of mesoporous silica with metal nanoparticles like palladium (Pd) or copper (Cu) also improves the catalytic properties, leading to better reduction and detection of anthraquinone derivatives in environmental samples. The catalytic efficiency, particularly for Pd-Cu nanocomposites, reached over 85% for anthraquinone reduction. Furthermore, these nanocomposites have high recyclability, retaining over 90% of their sensitivity after five cycles of use, highlighting their robustness for long-term applications in environmental monitoring [122].

## *Quantum Dots*

Semiconductor quantum dots can be used as fluorescent markers to tag anthraquinone compounds. The strong and stable fluorescence of quantum dots (QD) helps in tracking the distribution and concentration of anthraquinones in biological and environmental systems [123]. In the case of anthraquinones, a synthetic derivative, mitoxantrone, was coupled with cadmium telluride (CdTe) quantum dots to create a fluorescent biosensor. This system was able to detect HIV dsDNA through a mechanism involving the interaction between mitoxantrone and DNA, which resulted in fluorescence quenching at 599 nm. This interaction is crucial because anthraquinones like mitoxantrone are known to intercalate with DNA, influencing their detection potential. The biosensor showed strong sensitivity in detecting specific viral markers, illustrating the potential for quantum dot-based detection in anthraquinone-related applications. By utilizing the fluorescence quenching and recovery mechanism of quantum dots, particularly with anthraquinone derivatives, QD-based biosensors provide a powerful tool for the sensitive and specific detection of biological molecules, paving the way for advancements in diagnostics, particularly in detecting viral infections or biomolecular interactions in cancer therapies [123].

## *Degradation of Anthraquinone Dyes in Environmental Samples*

Studies have utilized nanomaterials like titanium dioxide ($TiO_2$) nanoparticles for photocatalytic degradation of anthraquinone pollutants, C.I. Acid Green 25, in water [124]. The $TiO_2$ nanoparticles, with a crystallite size of 8 nm and a surface area of 320.76 $m^2$/g, were immobilized on glass plates through a sol-gel dip-coating method, enhancing their reusability and mechanical stability. This immobilization method addresses challenges in separating suspended nanoparticles after treatment, making it more practical for environmental applications. The study optimized key parameters like dye concentration, UV light intensity, flow rate, and reaction time using Response Surface Methodology (RSM), and achieved a maximum decolorization efficiency of 79.43% under optimal conditions [124]. These findings were validated experimentally, with the results closely matching the model predictions.

## *Biomedical Applications*

In biomedical research, nanomaterials are used to enhance the imaging and detection of anthraquinone derivatives used as therapeutic agents, aiding in the study of their pharmacokinetics and distribution in tissues [125]. Nanomaterials, particularly carbon quantum dots (C-dots), have emerged as powerful tools in theranostics, blending diagnostic imaging with therapeutic functions. In the context of anthraquinones, these materials demonstrate significant potential due to their biocompatibility and unique optical properties. C-dots synthesized using large amino acids and tetramino-anthraquinone (TAAQ) derivatives exhibit strong light absorption at 230, 280, and 650 nm, making them highly effective in tumor targeting through LAT-1-mediated pathways. These nanomaterials have shown potential in near-infrared (NIR) fluorescence imaging, a critical feature for non-invasive tumor detection and phototherapy. Compared to traditional metal-based quantum dots, the anthraquinone-functionalized C-dots offer better photostability and reduced toxicity, positioning them as promising candidates for cancer theranostics [125].

Despite these advantages, challenges remain in ensuring the stability and consistency of anthraquinone-based nanomaterials in biological environments. One of the key goals for future research is to enhance the delivery capabilities of these nanomaterials, minimizing off-target effects while maximizing therapeutic efficacy. Additionally, improving the scalability of their synthesis while maintaining the same level of performance is crucial for their broader application in clinical settings. As this field evolves, anthraquinone-based C-dots hold significant promise for advancing targeted cancer therapies and precise diagnostic techniques.

## *Molecular Imprinting Techniques*

Molecular imprinting techniques involve creating polymer matrices with specific binding sites that are complementary in shape, size, and functional groups to the target molecule, in this case, anthraquinone. This approach allows for highly selective and sensitive detection of anthraquinones, making it a valuable tool in various applications, including environmental monitoring, food safety, and pharmaceutical analysis.

## *Environmental Monitoring*

Molecularly imprinted polymers (MIPs) are used to detect anthraquinone derivatives in environmental samples such as water and soil. These MIPs can selectively bind to anthraquinone pollutants, facilitating their detection even in complex matrices and at low concentrations [126]. The polymers, known as molecularly imprinted polymers (MIPs), are synthesized through a co-polymerization process involving functional monomers, cross-linkers, and the template molecule. For anthraquinone detection, functional monomers like methacrylic acid (MAA) or acrylamide are often used because of their strong hydrogen bonding with the hydroxyl and carbonyl groups present in anthraquinones. Upon removal of the template molecule, the polymer retains specific cavities that match the shape and chemical functionality of the target molecule, allowing for highly selective detection in complex environmental matrices such as soil or water [126].

Recent advancements in MIP design for anthraquinone environmental monitoring include the use of computational modeling to optimize the interactions between the template and monomer. Computational approaches, like density functional theory (DFT) and molecular dynamics simulations, help predict the binding energies and select the most suitable monomers and solvents for the synthesis process. In addition, nanostructured MIPs, particularly those based on multi-walled carbon nanotubes (MWCNTs), significantly enhance the sensitivity and binding efficiency of the polymers by increasing the surface area available for binding. MIPs fabricated on the surface of MWCNTs exhibit faster binding kinetics and higher binding capacity due to improved accessibility of binding sites. These innovations have made MIPs an effective tool for detecting trace levels of anthraquinones in environmental samples, with applications ranging from pollution monitoring to assessing contamination in industrial effluents and agricultural soils [126].

## *Food Safety Analysis*

In the food industry, MIPs assist in the identification and quantification of anthraquinone dyes and contaminants, ensuring food safety and compliance with regulations [127]. Molecularly imprinted polymers (MIPs) are synthesized by polymerizing around a template molecule, in this case, anthraquinones or their derivatives, creating highly specific binding sites. After the template is removed, the resulting cavities in the polymer are shaped to selectively bind anthraquinones. The addition of magnetic nanoparticles (MNPs) enables rapid magnetic separation, simplifying the extraction process and significantly improving the sensitivity and specificity of detection in complex food matrices, such as herbal products and teas, where anthraquinones naturally occur [127]. This method offers several advantages over conventional analytical techniques like chromatography, including being faster, more cost-effective, and environmentally friendly, with reduced solvent use. MIP-based sensors exhibit high selectivity, ensuring accurate detection even in the presence of other food components. Their application to monitor anthraquinone levels ensures that concentrations stay within safe limits, protecting consumer health. As the field advances, ongoing research focuses on enhancing the selectivity and capacity of MIPs, as well as exploring other nanoparticle-based materials to further improve detection efficiency. These technologies are expected to play an increasingly important role in routine food safety assessments [127].

## *Pharmaceutical Quality Control*

MIPs are employed in the pharmaceutical industry to detect and quantify anthraquinone compounds in drug formulations, helping in quality control and ensuring the efficacy and safety of pharmaceutical products [128]. The study on molecular imprinting techniques (MIPs) for anthraquinone pharmaceutical quality control focused on developing selective adsorbents to remove toxic compounds like emodin and physcion from *Polygonum multiflorum* extracts [128]. By using these anthraquinones as template molecules, the researchers synthesized MIPs that exhibited specific recognition and binding capacity, leaving the beneficial bioactive compound, THSG (tetrahydroxystilbene-O-glucoside), intact. The MIPs demonstrated excellent adsorption performance with a maximum capacity of 48.87 µmol/g for emodin and 32.00 µmol/g for physcion, fitting the Langmuir isotherm model. The kinetic analysis showed the adsorption followed pseudo-first-order dynamics, with high selectivity towards the targeted anthraquinones compared to non-imprinted polymers (NIPs). This selective removal process effectively eliminated potential toxicity risks while preserving THSG, a key active ingredient in pharmaceutical formulations derived from *Polygonum multiflorum*. Furthermore, the tandem use of emodin-MIPs and physcion-MIPs allowed for efficient removal of over 96% of the anthraquinones while recovering more than 90% of THSG, demonstrating the system's potential in pharmaceutical quality control. The MIPs were also shown to be reusable, maintaining high removal efficiency for five consecutive cycles, making them a cost-effective solution for industrial applications. This approach provides a practical method to improve the safety profile of plant-based pharmaceuticals by targeting and removing specific harmful compounds while retaining the therapeutic agents essential for their efficacy [128].

The analysis of anthraquinones requires a multifaceted approach due to the complexity of their chemical nature and the diversity of their matrices. While traditional methods like HPLC and MS remain staples in the analytical repertoire, emerging technologies offer new avenues for enhanced specificity and sensitivity. Each method has its strengths and limitations, and the choice of technique often depends on the specific requirements of the analysis, such as sensitivity, selectivity, speed, and cost-effectiveness. As research and technology evolve, the analytical methodology for anthraquinones continues to advance, promising more efficient, accurate, and sophisticated means of analysis.

## APPLICATIONS

Anthraquinones are versatile compounds with a broad range of applications across various industries due to their unique chemical properties. These applications include their use in foods, dyes, medical treatments, and functional materials.

### Food Color

Anthraquinones, such as *emodin, aloe-emodin, rhein, chrysophanol*, and *physcion*, are widely explored as natural pigments with vibrant red hues. These compounds are derived from various plant and microbial sources, making them attractive alternatives to synthetic colorants in the food industry. For example, *emodin* and *aloe-emodin* are commonly found in plants like *Rheum palmatum* (rhubarb) and *Aloe vera*, while *rhein* and *chrysophanol* are abundant in *Cassia obtusifolia* and *Rhamnus frangula* [129]. In addition to plant sources, fungi are emerging as a promising source of anthraquinones. Marine-derived fungi such as *Talaromyces purpureogenus* have been identified as prolific producers of anthraquinones, including *bostrycin* and *catenarin*. These fungal anthraquinones not only offer a bright red color for food applications but also exhibit antimicrobial and antioxidant properties. For instance, *bostrycin* has shown potential as a natural preservative by inhibiting bacterial growth in processed meat products like beef burgers [130]. *Herqueinone* and *austrocortinin*, two lesser-known fungal anthraquinones, have been isolated from *Aspergillus* and *Penicillium* species. These compounds are stable across a range of pH levels and demonstrate strong solubility, making them suitable for various food matrices [131]. Such stability is a significant advantage over plant-derived anthraquinones, which often degrade under varying environmental conditions [130,131]. Thus, the diverse sources of anthraquinones, from plants like rhubarb and aloe to marine-derived fungi, offer a sustainable and bioactive solution for replacing synthetic food colorants and preservatives, addressing both consumer demand and industry challenges [129-131].

## Dye and Pigment Industry

Anthraquinones are extensively used as dyes and pigments in the textile and paper industries due to their brilliant colors and excellent fastness properties. For example, alizarin, one of the oldest known anthraquinone dyes, is used to produce red and pink hues on fabrics. Synthetic derivatives like anthraquinone blue and green are used in high-performance inks and coatings. These dyes are valued for their stability, which is essential for long-lasting color in various materials [132].

## Pharmaceutical Applications

In medicine, anthraquinone derivatives have therapeutic significance. Emodin, for example, is researched for its potential in cancer therapy due to its ability to inhibit the proliferation of cancer cells. Another compound, doxorubicin, is a widely used anthraquinone-based chemotherapy agent effective against a range of cancers. These compounds are studied for their pharmacological effects, including anti-inflammatory, antibacterial, and antiviral activities [133–135].

## Functional Materials

In materials science, anthraquinones are employed in advanced applications such as organic semiconductors and photovoltaic cells [136]. Their solid-state properties are harnessed in designing organic electronic devices. Furthermore, anthraquinone compounds are used in energy storage systems, particularly in redox flow batteries, where their electrochemical properties facilitate efficient energy storage and release [137].

## CONCLUSION

Anthraquinones (AQs) represent a diverse class of aromatic organic compounds pivotal to both natural ecosystems and synthetic applications, showcasing a broad spectrum of biological and industrial functionalities. Originating from benzoquinones and naphthoquinones, these compounds, particularly noted for their structural variety, are synthesized chemically and occur naturally across various species, including bacteria, fungi, plants, animals, and insects. The dual synthesis pathways in nature, namely the polyketide and shikimate pathways, underscore the intricate biochemical processes AQs undergo, reflecting their ecological significance and evolutionary adaptability. Synthetic methodologies, like the Friedel-Crafts and Diels-Alder reactions, mirror this complexity, providing a framework for producing diverse AQ derivatives. Analytical advancements in AQ research, leveraging techniques like high-performance liquid chromatography and mass spectrometry, have propelled our understanding of these compounds. These methodologies not only facilitate the isolation and characterization of AQs but also ensure the refinement of their extraction and synthesis processes, enhancing purity and yield. Future research should focus on optimizing AQ biosynthesis, exploring the ecological roles of these compounds, and refining synthetic methods to improve their accessibility and sustainability. The potential for AQs in emerging fields, such as nanotechnology and advanced material science, invites further investigation, promising innovative applications that leverage their unique chemical properties. The medical industry's interest in AQs stems from their potent bioactivities, particularly in oncology. The anticancer properties of anthracycline antibiotics, derived from AQs, have led to substantial therapeutic advancements. However, the dual nature of AQs is evident in their toxicity profile, necessitating careful evaluation and management in pharmaceutical development.

## LIST OF ABBREVIATIONS

| | |
|---|---|
| **AQs** | Anthraquinones |
| **NOAQ** | Naturally Occurring Anthraquinones |
| **UAE** | Ultrasound-Assisted Extraction |
| **SCFE** | Super Critical Fluid Extraction |
| **HPLC** | High-Performance Liquid Chromatography |
| **GC** | Gas Chromatography |
| **MS** | Mass Spectrometry |
| **CE** | Capillary Electrophoresis |
| **ATPS** | Aqueous Two Phase Extraction System |
| **HFIP-DES** | Hexafluoroisopropanol-Based Deep Eutectic Solvent |
| **PHWE** | Pressurized Hot Water Extraction |
| **NMR** | Nuclear Magnetic Resonance |
| **XRD** | X-ray Diffraction |
| **SERS** | Surface-Enhanced Raman Spectroscopy |

## REFERENCES

[1] Diaz-Muñoz G, Miranda IL, Sartori SK, de Rezende DC, Diaz MAN. Chapter 11 - anthraquinones: An overview. In: Atta-ur-Rahman, editor Stud Nat Prod Chem. Elsevier 2018; 58: pp. 313-8.
http://dx.doi.org/10.1016/B978-0-444-64056-7.00011-8

[2] Duval J, Pecher V, Poujol M, Lesellier E. Research advances for the extraction, analysis and uses of anthraquinones: A review. Ind Crops Prod 2016; 94: 812-33.
http://dx.doi.org/10.1016/j.indcrop.2016.09.056

[3] Kim K, Min M, Hong S. Efficient synthesis of anthraquinones from diaryl carboxylic acids *via* palladium(ii)-catalyzed and visible light-mediated transformations. Adv Synth Catal 2017; 359(5): 848-52.
http://dx.doi.org/10.1002/adsc.201601057

[4] Mund NK, Čellárová E. Recent advances in the identification of biosynthetic genes and gene clusters of the polyketide-derived pathways for anthraquinone biosynthesis and biotechnological applications. Biotechnol Adv 2023; 63: 108104.

http://dx.doi.org/10.1016/j.biotechadv.2023.108104 PMID: 36716800

[5]     Liu C, Wang R, Wang S, *et al.* A prenyltransferase participates in the biosynthesis of anthraquinones in *Rubia cordifolia*. Plant Physiol 2024; 195(4): 2860-76.

http://dx.doi.org/10.1093/plphys/kiae171 PMID: 38502063

[6]     Schmalhofer M, Vagstad AL, Zhou Q, Bode HB, Groll M. Polyketide trimming shapes dihydroxynaphthalene-melanin and anthraquinone pigments. Adv Sci (Weinh) 2024; 11(22): 2400184.

http://dx.doi.org/10.1002/advs.202400184 PMID: 38491909

[7]     Xu H, Yuan Z, Yang S, *et al.* Discovery of a fungal P450 with an unusual two-step mechanism for constructing a bicyclo [3.2.2] nonane Skeleton. J Am Chem Soc 2024: jacs. 4c01284.

http://dx.doi.org/10.1021/jacs.4c01284

[8]     Qiang He HD Yuqing Miao, Xinyuan Zheng, Yaru Wang, *et al.*, The near-complete genome assembly of Reynoutria multiflora reveals the genetic basis of stilbenes and anthraquinones biosynthesis. J Syst Evol n.d.:0.

[9]     Zhao X, Yan F, Li Y, *et al.* Comparative transcriptome analysis and identification of candidate R2R3-MYB genes involved in anthraquinone biosynthesis in Rheum palmatum L. Chin Med 2024; 19(1): 23.

http://dx.doi.org/10.1186/s13020-024-00891-4 PMID: 38317158

[10]    Chien SC, Wu YC, Chen ZW, Yang WC. Naturally occurring anthraquinones: chemistry and therapeutic potential in autoimmune diabetes. Evid Based Complement Alternat Med 2015; 2015: 1-13.

http://dx.doi.org/10.1155/2015/357357 PMID: 25866536

[11]    Akev N, Candoken E, Erdem Kuruca S. Comparative study on the anticancer drug potential of a lectin purified from aloe vera and aloe-emodin. Asian Pac J Cancer Prev 2020; 21(1): 99-106.

http://dx.doi.org/10.31557/APJCP.2020.21.1.99 PMID: 31983171

[12]    Liu W, Qaed E, Zhu Y, *et al.* Research progress and new perspectives of anticancer effects of emodin. Am J Chin Med 2023; 51(7): 1751-93.

http://dx.doi.org/10.1142/S0192415X23500787 PMID: 37732372

[13]    Prateeksha, Yusuf MA, Singh BN, *et al.* Chrysophanol: A natural anthraquinone with multifaceted biotherapeutic potential. biomolecules 2019; 9:68.

http://dx.doi.org/10.3390/biom9020068

[14]    Shang XF, Zhao ZM, Li JC, *et al.* Insecticidal and antifungal activities of Rheum palmatum L. anthraquinones and structurally related compounds. Ind Crops Prod 2019; 137: 508-20.

http://dx.doi.org/10.1016/j.indcrop.2019.05.055

[15]    Khattak AK, Hassan SM, Mughal SS. general overview of phytochemistry and pharmacological potential of rheum palmatum (chinese rhubarb). Innovare Journal of Ayurvedic Sciences 2020; 5–9: 5-9.

http://dx.doi.org/10.22159/ijas.2020.v8i6.39192

[16]    Pham DQ, Ba DT, Dao NT, *et al.* Antimicrobial efficacy of extracts and constituents fractionated from Rheum tanguticum Maxim. ex Balf. rhizomes against phytopathogenic fungi and bacteria. Ind Crops Prod 2017; 108: 442-50.
        http://dx.doi.org/10.1016/j.indcrop.2017.06.067

[17]    Malik EM, Müller CE. Anthraquinones as pharmacological tools and drugs. Med Res Rev 2016; 36(4): 705-48.
        http://dx.doi.org/10.1002/med.21391 PMID: 27111664

[18]    Capasso F, Gaginella TS. Laxatives. Milano: Springer Milan 1997.
        http://dx.doi.org/10.1007/978-88-470-2227-0

[19]    Younes M, Aggett P, Aguilar F, *et al.* EFSA panel on food additives and nutrient sources added to food (ANS), safety of hydroxyanthracene derivatives for use in food. EFSA J 2018; 16.
        http://dx.doi.org/10.2903/j.efsa.2018.5090

[20]    Shukla V, Asthana S, Gupta P, Dwivedi PD, Tripathi A, Das M. Toxicity of naturally occurring anthraquinones Adv Mol Toxicol. Elsevier 2017; Vol. 11: pp. 1-50.
        http://dx.doi.org/10.1016/B978-0-12-812522-9.00001-4

[21]    Fouillaud M, Caro Y, Venkatachalam M, Grondin I, Dufossé L. Anthraquinones. In: Nollet LML, Gutierrez-Uribe JA, editors. Phenolic Compd. Food. 1st ed., Boca Raton : CRC Press, Taylor & Francis Group, 2018.: CRC Press; 2018, p. 131–72. In: 2018.
        http://dx.doi.org/10.1201/9781315120157-9

[22]    Do KL, Su M, Zhao F. From historical dye to bio-colourant: Processing, identification in historical textiles and potential applications of anthraquinone-based morindone. Dyes Pigments 2022; 205: 110482.
        http://dx.doi.org/10.1016/j.dyepig.2022.110482

[23]    Cheemalamarri C, Batchu UR, Thallamapuram NP, Katragadda SB, Reddy Shetty P. A review on hydroxy anthraquinones from bacteria: crosstalk's of structures and biological activities. Nat Prod Res 2022; 36(23): 6186-205.
        http://dx.doi.org/10.1080/14786419.2022.2039920 PMID: 35175877

[24]    Cheng MM, Tang XL, Sun YT, *et al.* Biological and chemical diversity of marine sponge-derived microorganisms over the last two decades from 1998 to 2017. Molecules 2020; 25(4): 853.
        http://dx.doi.org/10.3390/molecules25040853 PMID: 32075151

[25]    El-Beih AA, Kawabata T, Koimaru K, Ohta T, Tsukamoto S. Monodictyquinone A: A new antimicrobial anthraquinone from a sea urchin-derived fungus Monodictys sp. Chem Pharm Bull (Tokyo) 2007; 55(7): 1097-8.
        http://dx.doi.org/10.1248/cpb.55.1097 PMID: 17603212

[26]    Sottorff I, Künzel S, Wiese J, *et al.* Antitumor anthraquinones from an easter island sea anemone: animal or bacterial origin? Mar Drugs 2019; 17(3): 154.
        http://dx.doi.org/10.3390/md17030154 PMID: 30841562

[27]    Hafez Ghoran S, Taktaz F, Ayatollahi SA, Kijjoa A. Anthraquinones and their analogues from marine-derived fungi: chemistry and biological activities. Mar Drugs 2022; 20(8): 474.
        http://dx.doi.org/10.3390/md20080474 PMID: 35892942

[28]   Masi M, Evidente A. Fungal bioactive anthraquinones and analogues. Toxins (Basel) 2020; 12(11): 714.
http://dx.doi.org/10.3390/toxins12110714 PMID: 33198270

[29]   Shamim G, Ranjan SK, Pandey DM, Ramani R. Biochemistry and biosynthesis of insect pigments. Eur J Entomol 2014; 111(2): 149-64.
http://dx.doi.org/10.14411/eje.2014.021

[30]   Mohammad F. Anthraquinone-based natural colourants from insects.Text Cloth Sustain. Singapore: Springer Singapore 2017; pp. 81-97.
http://dx.doi.org/10.1007/978-981-10-2185-5_3

[31]   Usman M, Rehman F, Afzal M, *et al.* Sustainable appraisal of lac (*Kerria Lacca*) based anthraquinone natural dye for chemical and bio-mordanted viscose and silk dyeing. Sci Prog 2023; 106(4): 00368504231215944.
http://dx.doi.org/10.1177/00368504231215944 PMID: 37993992

[32]   Amin N, Rehman F, Adeel S, Ahamd T, Muneer M, Haji A. Sustainable application of cochineal-based anthraquinone dye for the coloration of bio-mordanted silk fabric. Environ Sci Pollut Res Int 2020; 27(7): 6851-60.
http://dx.doi.org/10.1007/s11356-019-06868-3 PMID: 31879870

[33]   Westendorf J, Marquardt H, Poginsky B, Dominiak M, Schmidt J, Marquardt H. Genotoxicity of naturally occurring hydroxyanthraquinones. Mutat Res Genet Toxicol Test 1990; 240(1): 1-12.
http://dx.doi.org/10.1016/0165-1218(90)90002-J PMID: 2294411

[34]   Naeimi H, Namdari R. Rapid, efficient and one pot synthesis of anthraquinone derivatives catalyzed by Lewis acid/methanesulfonic acid under heterogeneous conditions. Dyes Pigments 2009; 81(3): 259-63.
http://dx.doi.org/10.1016/j.dyepig.2008.10.019

[35]   Petrov M, Chikin D, Abunaeva L, *et al.* Mixture of anthraquinone sulfo-derivatives as an inexpensive organic flow battery negolyte: optimization of battery cell. Membranes (Basel) 2022; 12(10): 912.
http://dx.doi.org/10.3390/membranes12100912 PMID: 36295671

[36]   Mattioli R, Ilari A, Colotti B, Mosca L, Fazi F, Colotti G. Doxorubicin and other anthracyclines in cancers: Activity, chemoresistance and its overcoming. Mol Aspects Med 2023; 93: 101205.
http://dx.doi.org/10.1016/j.mam.2023.101205 PMID: 37515939

[37]   Gu D, Wang H, Li Z, *et al.* Green synthesis of anthraquinone by one-pot method with Ni-modified Hβ Zeolite. Molecular Catalysis 2023; 538: 112969.
http://dx.doi.org/10.1016/j.mcat.2023.112969

[38]   Xu H, Lu Y, Zhang T, *et al.* Characterization of binding interactions of anthraquinones and bovine β-lactoglobulin. Food Chem 2019; 281: 28-35.
http://dx.doi.org/10.1016/j.foodchem.2018.12.077 PMID: 30658758

[39]   Li M, Zhou D, Li Y, Li Q, Geng F, Wu D. Insights into the interaction and the influence of the antioxidant activity between beta-lactoglobulin and two hydroxyanthracene derivatives from aloe 2023.

http://dx.doi.org/10.2139/ssrn.4612616

[40] Kishikawa N, Kuroda N. Analytical techniques for the determination of biologically active quinones in biological and environmental samples. J Pharm Biomed Anal 2014; 87: 261-70.
http://dx.doi.org/10.1016/j.jpba.2013.05.035 PMID: 23791303

[41] Arvindekar AU, Laddha KS. An efficient microwave-assisted extraction of anthraquinones from Rheum emodi: Optimisation using RSM, UV and HPLC analysis and antioxidant studies. Ind Crops Prod 2016; 83: 587-95.
http://dx.doi.org/10.1016/j.indcrop.2015.12.066

[42] Mustafa A, Turner C. Pressurized liquid extraction as a green approach in food and herbal plants extraction: A review. Anal Chim Acta 2011; 703(1): 8-18.
http://dx.doi.org/10.1016/j.aca.2011.07.018 PMID: 21843670

[43] Zhang QW, Lin LG, Ye WC. Techniques for extraction and isolation of natural products: a comprehensive review. Chin Med 2018; 13(1): 20.
http://dx.doi.org/10.1186/s13020-018-0177-x PMID: 29692864

[44] Zhao LC, Liang J, Li W, *et al.* The use of response surface methodology to optimize the ultrasound-assisted extraction of five anthraquinones from Rheum palmatum L. Molecules 2011; 16(7): 5928-37.
http://dx.doi.org/10.3390/molecules16075928 PMID: 21765390

[45] Chemat F, Zill-e-Huma , Khan MK. Applications of ultrasound in food technology: Processing, preservation and extraction. Ultrason Sonochem 2011; 18(4): 813-35.
http://dx.doi.org/10.1016/j.ultsonch.2010.11.023 PMID: 21216174

[46] Vinatoru M. An overview of the ultrasonically assisted extraction of bioactive principles from herbs. Ultrason Sonochem 2001; 8(3): 303-13.
http://dx.doi.org/10.1016/S1350-4177(01)00071-2 PMID: 11441615

[47] Yao JY, Lin L-Y, Yuan X-M, *et al.* Antifungal activity of rhein and aloe-emodin from *Rheum palmatum* on fish pathogenic *Saprolegnia* sp. J World Aquacult Soc 2017; 48(1): 137-44.
http://dx.doi.org/10.1111/jwas.12325

[48] Dai LX, Li JC, Miao XL, *et al.* Ultrasound-assisted extraction of five anthraquinones from Rheum palmatum water extract residues and the antimicrobial activities. Ind Crops Prod 2021; 162: 113288.
http://dx.doi.org/10.1016/j.indcrop.2021.113288

[49] Nortjie E, Basitere M, Moyo D, Nyamukamba P. Extraction methods, quantitative and qualitative phytochemical screening of medicinal plants for antimicrobial textiles: A review. Plants 2022; 11(15): 2011.
http://dx.doi.org/10.3390/plants11152011 PMID: 35956489

[50] Khaw KY, Parat MO, Shaw PN, Falconer JR. Solvent supercritical fluid technologies to extract bioactive compounds from natural sources: A review. Molecules 2017; 22(7): 1186.
http://dx.doi.org/10.3390/molecules22071186 PMID: 28708073

[51] Herrero M, Mendiola JA, Cifuentes A, Ibáñez E. Supercritical fluid extraction: Recent advances and applications. J Chromatogr A 2010; 1217(16): 2495-511.
http://dx.doi.org/10.1016/j.chroma.2009.12.019 PMID: 20022016

[52]   Girotra P, Singh SK, Nagpal K. Supercritical fluid technology: A promising approach in pharmaceutical research. Pharm Dev Technol 2013; 18(1): 22-38.
       http://dx.doi.org/10.3109/10837450.2012.726998 PMID: 23036159

[53]   Arumugham T, K R, Hasan SW, Show PL, Rinklebe J, Banat F. Supercritical carbon dioxide extraction of plant phytochemicals for biological and environmental applications – A review. Chemosphere 2021; 271: 129525.
       http://dx.doi.org/10.1016/j.chemosphere.2020.129525 PMID: 33445028

[54]   Aichner D, Ganzera M. Analysis of anthraquinones in rhubarb (Rheum palmatum and Rheum officinale) by supercritical fluid chromatography. Talanta 2015; 144: 1239-44.
       http://dx.doi.org/10.1016/j.talanta.2015.08.011 PMID: 26452953

[55]   Alwi RS, Garlapati C, Tamura K. Solubility of Anthraquinone derivatives in supercritical carbon dioxide: New correlations. Molecules 2021; 26(2): 460.
       http://dx.doi.org/10.3390/molecules26020460 PMID: 33477249

[56]   Wang Y, Wang S, Liu L. Recovery of natural active molecules using aqueous two-phase systems comprising of ionic liquids/deep eutectic solvents. Green Chemical Engineering 2022; 3(1): 5-14.
       http://dx.doi.org/10.1016/j.gce.2021.07.007

[57]   Tan Z, Li F, Xu X. Isolation and purification of aloe anthraquinones based on an ionic liquid/salt aqueous two-phase system. Separ Purif Tech 2012; 98: 150-7.
       http://dx.doi.org/10.1016/j.seppur.2012.06.021

[58]   Deng WW, Zong Y, Xiao YX. Hexafluoroisopropanol-based deep eutectic solvent/salt aqueous two-phase systems for extraction of anthraquinones from rhei radix et rhizoma samples. ACS Sustain Chem& Eng 2017; 5(5): 4267-75.
       http://dx.doi.org/10.1021/acssuschemeng.7b00282

[59]   Wang J, Feng J, Xu L, *et al.* Ionic liquid-based salt-induced liquid-liquid extraction of polyphenols and anthraquinones in Polygonum cuspidatum. J Pharm Biomed Anal 2019; 163: 95-104.
       http://dx.doi.org/10.1016/j.jpba.2018.09.050 PMID: 30286440

[60]   Sun Q, Du B, Wang C, *et al.* Ultrasound-assisted ionic liquid solid–liquid extraction coupled with aqueous two-phase extraction of naphthoquinone Pigments in arnebia euchroma (Royle) Johnst. Chromatographia 2019; 82(12): 1777-89.
       http://dx.doi.org/10.1007/s10337-019-03804-y

[61]   Buarque F, Gautério G, Coelho M, Lemes A, Ribeiro B. Aqueous two-phase systems based on ionic liquids and deep eutectic solvents as a tool for the recovery of non-protein bioactive compounds—A Review. Processes (Basel) 2022; 11(1): 31.
       http://dx.doi.org/10.3390/pr11010031

[62]   Kalyniukova A, Holuša J, Musiolek D, Sedlakova-Kadukova J, Płotka-Wasylka J, Andruch V. Application of deep eutectic solvents for separation and determination of bioactive compounds in medicinal plants. Ind Crops Prod 2021; 172: 114047.
       http://dx.doi.org/10.1016/j.indcrop.2021.114047

[63]   Plastiras OE, Samanidou V. Applications of deep eutectic solvents in sample preparation and extraction of organic molecules. Molecules 2022; 27(22): 7699.

http://dx.doi.org/10.3390/molecules27227699 PMID: 36431799

[64]   Serna-Vázquez J, Ahmad MZ, Boczkaj G, Castro-Muñoz R. Latest insights on novel deep eutectic solvents (DES) for sustainable extraction of phenolic compounds from natural sources. Molecules 2021; 26(16): 5037.
       http://dx.doi.org/10.3390/molecules26165037 PMID: 34443623

[65]   Aduloju EI, Yahaya N, Mohammad Zain N, Anuar Kamaruddin M, Ariffuddin Abd Hamid M. An overview on the use of DEEP eutectic solvents for green extraction of some selected bioactive compounds from natural matrices. Adv J Chem 2023.
       http://dx.doi.org/10.22034/ajca.2023.389403.1356

[66]   Dheyab AS, Abu Bakar MF, AlOmar M, Sabran SF, Muhamad Hanafi AF, Mohamad A. Deep eutectic solvents (DESs) as green extraction media of beneficial bioactive phytochemicals. Separations 2021; 8(10): 176.
       http://dx.doi.org/10.3390/separations8100176

[67]   Zhang M, Zhang Z, Gul Z, *et al.* Advances of responsive deep eutectic solvents and application in extraction and separation of bioactive compounds. J Sep Sci 2023; 46(15): 2300098.
       http://dx.doi.org/10.1002/jssc.202300098 PMID: 37246933

[68]   Pongnaravane B, Goto M, Sasaki M, Anekpankul T, Pavasant P, Shotipruk A. Extraction of anthraquinones from roots of Morinda citrifolia by pressurized hot water: Antioxidant activity of extracts. J Supercrit Fluids 2006; 37(3): 390-6.
       http://dx.doi.org/10.1016/j.supflu.2005.12.013

[69]   Shotipruk A, Kiatsongserm J, Pavasant P, Goto M, Sasaki M. Pressurized hot water extraction of anthraquinones from the roots of Morinda citrifolia. Biotechnol Prog 2004; 20(6): 1872-5.
       http://dx.doi.org/10.1021/bp049779x PMID: 15575725

[70]   Vázquez MFB, Comini LR, Milanesio JM, *et al.* Pressurized hot water extraction of anthraquinones from Heterophyllaea pustulata Hook f. (Rubiaceae). J Supercrit Fluids 2015; 101: 170-5.
       http://dx.doi.org/10.1016/j.supflu.2015.02.029

[71]   Machatová Z, Barbieriková Z, Poliak P, Jančovičová V, Lukeš V, Brezová V. Study of natural anthraquinone colorants by EPR and UV/vis spectroscopy. Dyes Pigments 2016; 132: 79-93.
       http://dx.doi.org/10.1016/j.dyepig.2016.04.046

[72]   Tissier RC, Rigaud B, Thureau P, Huix-Rotllant M, Jaber M, Ferré N. Stressing the differences in alizarin and purpurin dyes through UV-visible light absorption and [1] H-NMR spectroscopies. Phys Chem Chem Phys 2022; 24(32): 19452-62.
       http://dx.doi.org/10.1039/D2CP00520D PMID: 35924547

[73]   Anouar EH, Osman CP, Weber JFF, Ismail NH. UV/Visible spectra of a series of natural and synthesised anthraquinones: experimental and quantum chemical approaches. Springerplus 2014; 3(1): 233.
       http://dx.doi.org/10.1186/2193-1801-3-233 PMID: 24851199

[74] Tolkou AK, Mitropoulos AC, Kyzas GZ. Removal of anthraquinone dye from wastewaters by hybrid modified activated carbons. Environ Sci Pollut Res Int 2023; 30(29): 73688-701.
http://dx.doi.org/10.1007/s11356-023-27550-9 PMID: 37195607

[75] Mohanty SS, Kumar A. Enhanced degradation of anthraquinone dyes by microbial monoculture and developed consortium through the production of specific enzymes. Sci Rep 2021; 11(1): 7678.
http://dx.doi.org/10.1038/s41598-021-87227-6 PMID: 33828207

[76] Guin PS, Mandal PC, Das S. The Binding of a Hydroxy-9,10-anthraquinone Cu $^{II}$ complex to calf thymus DNA: Electrochemistry and UV/V is spectroscopy. ChemPlusChem 2012; 77(5): 361-9.
http://dx.doi.org/10.1002/cplu.201100046

[77] Valarmathi T, Premkumar R, James Jebaseelan Samuel E, Benial AMF. Spectroscopic characterization, quantum chemical and molecular docking studies on 1-chloroanthraquinone: a novel oral squamous cell carcinoma drug. Polycycl Aromat Compd 2023; 44(3): 1816-34.
http://dx.doi.org/10.1080/10406638.2023.2209249

[78] Current achievement and future potential of fluorescence Spectroscopy. Macro Nano Spectrosc. InTech 2012.
http://dx.doi.org/10.5772/48034

[79] Gao J, Guo Y, Wang J, *et al.* Spectroscopic analysis of the interactions of anthraquinone derivatives (Alizarin, Alizarin-DA and Alizarin-DA-Fe) with bovine serum albumin (BSA). J Solution Chem 2011; 40(5): 876-88.
http://dx.doi.org/10.1007/s10953-011-9692-4

[80] Bi S, Song D, Kan Y, *et al.* Spectroscopic characterization of effective components anthraquinones in Chinese medicinal herbs binding with serum albumins. Spectrochim Acta A Mol Biomol Spectrosc 2005; 62(1-3): 203-12.
http://dx.doi.org/10.1016/j.saa.2004.12.049 PMID: 16257715

[81] Fu Z, Cui Y, Cui F, Zhang G. Modeling techniques and fluorescence imaging investigation of the interactions of an anthraquinone derivative with HSA and ctDNA. Spectrochim Acta A Mol Biomol Spectrosc 2016; 153: 572-9.
http://dx.doi.org/10.1016/j.saa.2015.09.011 PMID: 26436845

[82] Junaedi EC, Lestari K, Muchtaridi M. Infrared spectroscopy technique for quantification of compounds in plant-based medicine and supplement. J Adv Pharm Technol Res 2021; 12(1): 1-7.
http://dx.doi.org/10.4103/japtr.JAPTR_96_20 PMID: 33532347

[83] Gribov LA, Zubkova OB, Sigarev AA. Theoretical analysis of infrared spectrum of 9,10-anthraquinone molecule. J Struct Chem 1993; 34(1): 147-54.
http://dx.doi.org/10.1007/BF00745414

[84] Chumbalov TK, Chanysheva IS, Muzychkina RA. UV and IR spectra of anthraquinone and chrysophanol derivatives. J Appl Spectrosc 1967; 6(6): 570-4.
http://dx.doi.org/10.1007/BF00606188

[85] Zhan H, Fang J, Wu H, *et al.* Rapid Determination of total content of five major anthraquinones in rhei radix et rhizoma by NIR spectroscopy. Chin Herb Med 2017; 9(3): 250-7.
http://dx.doi.org/10.1016/S1674-6384(17)60101-1

[86] Shahid M, Wertz J, Degano I, Aceto M, Khan MI, Quye A. Analytical methods for determination of anthraquinone dyes in historical textiles: A review. Anal Chim Acta 2019; 1083: 58-87.
http://dx.doi.org/10.1016/j.aca.2019.07.009 PMID: 31493810

[87] Li M, Feng Y, Yu Y, *et al.* Quantitative analysis of polycyclic aromatic hydrocarbons in soil by infrared spectroscopy combined with hybrid variable selection strategy and partial least squares. Spectrochim Acta A Mol Biomol Spectrosc 2021; 257: 119771.
http://dx.doi.org/10.1016/j.saa.2021.119771 PMID: 33853000

[88] Wcisło A, Niedziałkowski P, Wnuk E, Zarzeczańska D, Ossowski T. Influence of different amino substituents in position 1 and 4 on spectroscopic and acid base properties of 9,10-anthraquinone moiety. Spectrochim Acta A Mol Biomol Spectrosc 2013; 108: 82-8.
http://dx.doi.org/10.1016/j.saa.2013.01.085 PMID: 23466318

[89] Maļeckis A, Cvetinska M, Kirjušina M, *et al.* A comparative study of new fluorescent anthraquinone and benzanthrone α-aminophosphonates: Synthesis, spectroscopy, toxicology, X-ray crystallography, and microscopy of *Opisthorchis felineus.* Molecules 2024; 29(5): 1143.
http://dx.doi.org/10.3390/molecules29051143 PMID: 38474655

[90] Dong JW, Cai L, Fang YS, Duan WH, Li ZJ, Ding ZT. Simultaneous, simple and rapid determination of five bioactive free anthraquinones in radix et rhizoma rhei by quantitative [1] H NMR. J Braz Chem Soc 2016.
http://dx.doi.org/10.5935/0103-5053.20160103

[91] Kalidhar SB. Structural elucidation in anthraquinones using 1H NMR glycosylation and alkylation shifts. Phytochemistry 1989; 28(12): 3459-63.
http://dx.doi.org/10.1016/0031-9422(89)80364-4

[92] Arnone A, Fronza G, Mondelli R, Pyrek JS. 13C NMR analysis of anthraquinones as models for anthracycline Antibiotics. Journal of Magnetic Resonance (1969) 1977; 28(1): 69-79.
http://dx.doi.org/10.1016/0022-2364(77)90257-8

[93] Xia XK, Huang HR, She ZG, *et al.* [1] H and [13] C NMR assignments for five anthraquinones from the mangrove endophytic fungus *Halorosellinia* sp. (No. 1403). Magn Reson Chem 2007; 45(11): 1006-9.
http://dx.doi.org/10.1002/mrc.2078 PMID: 17894425

[94] Luo P, Su J, Zhu Y, Wei J, Wei W, Pan W. A new anthraquinone and eight constituents from *Hedyotis caudatifolia* Merr. et Metcalf: isolation, purification and structural identification. Nat Prod Res 2016; 30(19): 2190-6.
http://dx.doi.org/10.1080/14786419.2016.1160231 PMID: 27027701

[95] Di Tullio V, Doherty B, Capitani D, *et al.* NMR spectroscopy and micro-analytical techniques for studying the constitutive materials and the state of conservation of an ancient Tapa barkcloth from Polynesia, is. Wallis. J Cult Herit 2020; 45: 379-88.

http://dx.doi.org/10.1016/j.culher.2020.02.009

[96]    Celik S, Ozkok F, Ozel AE, *et al.* Synthesis, FT-IR and NMR characterization, antimicrobial activity, cytotoxicity and DNA docking analysis of a new anthraquinone derivate compound. J Biomol Struct Dyn 2020; 38(3): 756-70.
http://dx.doi.org/10.1080/07391102.2019.1587513 PMID: 30890106

[97]    El-Gogary TM. Molecular complexes of some anthraquinone anti-cancer drugs: experimental and computational study. Spectrochim Acta A Mol Biomol Spectrosc 2003; 59(5): 1009-15.
http://dx.doi.org/10.1016/S1386-1425(02)00283-4 PMID: 12633717

[98]    Al-Otaibi JS, Teesdale Spittle P, El Gogary TM. Interaction of anthraquinone anti-cancer drugs with DNA: Experimental and computational quantum chemical study. J Mol Struct 2017; 1127: 751-60.
http://dx.doi.org/10.1016/j.molstruc.2016.08.007

[99]    Köseoğlu Yılmaz P, Kolak U. Development and validation of a SPE–HPLC method for quantification of rhein, emodin, chrysophanol and physcion in *Rhamnus petiolaris* Boiss. & Balansa. J Chromatogr Sci 2024; 62(9): 872-77.
http://dx.doi.org/10.1093/chromsci/bmad053 PMID: 37501520

[100]   Luo H, Qin W, Zhang H, *et al.* Anthraquinones from the aerial parts of *Rubia cordifolia* with their NO inhibitory and antibacterial activities. Molecules 2022; 27(5): 1730.
http://dx.doi.org/10.3390/molecules27051730 PMID: 35268830

[101]   Kishikawa N, El-Maghrabey M, Kawamoto A, Ohyama K, Kuroda N. Determination of anthraquinone-tagged amines using high-performance liquid chromatography with online uv irradiation and luminol chemiluminescence detection. Molecules 2023; 28(5): 2146.
http://dx.doi.org/10.3390/molecules28052146 PMID: 36903390

[102]   Furuya T, Shibata S, Iizuka H. Gas—liquid chromatography of anthraquinones. J Chromatogr A 1966; 21(1): 116-8.
http://dx.doi.org/10.1016/S0021-9673(01)91268-3 PMID: 5940122

[103]   Pitoi MM, Ariyani M, Koesmawati TA, Yusiasih R. Preliminary study for 9,10-anthraquinone residue analysis in tea-based functional beverage: GC-ECD optimization and method development. IOP Conf Ser Earth Environ Sci 2019; 277(1): 012020.
http://dx.doi.org/10.1088/1755-1315/277/1/012020

[104]   Terrill JB, Jacobs ES. Application of Gas-Liquid Chromatography to the analysis of anthraquinone dyes and intermediates. J Chromatogr Sci 1970; 8(10): 604-7.
http://dx.doi.org/10.1093/chromsci/8.10.604

[105]   Egan JM, Rickenbach M, Mooney KE, Palenik CS, Golombeck R, Mueller KT. Bank security dye packs: synthesis, isolation, and characterization of chlorinated products of bleached 1-(methylamino)anthraquinone. J Forensic Sci 2006; 51(6): 1276-83.
http://dx.doi.org/10.1111/j.1556-4029.2006.00295.x PMID: 17199613

[106]   Kučera L, Kurka O, Golec M, Bednář P. Study of Tetrahydroxylated anthraquinones—potential tool to assess degradation of anthocyanins Rich Food. Molecules 2020; 26(1): 2.
http://dx.doi.org/10.3390/molecules26010002 PMID: 33374941

[107]  Silberstein KE, Pastore JP, Zhou W, *et al.* Electrochemical lithiation-induced polymorphism of anthraquinone derivatives observed by operando X-ray diffraction. Phys Chem Chem Phys 2015; 17(41): 27665-71.
http://dx.doi.org/10.1039/C5CP04201A PMID: 26427626

[108]  Hu T, Wang Y, Dong M, *et al.* Systematical investigation of chain length effect on the melting point of a series of bifunctional anthraquinone derivatives *via* X-ray diffraction and scanning tunneling microscopy. J Phys Chem C 2020; 124(2): 1646-54.
http://dx.doi.org/10.1021/acs.jpcc.9b08710

[109]  Xie W, Huang C, Hong D, Hou J, Deng X, Han C. Determination of anthraquinone in tea by stable isotope dilution assay-gas chromatography-tandem mass spectrometry. SN Applied Sciences 2020; 2(6): 1104.
http://dx.doi.org/10.1007/s42452-020-2919-5

[110]  Dai H, Chen Z, Shang B, Chen Q. Identification and quantification of four anthraquinones in rhubarb and its preparations by gas chromatography–Mass spectrometry. J Chromatogr Sci 2018; 56(3): 195-201.
http://dx.doi.org/10.1093/chromsci/bmx103 PMID: 29206919

[111]  Wei S, Yao W, Ji W, Wei J, Peng S. Qualitative and quantitative analysis of anthraquinones in rhubarbs by high performance liquid chromatography with diode array detector and mass spectrometry. Food Chem 2013; 141(3): 1710-5.
http://dx.doi.org/10.1016/j.foodchem.2013.04.074 PMID: 23870882

[112]  Díaz-Galiano FJ, Murcia-Morales M, Gómez-Ramos MM, Ferrer C, Fernández-Alba AR. Presence of anthraquinone in coffee and tea samples. An improved methodology based on mass spectrometry and a pilot monitoring programme. Anal Methods 2021; 13(1): 99-109.
http://dx.doi.org/10.1039/D0AY01962C PMID: 33305763

[113]  Yang W, Su Y, Dong G, *et al.* Liquid chromatography–mass spectrometry-based metabolomics analysis of flavonoids and anthraquinones in Fagopyrum tataricum L. Gaertn. (tartary buckwheat) seeds to trace morphological variations. Food Chem 2020; 331: 127354.
http://dx.doi.org/10.1016/j.foodchem.2020.127354 PMID: 32569973

[114]  Sousa ET, Cardoso MP, Silva LA, de Andrade JB. Direct determination of quinones in fine atmospheric particulate matter by GC–MS. Microchem J 2015; 118: 26-31.
http://dx.doi.org/10.1016/j.microc.2014.07.013

[115]  Ling C, Shi Q, Wei Z, Zhang J, Hu J, Pei J. Rapid analysis of quinones in complex matrices by derivatization-based wooden-tip electrospray ionization mass spectrometry. Talanta 2022; 237: 122912.
http://dx.doi.org/10.1016/j.talanta.2021.122912 PMID: 34736649

[116]  Feilcke R, Arnouk G, Raphane B, *et al.* Biological activity and stability analyses of knipholone anthrone, a phenyl anthraquinone derivative isolated from Kniphofia foliosa Hochst. J Pharm Biomed Anal 2019; 174: 277-85.
http://dx.doi.org/10.1016/j.jpba.2019.05.065 PMID: 31185339

[117]  Gong YX, Li SP, Wang YT, Li P, Yang FQ. Simultaneous determination of anthraquinones in Rhubarb by pressurized liquid extraction and capillary zone electrophoresis. Electrophoresis 2005; 26(9): 1778-82.

http://dx.doi.org/10.1002/elps.200400001 PMID: 15800969

[118] Ahmadi S, Absalan G, Craig D, Goltz D. Photochemical properties of purpurin and its implications for capillary electrophoresis with laser induced fluorescence detection. Dyes Pigments 2014; 105: 57-62.
http://dx.doi.org/10.1016/j.dyepig.2013.12.011

[119] Ali I, Alharbi OML, Marsin Sanagi M. Nano-capillary electrophoresis for environmental analysis. Environ Chem Lett 2016; 14(1): 79-98.
http://dx.doi.org/10.1007/s10311-015-0547-x PMID: 32214934

[120] Koyama J, Morita I, Tagahara K, Bakari J, Aqil M. Capillary electrophoresis of anthraquinones from Cassia siamea. Chem Pharm Bull (Tokyo) 2002; 50(8): 1103-5.
http://dx.doi.org/10.1248/cpb.50.1103 PMID: 12192145

[121] Zhu J, Liu J, Fan Y, *et al.* SERS detection of anthraquinone dyes: Using solvothermal silver colloid as the substrate. Spectrochim Acta A Mol Biomol Spectrosc 2022; 282: 121646.
http://dx.doi.org/10.1016/j.saa.2022.121646 PMID: 35926284

[122] Rando G, Sfameni S, Galletta M, Drommi D, Cappello S, Plutino MR. Functional nanohybrids and nanocomposites development for the removal of environmental pollutants and Bioremediation. Molecules 2022; 27(15): 4856.
http://dx.doi.org/10.3390/molecules27154856 PMID: 35956804

[123] Mousavi SM, Hashemi SA, Yari Kalashgrani M, *et al.* The pivotal role of quantum dots-based biomarkers integrated with ultra-sensitive probes for multiplex detection of human viral infections. Pharmaceuticals (Basel) 2022; 15(7): 880.
http://dx.doi.org/10.3390/ph15070880 PMID: 35890178

[124] Khataee AR, Zarei M, Fathinia M, Jafari MK. Photocatalytic degradation of an anthraquinone dye on immobilized TiO2 nanoparticles in a rectangular reactor: Destruction pathway and response surface approach. Desalination 2011; 268(1-3): 126-33.
http://dx.doi.org/10.1016/j.desal.2010.10.008

[125] Paramasivam G, Palem VV, Sundaram T, Sundaram V, Kishore SC, Bellucci S. Nanomaterials: Synthesis and applications in theranostics. Nanomaterials (Basel) 2021; 11(12): 3228.
http://dx.doi.org/10.3390/nano11123228 PMID: 34947577

[126] Sajini T, Mathew B. A brief overview of molecularly imprinted polymers: Highlighting computational design, nano and photo-responsive imprinting. Talanta Open 2021; 4: 100072.
http://dx.doi.org/10.1016/j.talo.2021.100072

[127] Ramin NA, Asman S, Ramachandran MR, Saleh NM, Mat Ali ZM. Magnetic nanoparticles molecularly imprinted polymers: A review. Curr Nanosci 2023; 19(3): 372-400.
http://dx.doi.org/10.2174/1573413718666220727111319

[128] Liu S, Zhang J, Sun T, *et al.* Synthesis and application of molecularly imprinted polymers for preferential removal of emodin and physcion from Polygonum multiflorum stem extract. Ind Crops Prod 2022; 178: 114659.
http://dx.doi.org/10.1016/j.indcrop.2022.114659

[129] Vega EN, Ciudad-Mulero M, Fernández-Ruiz V, Barros L, Morales P. Natural sources of food colorants as potential substitutes for artificial additives. Foods 2023; 12(22): 4102.

http://dx.doi.org/10.3390/foods12224102 PMID: 38002160

[130]   Soliman IA, Hasanien YA, Zaki AG, Shawky HA, Nassrallah AA. Irradiation impact on biological activities of anthraquinone pigment produced from talaromyces purpureogenus and its evaluation, characterization and application in beef burger as natural preservative. BMC Microbiol 2022; 22(1): 325.

http://dx.doi.org/10.1186/s12866-022-02734-4 PMID: 36581795

[131]   Fouillaud M. Mekala Venkatachalam, Yanis Caro, Dufossé L. Marine-derived fungi producing red anthraquinones: new resources for natural colors?2016.

http://dx.doi.org/10.13140/RG.2.2.27618.53446

[132]   Dulo B, Phan K, Githaiga J, Raes K, De Meester S. Natural Quinone Dyes: A review on structure, extraction techniques, analysis and application potential. Waste Biomass Valoriz 2021; 12(12): 6339-74.

http://dx.doi.org/10.1007/s12649-021-01443-9

[133]   Shafiq N, Zareen G, Arshad U, *et al.* A mini review on the chemical and bio-medicinal aspects along with energy storage applications of anthraquinone and its analogues. Mini Rev Org Chem 2024; 21(2): 134-50.

http://dx.doi.org/10.2174/1570193X19666220512141411

[134]   Trung NQ, Thong NM, Cuong DH, *et al.* Radical scavenging activity of natural anthraquinones: a Theoretical insight. ACS Omega 2021; 6(20): 13391-7.

http://dx.doi.org/10.1021/acsomega.1c01448 PMID: 34056486

[135]   Zhao L, Zheng L. A Review on Bioactive Anthraquinone and derivatives as the regulators for ROS. Molecules 2023; 28(24): 8139.

http://dx.doi.org/10.3390/molecules28248139 PMID: 38138627

[136]   Li C, Yang X, Chen R, *et al.* Anthraquinone dyes as photosensitizers for dye-sensitized solar cells. Sol Energy Mater Sol Cells 2007; 91(19): 1863-71.

http://dx.doi.org/10.1016/j.solmat.2007.07.002

[137]   Park G, Jeong H, Lee W, Han JW, Chang DR, Kwon Y. Scaled-up aqueous redox flow battery using anthraquinone negalyte and vanadium posilyte with inorganic additive. Appl Energy 2024; 353: 122171.

http://dx.doi.org/10.1016/j.apenergy.2023.122171

# CHAPTER 2

# An Overview of Chemistry and Biosynthesis of Anthraquinones

**Pooja Sharma**[1,*] and **Amrit Kaur**[1]

[1]*Department of Chemistry, Guru Nanak Dev University, Amritsar, Punjab-143005, India*

**Abstract:** Anthraquinones are a class of secondary metabolites, have garnered significant interest due to their diverse biological activities and various industrial applications. The derivatives of anthraquinones are widely distributed in nature, being found in numerous plants, fungi, and bacteria. The biosynthetic pathways leading to anthraquinones differ among various organisms, yet common underlying mechanisms can be observed. Enzymatic reactions play a pivotal role in the functionalization and diversification of anthraquinones. Cytochrome P450 monooxygenases, glycosyltransferases, and acyltransferases are key enzymes involved in modifying the basic anthraquinone skeleton, leading to a wide array of structurally distinct derivatives. Moreover, advances in genomic and proteomic technologies have facilitated the discovery of genes and enzymes responsible for anthraquinone biosynthesis. Genetic engineering and synthetic biology approaches have enabled the manipulation of biosynthetic pathways, paving the way for the production of novel anthraquinones with engineering and synthetic biology approaches have enabled the manipulation of biosynthetic pathways, paving the way for the production of novel anthraquinones with enhanced bioactivity and potential applications in pharmaceuticals, agrochemicals, and the dye industry. In the present work, we will focus on the different biosynthetic pathways for the biosynthesis of anthraquinones.

**Keywords:** Anthracene, Aromatic fused ring, Antioxidant properties, Dye industry, 9,10-anthraquinone, Mevalonate (MVA), Methyl erythritol phosphate (MEP) quinone, Polyketone pathway, Polygonaceae, Plant pigment, Polyketide biosynthesis, Shikimate pathway, Tricarboxylic acid (TCA).

## INTRODUCTION

Anthraquinone is a prominent organic compound, known for its vibrant colors and diverse activities, which has captivated the attention of chemists, scientists, and industries for centuries [1]. Its name is derived from "anthracene", a tricyclic aromatic hydrocarbon, and "quinone", referring to the presence of a carbonyl group

* **Corresponding author Pooja Sharma:** Department of Chemistry, Guru Nanak Dev University, Amritsar, Punjab-143005, India; E-mail: poojamukerian92@gmail.com

**Pardeep Kaur, Ajay Kumar, Robin, Tarunpreet Singh Thind & Kamaljit Kaur (Eds.)**

(C=O) in its chemical structure. Anthraquinone is characterized by a bicyclic core structure composed of three fused benzene rings with two ketone functional groups at adjacent carbon positions [2]. They are widely distributed in nature and play significant roles in various physiological activities. Along with their medicinal properties, natural anthraquinones are seeking attraction as an alternative to synthetic dyes which harm aquatic ecosystems [3, 4]. Anthraquinones are widely distributed in nature and can be found in various plants, fungi, and certain types of bacteria including Madder Root, Cinchona Bark, Rhubarb, and fungi [5, 6]. These are also found in various food sources of humans like cabbage, and beans, which provide around 0.04 to 36 mg of anthraquinone. The natural occurrence of anthraquinones highlights their significance in ecological and pharmacological contexts, as well as their potential as sources of bioactive compounds [7]. The chemical reactivity of anthraquinones is diverse, owing to their conjugated and electron-rich structure. They readily participate in a variety of chemical reactions (substitution, redox, cycloaddition, acid-base reactions, *etc.*), making them versatile building blocks for organic synthesis [8].

The versatility of anthraquinones extends to a wide range of applications across various fields, including chemistry (synthetic chemistry, analytical chemistry, and environmental chemistry), pharmaceuticals, dyes, biological research, and many more (Fig. **1**). Anthraquinones have garnered significant attention due to their diverse pharmacological properties, including anticancer, anti-inflammatory, antibacterial, and antioxidant effects [9, 10]. Anthraquinones are commonly found in many different organisms, ranging from bacteria and fungi to plants and some animals. In plants, anthraquinones are found in a wide range of species, especially in the families; *Rubiaceae, Polygonaceae, and Rhamnaceae* [11-13]. Understanding the biosynthetic pathways of anthraquinones is crucial for both natural product synthesis and biotechnological production. Anthraquinone biosynthesis primarily occurs in plants, fungi, and certain bacteria. The biosynthetic pathways are complex and involve multiple enzymatic reactions. In plants, the biosynthesis of anthraquinones typically begins with the shikimate pathway, a central metabolic route responsible for the synthesis of aromatic compounds. This pathway usually starts with the conversion of phosphoenolpyruvate and erythrose-4-phosphate into shikimate. The second pathway is chorismic acid conversion in which chorismic acid serves as a precursor for the formation of various aromatic compounds. In the context of anthraquinone biosynthesis, it undergoes a series of reactions involving enzymes like isochorismate synthase and isochorismate-pyruvate lyase to produce isochorismic acid, which is then converted into intermediates like 1,2-dihydroxyanthraquinone, which subsequently undergoes oxidation and cyclization reactions catalyzed by various enzymes. Once the anthraquinone skeleton is formed, various tailoring enzymes may further modify the structure [14-17]. These enzymes can introduce functional groups like hydroxyl, methyl, or glycosyl groups at specific positions on the anthraquinone ring system, yielding a wide range of anthraquinone derivatives with distinct properties. After biosynthesis, anthraquinones are often transported to specific cellular compartments, such as vacuoles, where they accumulate [18-20]. This compartmentalization helps prevent cellular damage from these often toxic compounds and also facilitates their storage for various purposes. In this context, this chapter provides an overview of the different biosynthetic pathways for the synthesis of anthraquinones.

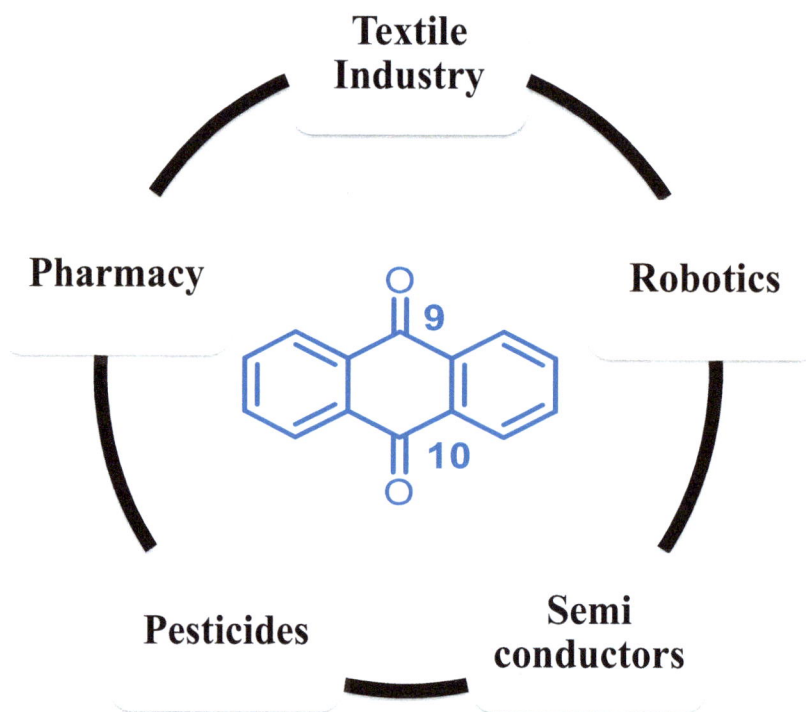

**Fig. (1).** Applications of anthraquinones in different fields.

## GENERAL STRUCTURE AND CHEMISTRY

Anthraquinones are a class of organic compounds known as quinones and are characterized by a cyclic conjugated structure with alternating double bonds and oxygen atoms. The core structure of anthraquinone consists of three fused benzene rings, forming a tricyclic aromatic system [21]. The aromaticity of anthraquinone is a crucial aspect of its chemistry. The general structure of anthraquinones can be represented by the molecular formula $C_{14}H_8O_2$, which corresponds to two fused benzene rings with an additional carbonyl group (C=O) attached to one of the rings. The carbonyl group is typically located at positions 9 and 10 of the anthraquinone ring system [22, 23]. This structural arrangement imparts unique chemical properties to anthraquinones, making them suitable for various reactions and applications. Chemically, anthraquinones can exist in different forms, depending on the substitution pattern of the benzene rings. Some common substitutions include hydroxyl (OH), methyl ($CH_3$), and methoxy ($OCH_3$) groups, among others [24]. These substitutions can greatly influence the chemical reactivity and biological activity of anthraquinones. Due to its three fused benzene rings, it exhibits a high degree of resonance stabilization, making it highly stable. The presence of alternating single and double bonds throughout the aromatic system contributes to its aromatic character [25].

Further, anthraquinones are known for their vibrant and diverse coloration. Many of these compounds are intensely colored, ranging from yellow and orange to red and purple. This property makes them valuable as natural dyes, and they have been used for centuries to color textiles, paints, and cosmetics. The chemistry of anthraquinones is characterized by their redox activity [26]. They can undergo both oxidation and reduction reactions, making them versatile in various chemical processes. Anthraquinones can be reduced to anthrahydroquinones by accepting electrons and protons, and they can be oxidized to quinones by donating electrons and protons (Fig. **2**).

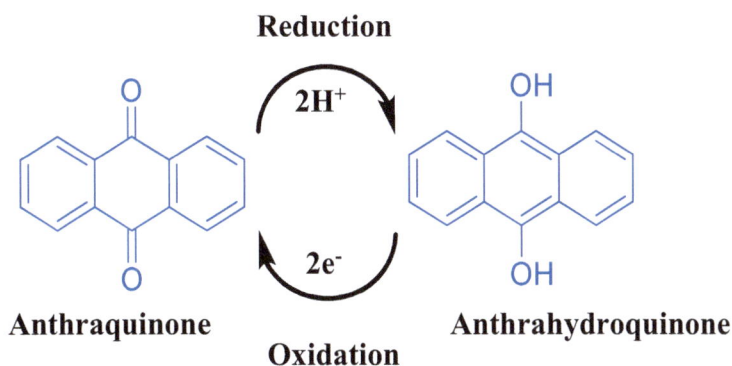

**Fig. (2).** Oxidized and reduced form of 9,10-anthraquinone.

These redox reactions are essential in the synthesis of various organic compounds and play a crucial role in biological processes such as respiration and photosynthesis. It can undergo reversible two-electron redox reactions, transitioning between the quinone (oxidized) and hydroquinone (reduced) forms. These redox reactions are essential in various applications, including energy storage systems like redox flow batteries. In nature, anthraquinones are widely distributed and can be found in various plants and organisms [27]. Some notable natural sources of anthraquinones include the roots of plants like madder (*Rubia tinctorum*), the bark of the cinchona tree (*Cinchona officinalis*), and certain species of fungi, such as the well-known medicinal fungus, Rhubarb (*Rheum* spp.), is another source of anthraquinones and has been used for its laxative properties for centuries. Additionally, some bacteria produce anthraquinones as secondary metabolites. Their natural occurrence in plants and microorganisms, as well as their versatility in synthetic chemistry, has led to their use in dyes, medicines, and various industrial processes [28, 29]. Understanding the general structure and chemistry of anthraquinones provides insight into their importance and potential in both the scientific and industrial worlds [30].

## Biosynthesis of Anthraquinone (AQs)

The chemical structure of AQs consists of an anthracene ring with keto groups at positions 9 and 10 and side group functionalised with various substituents such as -OH, -CH$_3$, -OCH$_3$, -CH$_2$OH, -CHO, -COOH, and others (Fig. 3). Anthraquinones are biosynthesised by plants belonging to families such as *Rubiaceae* (*Cinchona*, *Morinda*), *Fabaceae* or *Rhamnaceae* [31].

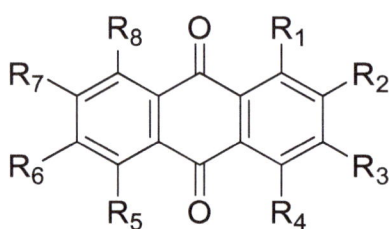

1. $R_1 = R_2 = R_3 = R_4 = R_5 = R_6 = H$

2. $R_1 = R_2 = R_3 = OH, R_4 = R_5 = R_6 = H$

3. $R_1 = OH, R_2 = CH_3, R_3 = R_4 = R_5 = R_6 = H$

4. $R_1 = R_2 = OH, R_3 = R_4 = R_5 = R_6 = H$

**Fig. (3).** The chemical structure of Anthraquinone **1.** Unsubstituted Anthraquinone, **2.** Pseudopurpurin, **3.** Quinzarin, **4.** Munjistin.

The literature reports suggest that the AQs can be synthesized by two pathways: 1) the polyketone pathway, and 2) the shikimate pathway (SA).

## Polyketone Pathway

As shown in Fig. (**4**), this pathway involves a series of condensation reactions using the substrates acetyl and malonyl CoA to extend the carbon chain. This reaction is catalysed by enzyme the chalcone synthase (CHS) resulting in the formation of an octa-ketide molecule that undergoes a series of different processes, such as cyclization, reduction, aldolization, dehydration, enolization, decarboxylation, oxidation, methylation, glycosylation, and radical coupling. These reactions finally, lead to the generation of anthraquinone and its derivatives such as emodin, aloe-emodin, aurantio-obtusin, obtusifolin, and rhein [32, 33].

## Shikimate (SA) Pathway

The various metabolites of the tricarboxylic acid (TCA), mevalonate (MVA), and methyl erythritol phosphate (MEP) pathways are used in plants to synthesize anthraquinones through the SA pathway [34-36]. The SA pathway is made up of the following three components.

The first component involves the generation of 1,4-dihyroxy-2-napthoic acid *via* aldol condensation between phosphoenol pyruvate (PEP) and erythrose-4-phosphate (E4P) catalysed by 3-deoxy-7-phosphoheptulonate synthase (DAHPS) [37]. The generated DAHPS is then dephosphorylated and cyclised to 3-dehydroquinate (DHQ) by enzyme 3-dehydroquinate synthase (DHQS). Further, the DHQ is dehydrated under the catalytic action of 3-dehydroquinate dehydratase (DHQD) to generate 3-dehydroshikimic acid (DHS). Next, in the presence of enzyme shikimate kinase, SA generates shikimic acid-3-Phosphate (S3P), followed by a condensation reaction between S3P and PEP, catalysed by 3-phosphoshikimate 1-carboxyvinyltransferase (EPSPS) leading to the formation of 5-enolpyruvylshikimate-3-phosphate (EPSP). EPSP is then dephosphorylated under the action of chorismate synthase (CS) generating chorismic acid (CHA). The generated CHA is converted into isochorismate (IC) under the catalytic action of isochorismate synthase (ICS) [38]. The next step of this module involves the reaction between IC generated in the previous step and $\alpha$-ketoglutarate generated in the TCA cycle, to generate O-succinyl benzoate (OSB) along with the liberation of $CO_2$ and PEP, in the presence of enzyme O-succinyl benzoate synthase (OSBS). Furthermore, the succinyl side chain of OSB is activated to O-succinyl benzoyl-CoA (OSB-CoA) under the action of O-succinyl benzoate-CoA ligase (MenE). The so-formed OSB-CoA is then cyclized to1,4-dihydroxy-2-naphthoyl-CoA (DHNA-CoA) by the action of 1,4-dihydroxy-2- naphthoyl-CoA synthase (MenB).

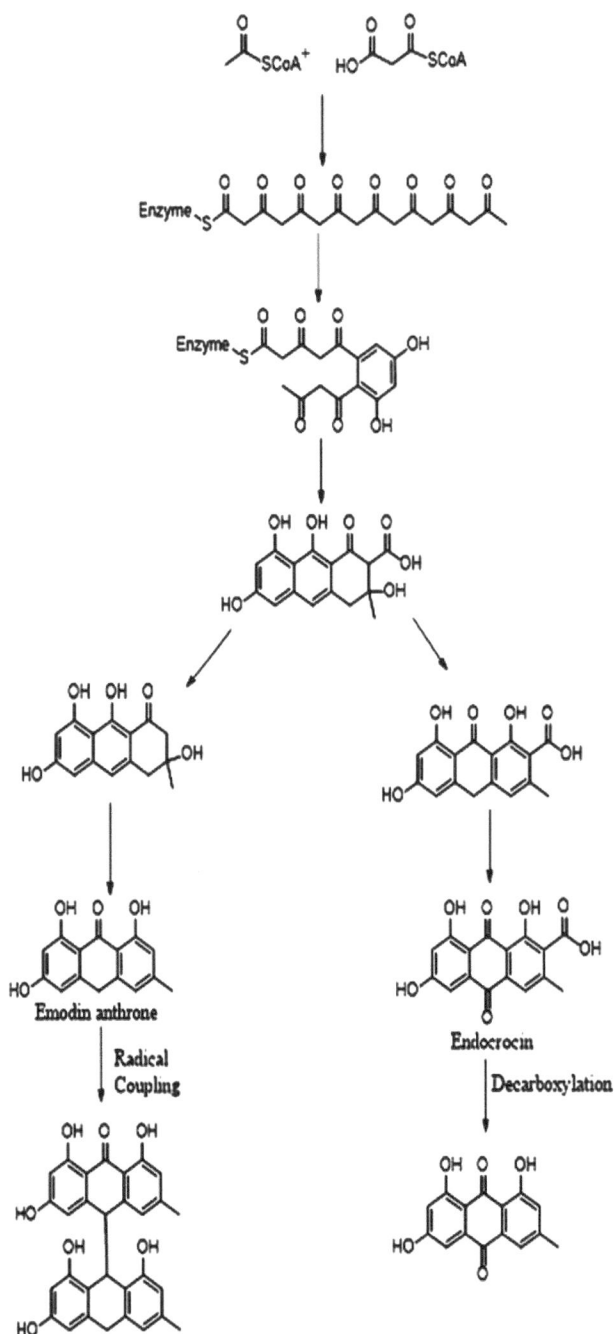

**Fig. (4).** The Schematic representation of the polyketone pathway for the synthesis of anthraquinones.

Finally, under the control enzymatic action of 1,4-dihydroxy-2-naphthoyl-CoA thioesterase (DHNAT), DHNA-CoA eventually produces DHNA which functions as the A and B rings of the anthraquinone nucleus (Fig. **5**) [39].

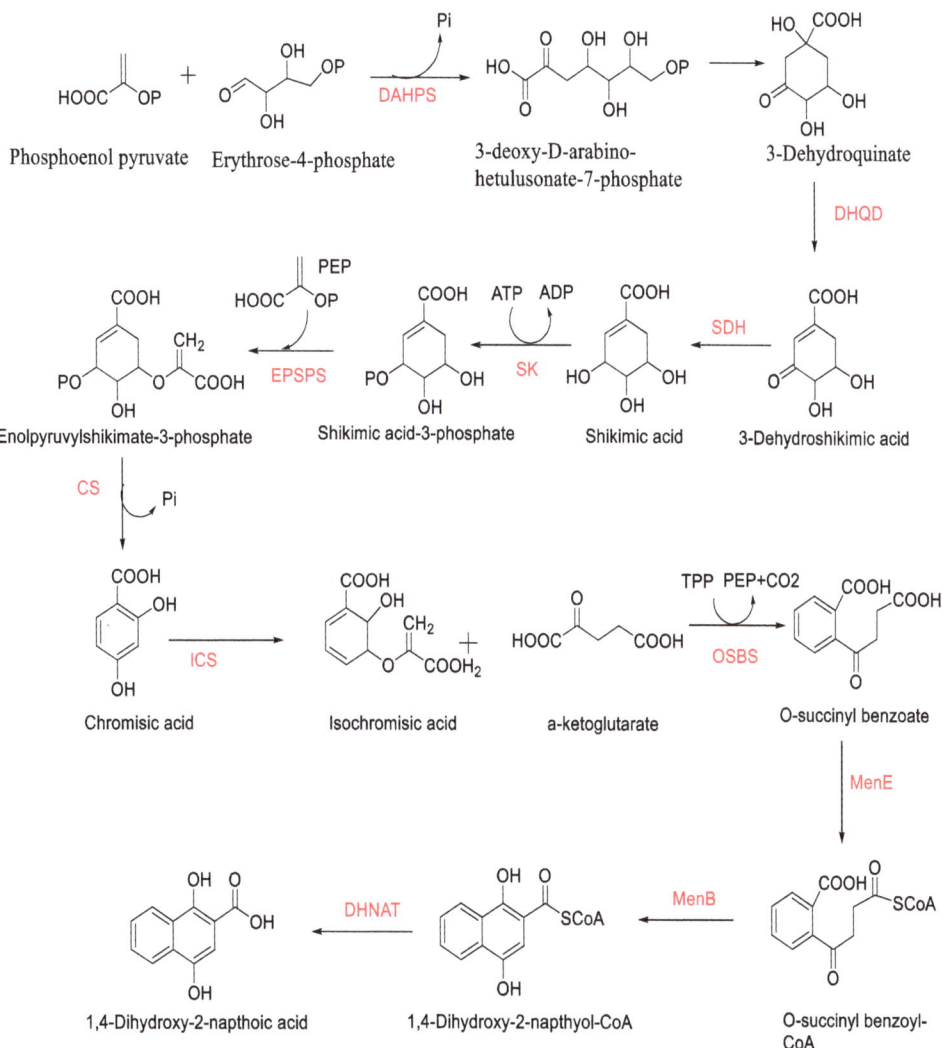

**Fig. (5).** The schematic representation of the first component of the shikimate Pathway showing the synthesis of two rings of Anthraquinones.

The second component involves the production of 3,3'-dimethylallyl diphosphate (DMAPP). The isoprenyl diphosphate (IPP) from the MVA pathway and 4-hydroxy-3-methylbutenyl-1 diphosphate (HMBPP) from the MEP pathway can both be used to make DMAPP [40, 41]. DMAPP is primarily generated in the plastid's MEP pathway and the cytoplasm's MVA pathway, which are both crucial precursors for the production of anthraquinones. As a result, the two synthetic pathways are covered separately.

In the MVA pathway, acetyl-CoA Carboxylase (ACCA) catalyses the initial condensation of two acetyl-CoA molecules to produce acetoacetyl-CoA. Further, under the control of the enzyme hydroxymethylglutaryl-CoA synthase (HMGS), acetoacetyl-CoA and acetyl-CoA are condensed to produce 3-hydroxyl-3-methylglutaryl-CoA (HMG-CoA). Next, the enzyme hydroxymethylglutaryl-CoA reductase (HMGR) catalyses the production of MVA from HMG-CoA. In the next step, mevalonate kinase (MVK) catalyzes, by utilizing 1 molecule of ATP by MVA to produce mevalonate-5-phosphate (MVA-P), which is then transformed by phosphomevalonate kinase (PMK), to produce mevalonate-5-diphosphate (MVA-DP) by the consumption of 1 molecule of ATP. The produced MVADP is converted into IPP by the action of the enzyme mevalonate disphosphate decarboxylase. Decarboxylase catalyzes to form an anthraquinone core structure, IPP cannot be cyclized with DHNA directly. Thus to synthesize DMAPP and subsequently cyclise it with DHNA, it is isomerised by isopentenyl diphosphate delta-isomerase (IDI) (Fig. **6**).

The first step in the MEP pathway involves the conversion of pyruvate and D-glyceraldehyde-3-phosphate into 1-deoxy-D-xylulose-5-phosphate (DXP) by the enzyme 1-deoxy-D-xylulose-5-phosphate synthase (DXS). The generated DXP is then reduced by NADPH, followed by isomerization to 1-deoxy-D-xylulose-5-phosphate-reductoisomerase (DXR), ultimately producing methylerythiol phosphate (MEP) and NADP$^+$. The produced MEP combines with cytidine 5'-triphosphate (CTP) to generate methylerythritol cytidyl diphosphate (CDP-ME) by enzyme 2-C-methyl-D-erythritol 4-phosphate cytidylyltransferase (CMS). CDP-ME is then converted into 4-diphosphocytidyl-2-C-methyl-D-erythritol-2-phosphate (CDP-MEP) by the action of enzyme 4-diphosphocytidyl-2-C-methyl-D-erythritol kinase (CMK). The CDP-MEP then undergoes a series of catalytic reactions to generate IPP and DMAPP (Fig. **7**).

The third and the final component of this pathway is the condensing steps. In this component, the DHNA generated by the SA pathway and DMAPP generated either by the MVA or MEP pathways unite and cyclize to form the C ring of the parent core skeleton of anthraquinone compounds. Further, a series of chemical modifications *via* methoxylation, hydroxylation, and glycosylation, generate a series of anthraquinone compounds. The rubidium anthraquinones are a major class of anthraquinones biosynthesized by this pathway (Fig. **8**).

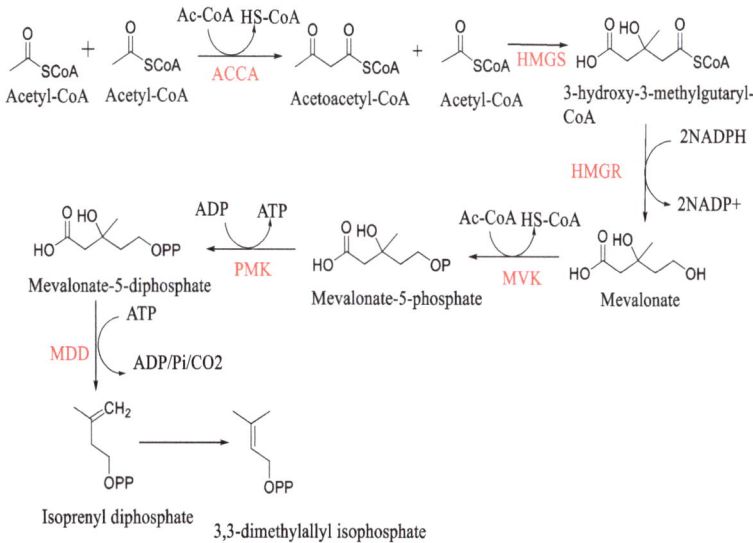

**Fig. (6).** The schematic representation of the second component of Shikimate Pathway; generation of 3,3-dimethylallyl diphosphate *via* mevalonate (MVA) pathway.

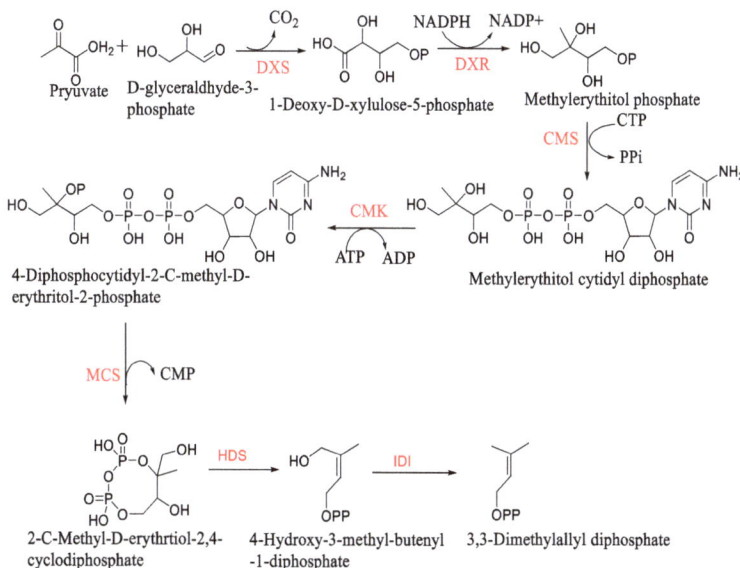

**Fig. (7).** The schematic representation of second component of Shikimate Pathway; generation of 3,3-dimethylallyl diphosphate *via* methyl erythritol phosphate (MEP) pathways.

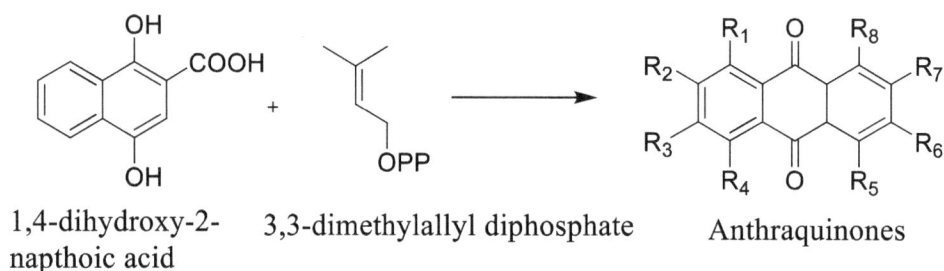

**Fig. (8).** The schematic representation of the third component of the Shikimate Pathway.

## Traditional Use of Anthraquinones

The majority of natural quinones are anthraquinones, of which more than 79 have been produced from natural sources [42]. The naturally occurring derivatives of anthraquinones have been used as therapeutic and dying agents for a long time in China and the USA.  Hydroxyanthraquinones, the active component of many traditional medicines are derived from *Rhamnus purshiana* (cascara sagrada), *Rhamnus.*

*Frangula, Cassia acutifolia, Rheum rhabarbarum, Aloe vera*, and *R. tinctorum,* which are used to treat kidney and bladder stones, as well as *Hypericum perforatum,* which has antidepressant and mild sedative properties [43].

Furthermore, the polyphenol anthraquinone derivative emodin (1,3,8-trihydroxy-6-methyl anthraquinone) that was extracted from the roots and rhizomes of *Rheum palmatum* is commonly used in traditional Chinese medicine. Three plant families namely, *Fabaceae* (*Cassia* spp.), *Polygonaceae* (*Rheum, Rumex,* and *Polygonum* spp.), and *Rhamnaceae* (*Rhamnus* and *Ventilago* spp.)- are the primary producers of this secondary metabolite [44]. Apart from its well-established antimutagenic, anti-inflammatory, and antibacterial properties, it also regulates the vasomotor system and has an impact on immunological and metabolic processes.

Traditionally used in Chinese folk medicine, rhubarb (*R. palmatum*) is a plant that contains anthraquinones and has been shown to have distinct and promising anticancer properties. These effects can be achieved through the induction of apoptosis, disruption of the cell cycle, inhibition of cell growth, or antimetastatic effects. This plant has been shown to contain emodin, the most prevalent secondary metabolite of this type in rhubarb, as well as aloe-emodin.

# BIOLOGICAL APPLICATION OF ANTHRAQUINONES

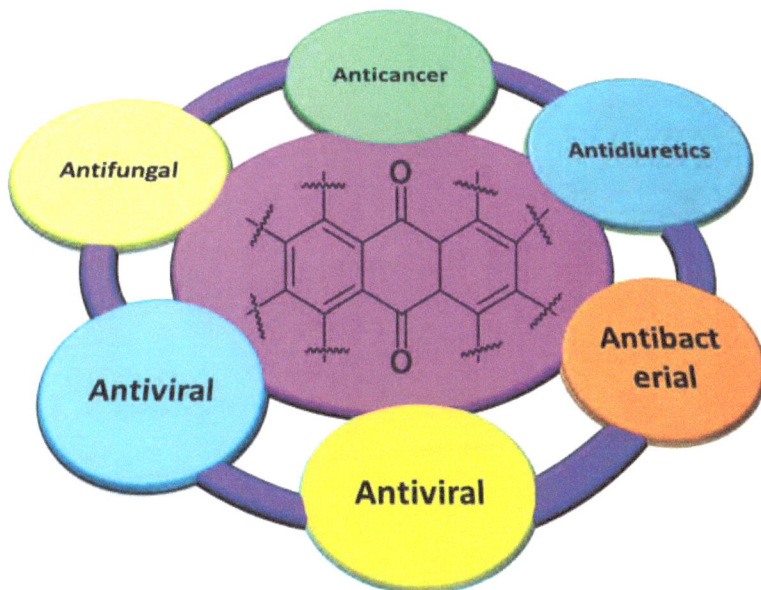

**Scheme 1.** The graphical representation of the various pharmacological applications of anthraquinone derivatives.

Anthraquinones are vital secondary metabolites of plants involved in defining crucial roles in numerous biological processes and environmental conditions. Numerous studies on these secondary metabolites (anthraquinones) have demonstrated their potential for usage as therapeutic agents (Table **1**) by humans for a variety of disorders. It is widely utilized as an anti-inflammatory, anti-cancer, antifungal, antiviral, antidiabetic, anti-arthritic, and antibacterial medication (Scheme **1**, Fig. **9**) [45].

## Anthraquinone as Anticancer Agent

Some anthraquinone derivatives, including emodin, physcion, and aloe-emodin, have anticancer properties that may encourage tumour apoptosis, cell proliferation, cell death, or mutation of tumour cells [49].

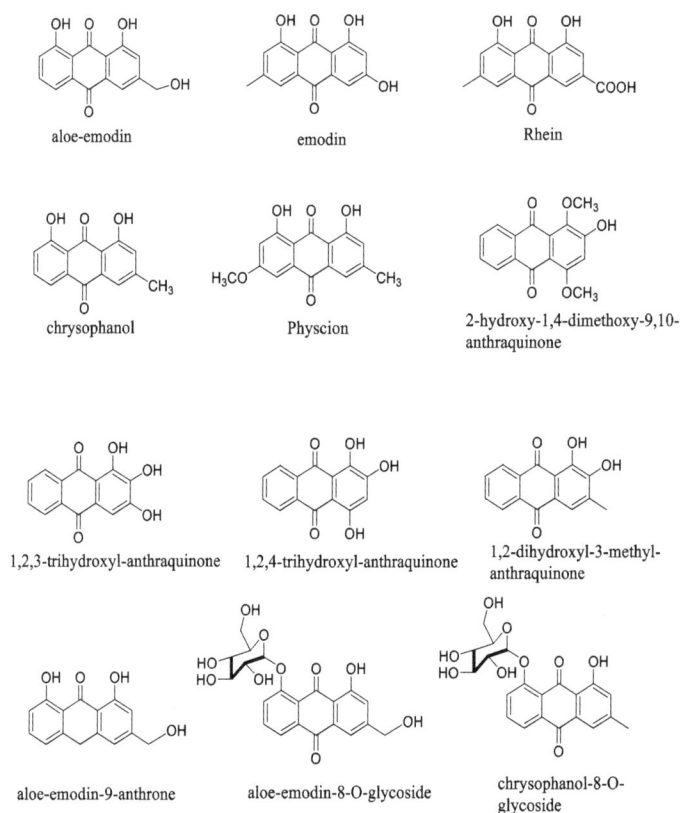

aloe-emodin          emodin          Rhein

chrysophanol          Physcion          2-hydroxy-1,4-dimethoxy-9,10-anthraquinone

1,2,3-trihydroxyl-anthraquinone   1,2,4-trihydroxyl-anthraquinone   1,2-dihydroxyl-3-methyl-anthraquinone

aloe-emodin-9-anthrone   aloe-emodin-8-O-glycoside   chrysophanol-8-O-glycoside

**Fig. (9).** The chemical structure of various pharmacologically active anthraquinone derivatives isolated from plants.

## Table 1. Representing the biological potential of Anthraquinones [46-48].

| Drug | Usage | Synthesis/Source | Route of Administration | Dosage |
|---|---|---|---|---|
| **Daunorubicin** | Antineoplastic | Natural (bacteria, *e.g.*, *Streptomyces* sp.), combinatorial biosynthesis (genetic engineering, strain improvement) | Intravenous | 1.2-3 mg/dL |

| Doxorubicin | Antineoplastic | Natural (bacteria, *e.g.*, *Streptomyces* sp.), combinatorial biosynthesis (genetic engineering, strain improvement) | Intravenous | <1.2 mg/dL |
|---|---|---|---|---|
| **Idarubicin** | Antineoplastic | Semisynthetic | Intravenous, oral | >5 mg/dL |
| **Epirubicin** | Antineoplastic | Semisynthetic, genetic engineering | Intravenous | >3 mg/dL |
| **Valrubicin** | Antineoplastic | Semisynthetic | Intravesical instillation into the bladder | 1.2 mg/dL |
| **Mitoxantrone** | Antineoplastic, multiple sclerosis | Synthetic | Intravenous | 12 mg/m$^2$ |
| **Pixantrone** | Antineoplastic | Synthetic | Intravenous | 50 mg/m$^2$ |
| **Senna (sennosides)** | Laxative | Herbal | Oral, rectal | 7.5 mg or 15 mg/day |
| **Cascara sagrada** | Laxative | Herbal | Oral | 300 mg/day |
| **Aloe (aloin)** | Laxative | Herbal | Oral | 1 mg/day |
| **Diacerein** | Antiarthritic | Semisynthetic | Oral | 100 mg/day |
| **Emodin** | laxative, antineoplastic agent | Herbal | oral | - |
| **Physcion** | antibacterial, and anticancer | Herbal | oral | 5 mg-10 mg/day |

*(Table 1) cont.....*

| Chrysophanol | Antidiabetic, Anti-inflammatory | Herbal | Intravenous, herbal | 5 mg-50 mg/day |
|---|---|---|---|---|
| **Rhein** | nephroprotective, | Herbal | oral | - |

## Anthraquinone as Antifungal, Antibacterial, and Antiviral Agent

The anthraquinone derivatives from *Polygonaceae, Asphodelaceae,* and *Rubiaceae* family can act as potential antifungal and antiviral agents by inhibiting biofilm formation, interfering with cell wall synthesis, inhibiting efflux pump activity, and viral replication, thereby stopping viral action. For example, anthraquinones such as *Rheum emodi*'s rhizome-extracted anthraquinones rhein, chrysophanol, physcion, and aloe-emodin demonstrated antifungal action against *Trichophyton mentagrophytes* and *Candida albicans* [50, 51]. Emodin has been isolated from the plants *Rumex abyssinicus, Ventilago madraspatana, Polygonum cuspidatum,* and *Rhamnus alaternus,* andexhibits antibacterial activity against bacteria, including strains of *Staphylococcus aureus* and methicillin-resistant (MRSA) and various other bacterial strains [52].

## Anthraquinone as Anti-arthritic, Anti-inflammatory and Antidiabetic Agent

Rheumatoid arthritis is an autoimmune disorder that is associated with inflammation of joints, chronic inflammation, and destruction of articular cartilage and bones. The 1,8-dihydroxyanthraquinone derivatives (1,8-DAD) isolated from a number of families, including *Rhamnaceae* (buckthorn, cascara), *Liliaceae* (aloe), *Polygonaceae* (rhubarbs), and *Caesalpiniaceae* (senna) exhibits anti-inflammatory activity. Aloe emodin, an anthraquinone glycoside displays the anti-inflammatory action by the inhibition of induced nitric oxide (iNO), prostaglandin E2, and prostaglandin E3 [53]. Emodin, rehin, and alalternin inhibit the enzyme protein tyrosine phosphatase 1B (PTP1B) and α-glucosidase involved in regulating blood sugar levels thereby, showing promising antidiabetic effects [54].

## Anthraquinone as Antioxidant

A possible antioxidant, anthraquinone derivatives derived from the plant *Paederia* might effectively scavenge free radicals such as hydroxyl radicals (OH), peroxide radicals (OOH), nitric oxide (NO), *etc.* [55].

Research into the potential of anthraquinones as an avian repellent led to the discovery of an additional significant use for them as an agrochemical in the 1940s. This group of substances is currently thought of as a biopesticide with a significant potential for application in insecticide and nonlethal pest management [56].

The first commercial formulations incorporating anthraquinone components, such as Morkit, were made possible by the patenting of the first anthraquinone-based repellents in 1943 and 1944. This finding sparked interest in creating novel crop protection formulas and methods for managing diverse agricultural pests.

## CONCLUSION AND FUTURE PERSPECTIVES

In conclusion, anthraquinones are an astonishing class of compounds having biological and industrial relevance. These quinone derivatives with diverse pharmacological properties are synthesized through a series of enzymatic reactions highlighting the complexity of nature's chemical factories. The biosynthetic pathways and complex chemistry of anthraquinones represent a fascinating journey for the production of natural products. The complete and deep insight into the biosynthetic pathways of anthraquinones has contributed to the applications in medicine, industry, and agriculture. The advancement in biotechnology and molecular biology offers promising tools for biosynthetic pathways in order to synthesize anthraquinones with tailored properties with enhanced yields. The biosynthesis of anthraquinones not only explores their chemical properties but also provides a gateway to understanding biological and ecological interactions. The research in this field continues to advance to uncover more insights into the regulation of these pathways, leading to the development of innovative strategies for harnessing the bioactive potential of anthraquinones. In essence, the biosynthesis of anthraquinones exemplifies the exquisite complexity of nature's chemical repertoire and underscores the immense potential for discovery and application in the world of natural product chemistry.

## REFERENCES

[1]    Duval J, Pecher V, Poujol M, Lesellier E. Research advances for the extraction, analysis and uses of anthraquinones: A review. Ind Crops Prod 2016; 94: 812-33.
       http://dx.doi.org/10.1016/j.indcrop.2016.09.056

[2]    Dave H, Ledwani L. A review on anthaquinones isolated from Cassia species and their applications. Indian J Nat Prod Resour 2012; 3: 291-319.

[3]    Murdock KC, Child RG, Fabio PF, *et al.* Antitumor agents. 1. 1,4-Bis[(aminoalkyl)amino]-9,10-anthracenediones. J Med Chem 1979; 22(9): 1024-30.
       http://dx.doi.org/10.1021/jm00195a002 PMID: 490545

[4]    Shrestha JP, Fosso MY, Bearss J, Chang CWT. Synthesis and anticancer structure activity relationship investigation of cationic anthraquinone analogs. Eur J Med Chem 2014; 77: 96-102.
       http://dx.doi.org/10.1016/j.ejmech.2014.02.060 PMID: 24631728

[5]    Davis R, Agnew P, Shapiro E. Natural sources of anthraquinones and their medicinal applications. J Am Podiat Med Assn 1986; 76: 61-6.
       http://dx.doi.org/10.7547/87507315-76-2-61

[6]    Sebak M, Molham F, Greco C, Tammam MA, Sobeh M, Demerdash AE. Chemical diversity, medicinal, potentialities, biosynthesis and pharmacokinetics of anthraquinones and their congeners derived from marine fungi: a comprehensive update 2022; 12: 24887-24921.
       http://dx.doi.org/10.1039/D2RA03610J

[7]    Gartman JA, Tambar UK. Recent total syntheses of anthraquinone-based natural products 2022; 105: 132501.
       http://dx.doi.org/10.1016/j.tet.2021.132501

[8]    Winter RW, Cornell K, Johnson LL, Ignatushchenko M, Hinrichs DJ, Riscoe MK. Antimicrobial activity of anthraquinone derivatives. Antimicrob agents chemother 1996; 40: 1408-11.
       http://dx.doi.org/10.1128/AAC.40.6.1408 PMID: 8726010

[9]    Iranshahi M, Amani M. The biosynthesis and metabolism of naturally occurring quinones. Compr Nat Prod Chem 1999; 3: 125-71.

[10]   Abe I, Morita H. Structure and function of the chalcone synthase superfamily of plant type III polyketide synthases. Nat Prod Rep 2010; 27(6): 809-38.
       http://dx.doi.org/10.1039/b909988n PMID: 20358127

[11]   Itkin M, Heinig U, Tzfadia O, *et al.* Biosynthesis of antinutritional alkaloids in solanaceous crops is mediated by clustered genes. Science 2013; 341(6142): 175-9.
       http://dx.doi.org/10.1126/science.1240230 PMID: 23788733

[12]   Wang D, Wang XH, Yu X, *et al.* Pharmacokinetics of anthraquinones from medicinal plants. Front Pharmacol 2021; 12(12): 638993.
       http://dx.doi.org/10.3389/fphar.2021.638993 PMID: 33935728

[13]   Zhang L, Zuo Z, Chen D, *et al.* Discovery of a new pathway for biosynthesis of 3,6-dihydroxyanthraquinone in the fungus Aspergillus terreus. Org Lett 2016; 18(15): 3794-7.
       PMID: 27410168

[14]   Mund NK, Čellárová E. Recent advances in the identification of biosynthetic genes and gene clusters of the polyketide-derived pathways for anthraquinone biosynthesis and biotechnological applications. Biotechnol Adv 2023; 63: 108104.
       http://dx.doi.org/10.1016/j.biotechadv.2023.108104 PMID: 36716800

[15]   Liu C, Wang R, Wang S, *et al.* A prenyltransferase participates in the biosynthesis of anthraquinones in Rubia cordifolia 2024; 195: 2860-2876.
       http://dx.doi.org/10.1093/plphys/kiae171

[16]   Oikawa H, Toyomasu T, Toshima H. Polyketide biosynthesis beyond the type I, II and III polyketide synthase paradigms. Curr Opin Chem Biol 2016; 31: 82-93.

[17]   Shrestha G, Shrestha A, Wang L, Zhang Z. Bioactive compounds and their derivatives from Ochrocarpus longifolius: A review. Fitoterapia 2018; 129: 235-42.

[18]   Liu Y, Yang X, Su Z, Li M. Unusual biosynthesis and structure of locillomycin A. J Am Chem Soc 2012; 134(39): 17323-6.
       PMID: 22953714

[19]   Beaudoin GA, Facchini PJ. Isoprenoid biosynthesis in the basal land plant Marchantia polymorpha. Plant Physiol 2013; 161(1): 375-81.

[20]   Das K, Roychoudhury A. Reactive oxygen species (ROS) and response of antioxidants as ROS-scavengers during environmental stress in plants. Front Environ Sci 2014; 2: 53.
       http://dx.doi.org/10.3389/fenvs.2014.00053

[21]   Hartmann T. Chemical ecology of pyrrolizidine alkaloids. Planta Med 1999; 65(2): 7-23.

[22]   Luo Y, Duan M, Zhao L, Zhang L. Anthraquinone compounds from the leaves of Morinda citrifolia (noni) exhibit quinone reductase induction and antitumor activity. J Agric Food Chem 2009; 57(10): 4343-8.

[23]   Kanokmedhakul K, Kanokmedhakul S, Phatchana R. Biological activity of anthraquinones and triterpenoids from Prismatomeris fragrans. J Ethnopharmacol 2005; 100(3): 284-8.
       http://dx.doi.org/10.1016/j.jep.2005.03.018 PMID: 15885942

[24]   Rodriguez E, Towers G, Mitchell J. Biological activities of lignans. Phytochemistry 1976; 15(11): 1573-80.
       http://dx.doi.org/10.1016/S0031-9422(00)97430-2

[25]   Zhao Y, Wu Q, Du J, *et al.* Anti-inflammatory and anti-oxidative effects of rhei radix et rhizoma and anthraquinones in LPS-induced RAW 264.7 cells. BMC Complement Altern Med 2017; 17(1): 111.
       PMID: 28202019

[26]   Bagewadi ZK, Mulla IS, Ravishankar B, Desai N, Derikvand F. Anthraquinones in tissue culture of Morinda elliptica. Phytochemistry 1997; 44(2): 339-42.

[27]   Proksa B, Fedoročko P. Plant metabolites as antioxidants and antimutagens.Antioxidants in Higher Plants. CRC Press 1999; pp. 137-46.

[28]   Bodas R, Prieto N, García-González R, Andrés S, Giráldez FJ, López S. Manipulation of rumen fermentation and methane production with plant secondary metabolites. Anim Feed Sci Technol 2012; 176(1-4): 78-93.
       http://dx.doi.org/10.1016/j.anifeedsci.2012.07.010

[29]   Dholakiya BZ, Kapadi AH, Dodia MS, Dudhat CD, Panara D, Patel PS. Antioxidant and hepatoprotective activity of the crude drug of Operculina turpethum (Linn). Indian J Pharmacol 2010; 42(6): 394-8.

[30]   May G. Biochemistry and physiology of anthraquinone synthesis in higher plants 1977.

[31]   Wang P, Wei J, Hua X, *et al.* Plant anthraquinones: Classification, distribution, biosynthesis, and regulation. J Cell Physiol 2024; 239(10): e31063.
       http://dx.doi.org/10.1002/jcp.31063 PMID: 37393608

[32]   Bringmann G, Noll TF, Gulder TAM, *et al.* Different polyketide folding modes converge to an identical molecular architecture. Nat Chem Biol 2006; 2(8): 429-33.
       http://dx.doi.org/10.1038/nchembio805 PMID: 16829953

[33]    Chang P, Lee KH. Antitumor agents, 75. Synthesis of cytotoxic anthraquinones digiferruginol and morindaparvin-B. J Nat Prod 1985; 48(6): 948-51.
        http://dx.doi.org/10.1021/np50042a011 PMID: 3841556

[34]    Kang SH, Lee WH, Lee CM, *et al. De novo* transcriptome sequence of *Senna tora* provides insights into anthraquinone biosynthesis. PLoS One 2020; 15(5): e0225564.
        http://dx.doi.org/10.1371/journal.pone.0225564 PMID: 32380515

[35]    Kang SH, Lee WH, Sim JS, *et al. De novo* transcriptome assembly of Sienna occidentals sheds light on the anthraquinone biosynthesis pathway. Front Plant Sci 2022; 12: 773553.
        http://dx.doi.org/10.3389/fpls.2021.773553 PMID: 35046973

[36]    Kang SH, Pandey RP, Lee CM, *et al.* Genome-enabled discovery of anthraquinone biosynthesis in *Senna tora.* Nat Commun 2020; 11(1): 5875.
        http://dx.doi.org/10.1038/s41467-020-19681-1 PMID: 33208749

[37]    Santos-Sánchez NF, Salas-Coronado R, Hernández-Carlos B, *et al.* Shikimic acid pathway in biosynthesis of phenolic compounds. Plant Physiological Aspects of Phenolic Compounds 2019; pp. 1-15.

[38]    Dempsey DMA, Vlot AC, Wildermuth MC, Klessig DF. Salicylic Acid biosynthesis and metabolism. Arabidopsis Book 2011; 9: e0156.
        http://dx.doi.org/10.1199/tab.0156 PMID: 22303280

[39]    Gaid MM, Sircar D, Müller A, *et al.* Cinnamate:CoA ligase initiates the biosynthesis of a benzoate-derived xanthone phytoalexin in *Hypericum calycinum* cell cultures. Plant Physiol 2012; 160(3): 1267-80.
        http://dx.doi.org/10.1104/pp.112.204180 PMID: 22992510

[40]    Lv X, Wang F, Zhou P, *et al.* Dual regulation of cytoplasmic and mitochondrial acetyl-CoA utilization for improved isoprene production in *Saccharomyces cerevisiae.* Nat Commun 2016; 7(1): 12851.
        http://dx.doi.org/10.1038/ncomms12851 PMID: 27650330

[41]    Zebec Z, Wilkes J, Jervis AJ, Scrutton NS, Takano E, Breitling R. Towards synthesis of monoterpenes and derivatives using synthetic biology. Curr Opin Chem Biol 2016; 34: 37-43.
        http://dx.doi.org/10.1016/j.cbpa.2016.06.002 PMID: 27315341

[42]    Diaz-Muñoz G, Miranda IL, Sartori SK, de Rezende DC, Diaz MAN. Chapter 11 - anthraquinones: An overview. In: Atta-ur-Rahman, editor Stud Nat Prod Chem. Elsevier 2018; 58: pp. 313-8.
        http://dx.doi.org/10.1016/B978-0-444-64056-7.00011-8

[43]    Dong M, Ming X, Xiang T, *et al.* Recent research on the physicochemical properties and biological activities of quinones and their practical applications: a comprehensive review. Food Funct 2024; 15(18): 8973-97.
        http://dx.doi.org/10.1039/D4FO02600D PMID: 39189379

[44]    Ayele TM, Abebe EC, Muche ZT, *et al. In vivo* antidiarrheal activity of the crude extract and solvent fractions of *Rhamnus prinoides* (Rhamnaceae) leaves. Heliyon 2023; 9(6): e16654.
        http://dx.doi.org/10.1016/j.heliyon.2023.e16654 PMID: 37292287

[45]  Silva M, Caro V, Guzmán C, Perry G, Areche C, Cornejo A. Chapter 1 - α-Synuclein and tau, two targets for dementia. In: Atta-ur-Rahman, editor Stud Nat Prod Chem. Elsevier Science B.V. 2020; 67: pp. 1-25.
http://dx.doi.org/10.1016/B978-0-12-819483-6.00001-1

[46]  Adnan M, Rasul A, Hussain G, *et al.* Physcion and physcion 8-O-β-D-glucopyranoside: Natural anthraquinones with potential anti-cancer activities. Curr Drug Targets 2021; 22(5): 488-504.
http://dx.doi.org/10.2174/18735592MTEwDNjQiz PMID: 33050858

[47]  Faizan S, Mohammed Abdo Mohsen M, Amarakanth C, *et al.* Quinone scaffolds as potential therapeutic anticancer agents: Chemistry, mechanism of Actions, Structure-Activity relationships and future perspectives. Results Chem 2024; 7: 101432.
http://dx.doi.org/10.1016/j.rechem.2024.101432

[48]  Gupta PO, Sekar N. NLOphoric Anthraquinone Dyes – A Review. ChemistrySelect 2024; 9(20): e202400736.
http://dx.doi.org/10.1002/slct.202400736

[49]  Nsairat H, Khater D, Odeh F, Jaber AM, *et al.* Phytosomes: a modernistic approach to the delivery of herbal drugs. Advanced and Modern Approaches for Drug Delivery 2023; pp. 301-55.
http://dx.doi.org/10.1016/B978-0-323-91668-4.00029-0

[50]  Kim YM, Lee CH, Kim HG, Lee HS. Anthraquinones isolated from *Cassia tora* (Leguminosae) seed show an antifungal property against phytopathogenic fungi. J Agric Food Chem 2004; 52(20): 6096-100.
http://dx.doi.org/10.1021/jf049379p PMID: 15453672

[51]  Ntemafack A, Singh RV, Ali S, Kuiate JR, Hassan QP. Antiviral potential of anthraquinones from Polygonaceae, Rubiaceae and Asphodelaceae: Potent candidates in the treatment of SARS-COVID-19, A comprehensive review. S Afr J Bot 2022; 151: 146-55.
http://dx.doi.org/10.1016/j.sajb.2022.09.043 PMID: 36193345

[52]  Kshirsagar AD, Panchal PV, Harle UN, Nanda RK, Shaikh HM. Anti-inflammatory and antiarthritic activity of anthraquinone derivatives in rodents. Int J Inflamm 2014; 2014: 1-12.
http://dx.doi.org/10.1155/2014/690596 PMID: 25610704

[53]  Jung H, Ali M, Choi J. Promising inhibitory effects of anthraquinones, naphthopyrone, and naphthalene glycosides, from *cassia obtusifolia* on α-glucosidase and human protein tyrosine phosphatases 1B. Molecules 2016; 22(1): 28.
http://dx.doi.org/10.3390/molecules22010028 PMID: 28035984

[54]  Qun T, Zhou T, Hao J, *et al.* Antibacterial activities of anthraquinones: structure–activity relationships and action mechanisms RSC Med. Chem. 2023;14(8):1446-1471. Org Lett 2016; 18(15): 3794-7.
PMID: 27410168

[55]   Ben Ammar R, Miyamoto T, Chekir-Ghedira L, Ghedira K, Lacaille-Dubois MA. Isolation and identification of new anthraquinones from *Rhamnus alaternus* L and evaluation of their free radical scavenging activity. Nat Prod Res 2019; 33(2): 280-6.
http://dx.doi.org/10.1080/14786419.2018.1446135 PMID: 29533086

[56]   Heckmanns F, Meisenheimer M. M, Bird Repellents, Germany Patent 743517 1943.

# CHAPTER 3

# Anthraquinones as Bioactive Agents: Recent Trends and Developments in Phytotherapy

**Sukhvinder Dhiman[1], Gulshan Kumar[2], Ajay Kumar[3], Sapna Devi[4], Sonika[5] and Manoj Kumar[6],***

[1]*Institute of Nano Science and Technology, Mohali-140306, Punjab, India*

[2]*Department of Chemistry, Banasthali University, Banasthali Newai-304022, Rajasthan, India*

[3]*University Centre for Research and Development (UCRD), Chandigarh University, Mohali-140413, Punjab, India*

[4]*Department of Microbiology, DAV University, Jalandhar-144012, Punjab, India*

[5]*Department of Chemistry, Maharishi Markandeshwar Engineering College, Maharishi Markandeshwar (Deemed to be University), Mullana-133207, Haryana, India*

[6]*Department of Microbiology, Guru Nanak Dev University, Amritsar-143005, Punjab, India*

**Abstract:** Anthraquinones are organic compounds and members of the Quinone family comprising 9, 10-anthracenedione core. Anthraquinone is a chemical scaffold that has been employed for many years in a variety of therapeutic applications such as antimicrobial, anticancer, diuretic, anti-inflammatory, and phytoestrogen activities. Anthraquinones are commonly produced as secondary metabolites in various higher plant species (senna, buckthorn, yellow dock) or they can either be synthesized using chemical routes such as the condensation of 1, 4-naphthoquinone with butadiene, naphthalene oxidation, anthracene oxidation, *etc.* Anthraquinones are used in various traditional and ethnomedical processes for the treatment of acute as well as chronic illness and are nowadays employed in modern pharmaceutical markets as a key bioactive agent. Hence, due to these properties, anthraquinone-based compounds are widely used in phytotherapy. Anthraquinones are unique in terms of their structure, chemical stability, biological properties, and industrial applications among all reported quinones, making them valuable in a wide range of drug formulations. The in-depth studies regarding the role of anthraquinones using *in vitro* and *in vivo* models need to be extensively explored for bioactive and phytotherapy applications. In addition to this, the safety and toxicity assessment need to be thoroughly investigated. The knowledge regarding its biochemical structure can pave the way to understanding its physiological and toxicological properties. The chapter dispenses compact knowledge regarding anthraquinones aspotential bioactive agents and their use as a therapeutic/health product.

* **Corresponding author Manoj Kumar:** Department of Microbiology, Guru Nanak Dev University, Amritsar-143005, Punjab, India; E-mail: manojdutta27@gmail.com

**Keywords:** Anthraquinones, Bioactive agents, Drug formulations, *In vivo*, *In vitro*, Medicinal plants, Toxicity studies.

## INTRODUCTION

Anthraquinone, a captivating organic compound, has gained attention for its versatile applications in chemistry, pharmaceuticals, and the textile industry [1]. This aromatic compound is characterized by its unique structure, consisting of a fused aromatic ring system comprising three benzene rings and a ketone group. The term "anthraquinone" is derived from a Greek word, "anthos" meaning flower and "rhodo" meaning rose, signifying its initial discovery from the roots of madder plants, which have been employed for centuries in the production of vibrant red dyes. As time passed, the study of anthraquinone compounds expanded beyond their use in dyeing, revealing their presence in a diverse array of natural sources. Anthraquinones are frequently found in the roots, leaves, and stems of numerous plant species, including madder (*Rubia tinctorum*), rhubarb (*Rheum rhabarbarum*), and senna (*Senna alexandrina*) [2]. They are particularly abundant in plants belonging to families like *Polygonaceae*, *Rhamnaceae*, and *Juglandaceae*. More than 70 different types of anthraquinones have been identified. These compounds often serve as secondary metabolites in plants, playing essential roles in defense mechanisms against herbivores and pathogens [3]. Furthermore, anthraquinones have been identified in various fungi, notably in species belonging to the genera *Aspergillus* and *Penicillium*, where they contribute to the metabolic processes of these organisms [4]. These compounds have garnered significant interest from researchers due to their diverse biological functions, including hepatoprotective, antifungal, antibacterial, laxative, antioxidant, and anti-cancer properties. Anthraquinones have also been reported for their bioactive properties and use in phytotherapy [5-9]. In this exploration of anthraquinone, we will delve deeper into their origins and the diverse sources reported for their bioactive and phyto-therapeutic properties. Also, the study related to safety and toxicity assessment needs to be explored for its pharmaceutical applications. Beyond their role as natural products, anthraquinones have found applications in synthetic chemistry, pharmaceuticals, and as colorants in various industries [10]. This makes them a versatile and captivating class of compounds with a rich history and promising prospects for the future.

## ANTHRAQUINONES AS BIOACTIVE AGENT

Anthraquinones are a group of chemical compounds that are found in a wide diversity of plants and have been explored for a range of bioactive properties [11].

The various biological activities that anthraquinones exhibit make them excellent candidates for use in medicine and other applications (Table **1**).

## Antimicrobial Potential of Anthraquinone

Anthraquinones' antimicrobial abilities have been widely studied *in vitro* using both pure and crude forms [12]. Among the most studied anthraquinones found in nature include chrysophanol, aloe-emodin, emodin, rhein, and physcione reported for their *in vitro* antimicrobial activity. Anthraquinones, both extracted and isolated, were effective against various Gram-negative and Gram-positive bacteria including *Pseudomonas aeruginosa, Helicobacter pylori, Neisseria gonorrhoeae*, MRSA strains of *Staphylococcus aureus* and *S. epidermitis* [11]. In literature, Wang *et al.* [13], isolated a new anthraquinone, 2-(dimethoxymethyl)-1-hydroxyanthracene-9,10-dione from the fermentation of *Aspergillus versicolor* derived from sea sediment. This anthraquinone showed strong inhibitory activity against MRSA strains and moderate activity against *Vibrio campbellii* due to the inhibition of topoisomerase-IV and AmpC β-lactamase. In literature, Song *et al.* [14] isolated two anthraquinone compounds *viz.* 3,8-dihydroxy-1-methylanthraquinon-2-carboxylic acid and 3,6,8-trihydroxy-1-methylanthraquinone-2-carboxylic acid from an actinobacterial strain named *Kitasatospora albolonga* R62. These compounds disrupted preformed MRSA biofilms, potentially by either killing or dispersing the biofilm cells. Similarly, Shupeniuk *et al.* [15] synthesized amino derivative fragments of anthraquinone using a Ullmann coupling reaction exhibiting an inhibitory effect against a wide range of Gram-positive and Gram-negative strains of clinical isolates *viz. Staphylococcus* spp., *Streptococcus* spp., *Escherichia coli, Klebsiella pneumoniae, Providencia stuartii, Pseudomonas aeruginosa,* and fungal strain *Candida*. Antibacterial activity is due to the presence of benzoic acid inhibiting the active uptake of some amino and oxo acids [16]. Yirdaw and Kassa [17] reported the antibacterial activity of terpenoids and anthraquinones present in the methanol extracts from the root bark of *Ferula communis* (Apiaceae) against Gram-negative (*Salmonella typhi, E. coli, Klebsiella pneumoniae* and *Pseudomonas erogenous*), and Gram-positive bacteria (*Staphylococcus aureus*). Zhuravleva *et al.* [18] isolated different anthraquinones namely Acruciquinones A to C from marine fungus *Asteromyces cruciatus* KMM 4696, which showed significant antimicrobial effects against *Staphylococcus aureus* and *Staphylococcus aureus*-infected human HaCaT keratinocytes. The antimicrobial activity of Acruciquinones A to C is due to the inhibition of sortase A and urease activity. Recently, Adekunle *et al.* [19] isolated two anthraquinone molecules from the extracts of *Morinda lucida viz.* 2-hydroxy-1-methoxy anthraquinone and 1,2-dihydroxyanthraquinone (Alizarin), which were found to have potent antibacterial activity against *Staphylococcus aureus* and *Pseudomonas aeruginosa*, respectively. Therefore, anthraquinone is a versatile compound with a broad spectrum of antimicrobial activity against clinical isolates, making it a promising candidate for therapeutic applications.

## Anticancer Potential of Anthraquinone

Anthraquinone compounds are recognized for their anticancer properties, primarily achieved by inducing DNA damage, cell cycle arrest, and apoptosis. Recent studies have discovered that novel anthraquinone compounds can inhibit cancer through diverse mechanisms, including paraptosis, autophagy, radiosensitization, and overcoming chemoresistance [1]. In literature, Bajpai *et al.* [20] carried out a study on root extracts of *Rubia philippinensis*, where they identified various anthraquinone derivatives such as 2-methyl-1,3,6-trihydroxy-9,10-anthraquinone-3-O-(6'-O-acetyl)-α-rhamnosyl (1 → 2)-β-glucoside, 2-methyl-1,3,6-trihydroxy-9,10-anthraquinone, alizarin, xanthopurpurin and lucidin-ω-methyl ether. They evaluated the cytotoxicity of all the derivatives against four different cancer cell lines, including murine melanoma (B16F10), human melanoma (SK-MEL-5), and human breast adenocarcinoma (MCF7 and MDA-MB-231) using 3-(4,5-dimethylthiazol-2-yl)-2,5-diphenyltetrazolium bromide (MTT) assay. The results showed significant anticancer potential of all the anthraquinone compounds against various cancer cell lines *viz.* B16F10, SK-MEL-5, MCF7, and MDA-MB-231 but xanthopurpurin and lucidin-ω-methyl ether exhibited high selective toxicity against breast cancer cells (MCF7 and MDA-MB-231) at lower concentrations as compared to normal cell line *i.e.* normal kidney epithelial cells (MDCK). Similarly, Hasanien *et al.* [21] reported the potent anticancer activity of anthraquinone isolated from *Talaromyces purpureogenus* AUMC2603, against three human cancer cell lines *viz.* mammary adenocarcinoma (MCF-7), hepatocellular carcinoma (HepG2) and colorectal adenocarcinoma (HCT-116) with two normal cell lines *i.e.* epithelial breast (MCF-12) and fibroblast (BJ-1) with less cytotoxicity on normal cell lines MCF12F and BJ-1T. They also reported the application of purified anthraquinone in kidney cancer diagnosis using a kidney radio-imaging technique. Manhas *et al.* [22] isolated bisaryl anthraquinone named 4-acetyl-10-(1-acetyl-2,5,10-trihydroxy-2-methyl-4-oxo-1,3-dihydroanthracen-9-yl)-8,9-dihydroxy-3-methyl-4H-anthracen-1-one (Setomimycin) from a new *Streptomyces* strain from Shivalik region of NW Himalayas. The setomimycin antibiotic, which contains 9,9' bisanthraquinone, demonstrated significant anticancer and anti-migratory effects against cancer cell lines *i.e.* colorectal adenocarcinoma (HCT-116) and mammary adenocarcinoma (MCF-7). In addition to this, *in vivo* studies against the mouse mammary carcinoma model (4T1) led to a reduction in primary tumor weight by

76% and tumor volume by 90.5% within two weeks. Recently, Hwang *et al*. [23] identified 1,8-Dihydroxy-3-methoxy-anthraquinone (DMA) isolated from *Photorhabdus luminescens* bacterium found to be a potent suppressor of tumor angiogenesis as a promising candidate for preventing cancer. DMA was found to downregulate hypoxia-inducible factor-1α (HIF-1α) protein and PI3K/AKT and c-RAF/ERK pathway to prevent tumor growth. Therefore, further investigation is needed to understand the pathways underlying the anticancer mechanisms of anthraquinones [24].

## Antioxidative and Anti-Inflammatory Potential of Anthraquinone

Anthraquinones can help reduce oxidative stress and prevent cellular damage by neutralizing and deactivating free radicals and reactive oxygen species (ROS) [25]. Some anthraquinones can chelate or bind to metal ions like iron and copper. These metal ions can catalyse the production of harmful ROS, and by binding to them, anthraquinones inhibit their ability to generate oxidative stress. Anthraquinones have the potential to be used in the development of natural remedies for various inflammatory conditions. In literature, Nam *et al*. [26] assessed the antioxidative and anti-inflammatory effects of anthraquinone and its hydroxy derivatives such as anthrarufin (1,5-dihydroxyanthraquinone), purpurin (1,2,4-trihydroxyanthraquinone) and chrysazin (1,8-dihydroxyanthraquinone) derived from the roots of madder plant named *Rubia tinctorum* L. using mammalian cells (murine macrophage RAW 264.7 cells) and chemical assay. The results revealed the highest antioxidative property of purpurin against both cultured cells and chemical assay. They have also suggested that purpurin can be used to prevent oxidative damage to food as it can down-regulate NLRP3 inflammasome assembly. Gupta *et al*. [27] evaluated the antioxidative potential of natural anthraquinones namely chrysophanol and emodin as well as synthesized anthraquinone such as 1,8-dimethoxy-3-methylanthraquinone, 2-methylanthraquinone, 2-bromoanthraquinone and rubiadin. All of these synthetic anthraquinones and emodin among natural anthraquinones exhibited a high DPPH• radical scavenging capacity (IC50=<500μg/mL) proving the potent potential of these anthraquinones for the treatment of cancer. Trung *et al*. [28] evaluated the superoxide and hydroperoxide radical scavenging activities of six anthraquinones such as 2-hydroxy-1,4-dimethoxy-9,10-anthraquinone, 1,3-dihydroxy-2,4-dimethoxy-9,10-anthraquinone, 1-hydroxy-2 hydroxymethyl-9,10-anthraquinone, 1-methyl-2,4-dimethoxy-3-hydroxyanthraquinone, 1-methoxy-3-hydroxy-2-ethoxymethyl-anthraquinone and 1-methoxy-2-methoxymethyl-3-hydroxy-9,10-anthraquinone isolated from *Paederia* plants using a computational approach. The results indicate that these anthraquinones show promising $O_2^{•-}$ radical scavenging abilities in polar media. Marković *et al.* [29] investigated the various deactivation mechanisms out of which hydrogen atom abstraction (HAA) was found to be the most thermodynamically plausible mechanism of alizarin for the deactivation of ROS. Alizarin can place itself between polypeptide and ROS, playing an important role in protecting polypeptide from hydroperoxyl radical attack. As per these literature reports, anthraquinone and its derivatives were found to have significant antioxidative and anti-inflammatory making it a promising candidate as a therapeutic agent for diseases related to inflammatory and oxidative stress [30]. However, further studies are required to understand its mechanism and safety concerns for clinical applications.

**Table 1. Bioactive properties of different anthraquinones: structure, activities, and mode of action.**

| Anthraquinone Name and Structure | Activities | Mode of Action | References |
|---|---|---|---|
| 2-(dimethoxymethyl)-1-hydroxyanthracene-9,10-dione | Antibacterial activity against Methicillin Resistant *Staphylococcus aureus* (MRSA) strains and *Vibrio campbellii* | Antibacterial activity is due to the inhibition of topoisomerase-IV and AmpC β-lacta-mase enzyme | [13] |
| 3,8-dihydroxy-1- | | | |

*(Table 1) cont.....*

| | | | |
|---|---|---|---|
| methylanthraquinon-2-carboxylic acid 3,6,8-trihydroxy-1-methylanthraquinone-2-carboxylic acid | Antibacterial and antibiofilm activities against Methicillin Resistant *Staphylococcus aureus* (MRSA) | Anthraquinones hydroxyl group at the C-2 position responsible for inhibiting biofilm formation, while the carboxyl group at the same C-2 position have antibacterial activity, leading to killing or dispersing biofilm cells | [14] |
| Triazene 1-(3-(benzoic acid (triaz-1-en-1-ol(-4-(1H-imidazol-1-yl(-9,10-dioxo-9,10-dihydroanthracene-2-sulfonic acid | Antibacterial activity against a wide range of Gram-positive and Gram-negative strain clinical isolates *viz.* *Staphylococcus* spp., *Streptococcus* spp., *Escherichia coli*, *Klebsiella pneumoniae*, *Providencia stuartii*, *Pseudomonas* | Antibacterial activity is due to the presence of benzoic acid inhibiting the active uptake of some | [15, 16] |

(Table 1) cont.....

| | | amino and oxoacids | |
|---|---|---|---|
| *aeruginosa* and antifungal activity against *Candida* | | | |
| Anthracene-9,10-dione (Anthraquinone) | Antibacterial activity against Gram-negative (*Salmonella typhi, E. coli, Klebsiella pneumoniae* and *Pseudomonas erogenous*), and Gram-positive bacteria (*Staphylococcus aureus*) | N/A | [17] |
| A R= H  B R= OH  C  Acruciquinones (A, B and C) | Antimicrobial activity against *Staphylococcus aureus* and *Staphylococcus aureus*-infected human HaCaT keratinocytes | Antimicrobial activity is due to the inhibition of sortase A and urease activity. | [18] |

*(Table 1) cont.....*

| | | | |
|---|---|---|---|
| <br><br>2-hydroxy-1-methoxy<br>anthraquinone<br><br><br><br>1,2-dihydroxyanthraquinone<br>(Alizarin) | Antibacterial activity of 2-hydroxy-1-methoxy anthraquinone against *Staphylococcus aureus* and 1,2-dihydroxyanthraquinone (Alizarin) against *P aeruginosa* | N/A | [19] |
| <br><br>2-methyl-1,3,6-trihydroxy-9,10-anthraquinone-3-O-(6′-O-acetyl)-α-rhamnosyl (1 → 2)-β-glucoside<br><br> | Anticancer activity against cancer cell lines *viz.* B16F10, SK-MEL-5, MCF7 and MDA-MB-231 but xanthopurpurin and lucidin-ω-methyl ether exhibited high selective toxicity against breast cancer cells (MCF7 and MDA-MB-231) | N/A | [20] |

*(Table 1) cont.....*

| | | | |
|---|---|---|---|
| 2-methyl-1,3,6-trihydroxy-9,10-anthraquinone<br><br>1,2-dihydroxyanthraquinone<br>(Alizarin)<br><br>1,3-Dihydroxyanthracene-9,10-dione<br>(Xanthopurpurin)<br><br>Lucidin-ω-methyl ether | | | |
| Purified anthraquinone (anthracene-9,10-dione) | Anticancer activity against three human cancer cell lines *viz.* mammary adenocarcinoma (MCF-7), hepatocellular carcinoma (HepG2) and colorectal | N/A | [21] |

| | adenocarcinoma (HCT-116) and application in kidney cancer diagnosis. | | |
|---|---|---|---|
| <br>4-acetyl-10-(1-acetyl-2,5,10-trihydroxy-2-methyl-4-oxo-1,3-dihydroanthracen-9-yl)-8,9-dihydroxy-3-methyl-4H-anthracen-1-one (Setomimycin) | Anticancer against colorectal adenocarcinoma (HCT-116) and mammary adenocarcinoma (MCF-7) cancer cell lines and reduction in primary tumor weight by 76% and tumor volume by 90.5% against mouse mammary carcinoma model (4T1) | Setomimycin causes a reduction in the expression of both MEK/ERK pathways | [22] |
| <br>1,8-Dihydroxy-3-methoxy-anthraquinone (DMA) | Potent suppressor of tumor angiogenesis | Downregulate hypoxia-inducible factor-1α (HIF-1α) protein and PI3K/AKT and c-RAF/ERK pathway | [23] |
| | Antioxidative property of purpurin tested on cultured cells *i.e.* mammalian cells | It can down-regulate NLRP3 | [26] |

*(Table 1) cont.....*

| Anthracene-9,10-dione (Anthraquinone) | (murine macrophage RAW 264.7 cells) and using chemical assay. | inflammaso me assembly. | |
|---|---|---|---|
| 1,5-dihydroxyanthraquinone (Anthrarufin) 1,2,4-trihydroxyanthraquinone (Purpurin) 1,8-dihydroxyanthraquinone (Chrysazin) | | | |
| 1,8-Dihydroxy-3-methylanthracene-9,10-dione (Chrysophanol) | Anticancer and antibacterial properties | High DPPH• radical scavenging capacity | [27] |

*(Table 1) cont.....*

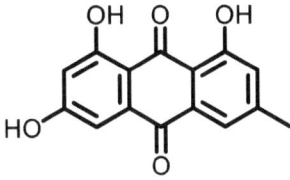

1,3,8-Trihydroxy-6-
methylanthracene-9,10-dione
(Emodin)

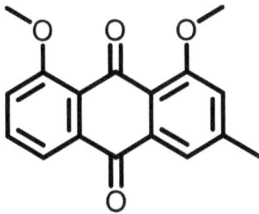

1,8-dimethoxy-3-
methylanthraquinone

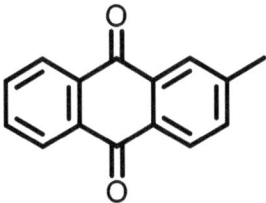

2-methylanthraquinone

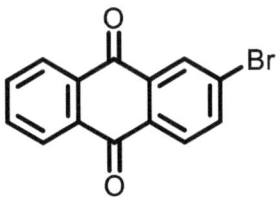

2-bromoanthraquinone

*(Table 1) cont.....*

| | | | |
|---|---|---|---|
| 1,3-dihydroxy-2-methyl anthraquinone (Rubiadin) | | | |
| 1,3-dihydroxy-2,4-dimethoxy-9,10-anthraquinone <br><br> 2-hydroxy-1,4-dimethoxy-9,10-anthraquinone <br><br> 1-methoxy-2-methoxymethyl-3-hydroxy-9,10-anthraquinone | Promising $O_2{}^{\bullet-}$ radical scavenging abilities | N/A | [28] |

*(Table 1) cont.....*

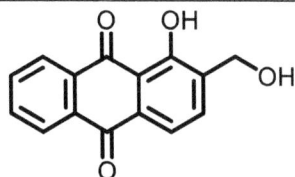

1-hydroxy-2hydroxymethyl-9,10-
anthraquinone

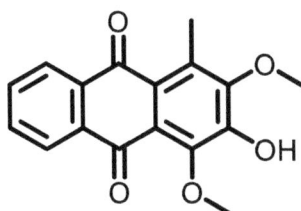

1-methyl-2,4-dimethoxy-3-
hydroxyanthraquinone

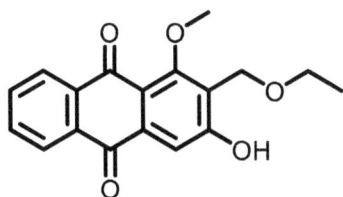

1-methoxy-3-hydroxy-2-
ethoxymethylanthraquinone

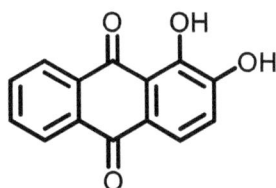

1,2-dihydroxyanthraquinone
(Alizarin)

| | | | |
|---|---|---|---|
| 1,2-dihydroxyanthraquinone (Alizarin) | Deactivation of ROS and protecting polypeptide from hydroperoxyl radical attack. | Deactivation of ROS by Hydrogen atom abstraction (HAA) mechanism. | [29] |

# RECENT DEVELOPMENTS IN PHYTOTHERAPY

Phytotherapy, often referred to as herbal medicine, entails the historical utilization of bioactive substances obtained from plants for therapeutic applications [31]. The term "phytotherapy" derives from the Greek terms, therapeia, denoting the concept of treatment, and phyton, signifying plants [32, 33]. The World Health Organization (WHO) has recognized the importance of medicinal plants as useful therapeutic resources. In addition, the WHO has been aggressively advocating for the investigation and use of traditional knowledge related to plants. This effort is aimed at broadening the field of phytotherapy and aiding the development of innovative medicines [34]. Among natural compounds, anthraquinones have attracted significant interest due to their notable bioactive characteristics and impact on the developing field of natural medicine [3]. The current resurgence of interest in these compounds within scientific investigations highlights their continued significance in contemporary healthcare practices [9]. Anthraquinones have exhibited a diverse array of pharmacological properties, rendering them vital constituents of herbal medicine. The functions of these substances encompass a wide range of applications, ranging from conventional laxatives to more contemporary uses in treating diverse health ailments such as inflammation, cancer, and microbial infections [10]. The extensive range of plant sources that contain anthraquinones, such as *Aloe vera* and *Rheum palmatum*, underscores their multifaceted nature and importance in the field of phytotherapy [35, 36]. The application of anthraquinones in phytotherapy has seen a period of significant change and development in recent years. Recent advancements in research and developing patterns have greatly contributed to our understanding of the therapeutic capabilities of anthraquinones [37, 38]. The advancements in extraction procedures and analytical approaches are significantly influencing the utilization of anthraquinone in natural medicine, leading to a transformation in its uses. The extraction of anthraquinones from botanical sources has traditionally relied on conventional methods such as maceration and Soxhlet extraction [39]. However, recent advancements have introduced novel extraction techniques that offer higher efficiency and selectivity. Supercritical fluid extraction (SFE), microwave-assisted extraction (MAE), and ultrasound-assisted extraction (UAE) are among the innovative approaches that have gained prominence. These methods enhance anthraquinone yield and minimize solvent use, aligning with the growing emphasis on green and sustainable practices in phytotherapy [40]. In literature, Su and Ferguson [41] developed a new method for extracting a pure fraction of anthraquinone aglycones and glycosides from cascara bark and senna leaflets using the aqueous/solvents extraction maceration method. Purification is achieved through recrystallization from solvents, with acetylated in benzene, before further purification. Cao and Zhao [42] investigated optimal extraction conditions for anthraquinone derivatives isolation from *Rheum officinale* Baill rhizomes powder with a particle size of 0.18 mm. The solvents (chloroform, glycerol, and sulfuric acid) are in the ratio of 4:1:1 with an extraction time of 110 min. being the optimal extraction condition. The study utilized natural deep eutectic solvent ultrasound-assisted extraction to extract total anthraquinones from *Rheum palmatum* L. Highly efficient natural deep eutectic solvents (NADES), such as lactic acid, glucose, and water (LGH), were found to be effective in this process. Optimization of extraction yields for aloe-emodin, rhein, emodin, chrysophanol, and physcione was conducted using DM130 macroporous resin, yielding recovery rates of 84.08%, 79.51%, 84.96%, 81.83%, and 78.35%, respectively. LGH was found to be environmentally friendly and highly efficient for extracting anthraquinones from *R. palmatum* L [43]. Two extraction methods (boiling extraction and Soxhlet extraction with 70%, 80%, and 95% ethanol solvents) were employed for the extraction of anthraquinones from three species of plant samples: noni roots, golden shower pods, and leaves of the Cassod tree. A high amount of anthraquinone glycoside is extracted from ethanolic extract (80%) from Cassod tree leaves [44].

Anthraquinones have been discovered to exhibit a diverse array of pharmacological effects, encompassing antioxidant, anti-inflammatory, anticancer, and laxative properties [26,45]. Moreover, the capacity to regulate many cellular pathways and selectively target disease processes renders them very prospective contenders for advancing innovative therapeutic medicines. Understanding the pharmacokinetics of anthraquinones is essential for optimizing dosing regimens and predicting their therapeutic efficacy. Recent research has employed sophisticated analytical methods to investigate anthraquinone absorption, distribution, metabolism, and excretion (ADME). This knowledge has contributed to developing more effective delivery systems, such as nanocarriers and prodrugs, to enhance anthraquinone bioavailability and tissue targeting. Anthraquinones are bioactive natural products used in Chinese medicines, with effects like purgation, anti-inflammation, and anti-cancer. They are mainly absorbed in the intestines and distributed throughout the body. They can be transformed into other anthraquinones, and their pharmacokinetics (PKs) may influence drug compatibility theory [46]. This study investigates diabetic nephropathy and acute liver injury in rats using the pharmacokinetic and pharmacodynamic properties of anthraquinones extracted from Rhubarb (*Rheum palmatum* L.). The rats were treated with different doses of the rhubarb extract (37.5, 75, and 150 mg/kg of rat weight) and pharmacokinetic markers were identified. While no significant pharmacokinetic differences were found between control and diabetic nephropathy rats, distinct variations were observed in liver injury rats. The extract exhibited therapeutic and preventive effects against diabetic nephropathy but did not ameliorate $CCl_4$-induced liver injury in rats. Further research is needed to understand its efficacy for liver injury [47].

In summary, phytotherapy serves as a testament to the enduring significance of these natural compounds in healthcare. Through this exploration, we aim to provide valuable insights into the evolving landscape of anthraquinones and their pivotal roles in shaping the future of phytotherapy.

## PHARMACEUTICAL APPLICATIONS OF ANTHRAQUINONES

Anthraquinones can be utilized as valuable compounds in pharmaceuticals (Table 2). They can serve as lead structures for future drug development. Anthraquinone derivatives exhibit a diverse range of pharmacological effects, encompassing laxative properties, anti-cancer potential, anti-inflammatory and anti-arthritic activities, antifungal and antibacterial capabilities, antiviral effects, and neuroprotective benefits [48-50]. Certain compounds derived from anthraquinone have been explored for their potential applications in photodynamic therapy (PDT) [51]. PDT employs light-sensitive substances to specifically target and eradicate cancer cells or undesirable tissues upon exposure to particular light wavelengths [52]. The clinical practice utilizes various anthraquinone derivatives, including anthracyclines, diacerein (also known as diacetylrhein), and natural hydroxyanthraquinones, primarily employed as laxatives, anticancer and anti-inflammatory medications [10].

Table 2. Different anthraquinone and their pharmaceutical applications.

| Anthraquinone | Brand Name | Pharmaceutical Application | References |
|---|---|---|---|
| Sennosides | Senokot Ex-Lax | Laxative and relief of constipation. | [53] |
| Aloe-emodin | Aloin | Treatment of constipation and as a laxative. | [54] |
| Daunorubicin | Cerubidine | Chemotherapy for certain cancers. | [55] |

*(Table 2) cont.....*

| Doxorubicin | Adriamycin, Doxil | Chemotherapy for various cancers. | [56] |
|---|---|---|---|
| Emodin | NA | Potential use in cancer treatment. | [57] |
| Rhein | Rheumate, Rheutin | Investigated for potential anti-inflammatory and anti-cancer properties. | [58] |
| Danthron | Chrysazin | Laxative for colon cleansing. | [59] |
| Anthralin | Psoriatec, Dritho- Scalp, Zithranol | Treatment of psoriasis and other skin conditions. | [60] |
| Epirubicin | Ellence | Cancer (antineoplastic) medication used to treat breast cancer. | [61] |
| Idarubicin | Idamycin | Chemotherapy drugs used for: acute myeloid leukaemia (AML) and acute lymphoblastic leukaemia (ALL). | [62] |
| Mitoxantrone | Novantrone® | Treats cancer by stopping the growth and spread of cancer cells. | [63] |
| Valrubicin | Valstar | Treatment of bladder cancer. | - |
| Hypericin | Kira® (USA), Jarsin 300®, | Antiviral and non-specific kinase inhibitor. | [64] |

*(Table 2) cont.....*

| | Hypericum 2000 plus® | | |
|---|---|---|---|
| Physcion | Physcion | Hepatoprotective, anti-inflammatory, laxative, anti-microbial and anti-proliferative properties. | [65] |

Anthracycline antibiotics represent a vital group of potent anti-cancer medications effective against a wide range of both solid and hematologic tumor in children and adult populations [66]. These drugs exert their therapeutic action due to the disruption of DNA within cancer cells, inhibiting their ability to undergo division and proliferation [67-69]. Additionally, anthracyclines generate free radicals, which have the potential to harm the DNA and proteins within these cancerous cells [67]. The Food and Drug Administration (FDA) in the United States has granted approval for the clinical utilization of daunorubicin, doxorubicin, epirubicin, idarubicin, and valrubicin [10]. Doxorubicin, commonly known by its brand name Adriamycin, is a widely employed anthracycline antibiotic utilized in the management of various cancers such as breast cancer, leukemia, and lymphoma. Doxorubicin works by inhibiting DNA replication and causing DNA damage in cancer cells [70]. Daunorubicin, another anthracycline, is applied in the treatment of leukemia and certain other cancer types under different brand names [71]. Epirubicin, marketed as Ellence, is employed in breast cancer treatment and is also used for other cancer types [72]. Idarubicin is primarily employed in the management of acute myeloid leukemia (AML) and its chemical structure bears resemblance to that of daunorubicin and doxorubicin [73]. Doxorubicin and epirubicin have demonstrated effectiveness in addressing solid tumors, while valrubicin is specifically utilized for managing bladder cancer [74]. Diacerein, also known as diacetylrhein, is a semi-synthetic compound derived from rhein, with acetylation of the hydroxyl groups at the 1 and 8 positions [75]. It is authorized for use in several European Union countries, including Austria, the Czech Republic, France, Greece, Italy, Portugal, Slovakia, and Spain, for managing osteoarthritis in the hip and knee joints. It can be prescribed either as an impartial treatment or in conjunction with other anti-inflammatory medications.

In addition to this, anthraquinone derivative hydroxyanthraquinones, such as emodin, rhein, physcion, chrysophanol, aloeemodin, and hypericin, along with glycosides like aloe-emodin-8-O-glycoside, chrysophanol-8-Oglycoside, sennosides, cascaroside, and aloins, are the active constituents found in various traditional medicinal plants [76]. These plants, including *Cassia acutifolia*, *Rhamnus purshiana* (Cascara sagrada), *Rhamnus frangula*, *Rheum rhabarbarum*, *Aloe vera*, *Rubia tinctorum*, and *Hypericum perforatum*, have a long history of use for their laxative effects, treatment of kidney and bladder stones, mild sedative properties, and antidepressant potential [76, 77]. Among these compounds, aloin, derived from aloe vera leaves, is notable for its traditional use as a laxative and its inclusion in over-the-counter laxative products. It causes irritation to the lining of the intestines, which in turn stimulates bowel movements. Emodin, a naturally occurring hydroxyanthraquinone, is found in a variety of plant species *i.e.*, rhubarb and Japanese knotweed. It has bioactive potential such as anti-inflammatory, antioxidant, and anticancer [78]. It is also explored to treat inflammatory conditions like arthritis and is used as a supplementary treatment in cancer therapy [79]. Similarly, Rhein, found in *Rhubarb* and *Aloe vera*, has anti-inflammatory and anticancer potential [80]. In laboratory conditions, Rhein has shown a potential to reduce inflammation and inhibit cancer cell growth [81]. Another synthetic hydroxyanthraquinone derivative dantron, was previously used as a laxative but is now less frequently used because of safety concerns and its potential to cause cancer. The Senna (*Cassia*) plant contains hydroxyanthraquinone glycosides, such as sennosides used in many pharmaceutical preparations as natural laxatives [82]. Senna extracts irritate the mucous membranes of the colon and lead to increased peristalsis and bowel movements. On the other hand, Cascara sagrada is a natural laxative made from the bark of the cascara tree and emodin is one of its hydroxyanthraquinones. Cascara sagrada has been used to alleviate constipation and promote regular bowel movements [83]. Anthranoid laxatives are a well-researched group of compounds derived from anthraquinones and are commonly used. They are primarily found in plants in a glycosylated form [82]. These compounds resist being broken down in the stomach when taken orally and are not affected by glucosidase enzymes in the small intestine because of the glycosidic bond between the sugar and the anthraquinone ring, bacterial enzymes, such as glucosidases and reductases, degrade the sugar component of these glycosylated substances in the large intestine, converting them into active aglycones. The epithelial cells of the large intestine are then affected locally by these aglycones, which alters the gut's absorption, excretion, and motility and finally causes diarrhoea. Therefore, anthraquinones is a versatile and pharmacologically important compound having a range of applications as laxatives to a potent anticancer agents. Still, more studies are required to uncover new therapeutic potentials and optimize their efficacy and safety profiles.

## TOXICITY & SAFETY REGULATIONS

Anthraquinone and its derivatives have a wide range of applications in the pharmaceutical sector due to its antioxidant, anti-inflammatory, and anticancer properties and use in industrial sectors for the production of dyes [26, 45]. Hence, it is important to know its toxicity as well as safety regulations so that it can be easily used in various pharmaceutical formulations. Safety regulations and guidelines for the use of anthraquinone typically focus on minimizing exposure and ensuring safe handling. The toxicity and safety assessment of any therapeutic agent or drug is the main concern before its use in any clinical setting. In addition to their usefulness in medicine, several anthraquinone derivatives have the potential to harm cells because of their similarity to the poisonous equivalent anthracene [84]. In literature, anthraquinone derivatives have been reported to clear *in vivo* genotoxicity in male mice colon and kidney after two oral doses of 500-2000 mg/kg body weight/day [10]. In addition to this, the clinical studies of anthraquinone show good therapeutic potential and tolerability in people of all ages including children and elderly people. The use of anthraquinone derivatives such as emodin is reported for their neuroprotective effects in cerebral ischemic stroke (CIS) [85, 86]. Emodin is also reported for the treatment of cancer, inflammation, peptic ulcers, and constipation. Rhein also known as cassic acid and chrysophanol which is one of the derivatives of anthraquinone, is reported for its neuroprotective effects due to its apoptosis and oxidative stress [87]. Whereas in some studies, anthraquinone obtained from *Rhamnus* species was reported for genotoxicity in mammalian cells, tumor promotion, mutation, diarrhoea, anorexia (long-term exposure), renal failure, and nephrotoxicity (regular intake). Also, the long-term use of anthraquinone laxative can cause melanosis coli and permanent damage to colonic tissue and cells causing Crohn's disease or ulcerative colitis [88]. Similarly, Liu *et al.* [89] studied 16 different anthraquinone compounds, out of which Rhein was found to be potentially hepatotoxic based on *in vitro* cytotoxicity and *in silico* modelling. He *et al.* [90] studied the toxic effects of emodin at a concentration of 0.25 µg/ml or above, which can have a negative effect on the hatching success and survival of zebrafish embryos. In addition to this, emodin can induce adverse effects on zebrafish embryos such as abnormal morphology, crooked trunk, and edema proving the toxicity of emodin at low concentrations. Similarly, Oshida *et al.* [91] reported the toxic effects of emodin on testicular gene expression of male mice models reporting testicular toxicity *via.* Insulin-like growth factors (IGF)-1 receptor signalling pathway. Emodin was also reported as a DNA intercalator causing chromosomal aberration [92]. Also, the use of emodin for its naturopathic effects is not approved by the US FDA. Anthraquinone and its derivatives are known to be toxic to aquatic life. They can have harmful effects on aquatic ecosystems and should be handled and disposed of in the environment in a responsible manner. Regulatory agencies, such as the Environmental Protection Agency (EPA) in the United States and the European Chemicals Agency (ECHA) in the European Union have established regulations and guidelines for the use and disposal of anthraquinone and its derivatives. Compliance with these regulations is essential for dealing with anthraquinone and its derivatives [3, 9-10]. Also, the European Food Safety Authority (EFSA) provided a scientific opinion on anthraquinones' long-term side effects on health but daily intake of anthraquinones is not defined for public safety [93]. The toxicity of anthraquinone may vary depending on its specific chemical form and the extent of exposure. It is important to conduct a thorough risk assessment when working with anthraquinone to determine the potential health risks and appropriate safety measures.

## CONCLUSION AND FUTURE PERSPECTIVE

Anthraquinone, an intriguing organic compound has attracted attention for its diverse applications in chemistry, pharmaceuticals, and the textile industry. It plays a significant role as a bioactive compound in the treatment of various diseases and as a promising candidate in phytotherapy applications. It has therapeutic potential with a range of properties such as antimicrobial, anti-inflammatory, antioxidant, anticancer, laxative, and antiviral agents. Their diverse biological activities highlight their value in natural medicine, but further research into their mechanisms, safety, and therapeutic efficacy is essential to fully harness their benefits in clinical applications. Also, the anthraquinone compounds or their derivatives can be used in combination with therapies in order to improve their effectiveness. The anthraquinone compounds must be thoroughly understood from a safety and toxicological perspective in both *in vitro* and *in vivo* systems. With these aspects, anthraquinones can be used as significant bioactive agents with diverse applications in phytotherapy, medicine, and related fields. The future prospects of anthraquinone studies relating to identifying novel compounds derived from unexplored natural sources and understanding the mode of action of anthraquinone in order to create safe and secure medical formulation are likely to be key areas of anthraquinone research in the future. As research advances, anthraquinones may become valuable assets in the development of drugs and therapies for a wide range of health conditions.

# REFERENCES

[1]   Diaz-Muñoz G, Miranda IL, Sartori SK, de Rezende DC, Diaz MAN. Anthraquinones: An overview. Studies in natural products chemistry 2018; 58: 313-38.
http://dx.doi.org/10.1016/B978-0-444-64056-7.00011-8

[2]   A M, Satdive R, Fulzele DP, *et al.* Enhanced production of anthraquinones by gamma-irradiated cell cultures of *Rubia cordifolia* in a bioreactor. Ind Crops Prod 2020; 145: 111987.
http://dx.doi.org/10.1016/j.indcrop.2019.111987

[3]   Duval J, Pecher V, Poujol M, Lesellier E. Research advances for the extraction, analysis and uses of anthraquinones: A review. Ind Crops Prod 2016; 94: 812-33.
http://dx.doi.org/10.1016/j.indcrop.2016.09.056

[4]   Gessler NN, Egorova AS, Belozerskaia TA. [Fungal anthraquinones (review)]. Prikl Biokhim Mikrobiol 2013; 49(2): 109-23.
PMID: 23795468

[5]   Rao GMM, Rao CV, Pushpangadan P, Shirwaikar A. Hepatoprotective effects of rubiadin, a major constituent of Rubia cordifolia Linn. J Ethnopharmacol 2006; 103(3): 484-90.
http://dx.doi.org/10.1016/j.jep.2005.08.073 PMID: 16213120

[6]   Watroly MN, Sekar M, Fuloria S, *et al.* Chemistry, biosynthesis, physicochemical and biological properties of rubiadin: A promising natural anthraquinone for new drug discovery and development. Drug Des Devel Ther 2021; 15: 4527-49.
http://dx.doi.org/10.2147/DDDT.S338548 PMID: 34764636

[7]   Anton R, Haag-Berrurier M. Therapeutic use of natural anthraquinone for other than laxative actions. Pharmacology 1980; 20(1): 104-12.
http://dx.doi.org/10.1159/000137404 PMID: 6246546

[8]   Khan NT. Anthraquinones-A naturopathic compound. Journal of New Developments in Chemistry 2019; 2(2): 25-8.
http://dx.doi.org/10.14302/issn.2377-2549.jndc-18-2569

[9]   Malik MS, Alsantali RI, Jassas RS, *et al.* Journey of anthraquinones as anticancer agents – a systematic review of recent literature. RSC Advances 2021; 11(57): 35806-27.
http://dx.doi.org/10.1039/D1RA05686G PMID: 35492773

[10]  Malik EM, Müller CE. Anthraquinones as pharmacological tools and drugs. Med Res Rev 2016; 36(4): 705-48.
http://dx.doi.org/10.1002/med.21391 PMID: 27111664

[11]  Gălăţanu ML, Panţuroiu M, Popescu M, Mihăilescu CM. Plant extracts with antibiotic effect.Handbook of Research on Advanced Phytochemicals and Plant-Based Drug Discovery. IGI Global 2022; pp. 49-72.
http://dx.doi.org/10.4018/978-1-6684-5129-8.ch004

[12]  Malmir M, Serrano R, Silva O. Anthraquinones as potential antimicrobial agents-A review. In: Antimicrobial Research: Novel Bioknowledge and Educational Programs; Mendez-Vilas,. A., Ed. 2017; pp. 55-61.

[13]    Wang W, Chen R, Luo Z, Wang W, Chen J. Antimicrobial activity and molecular docking studies of a novel anthraquinone from a marine-derived fungus *Aspergillus versicolor*. Nat Prod Res 2018; 32(5): 558-63.
http://dx.doi.org/10.1080/14786419.2017.1329732 PMID: 28511613

[14]    Song ZM, Zhang JL, Zhou K, *et al.* Anthraquinones as potential antibiofilm agents against methicillin-resistant *Staphylococcus aureus*. Front Microbiol 2021; 12: 709826.
http://dx.doi.org/10.3389/fmicb.2021.709826 PMID: 34539607

[15]    Shupeniuk V, Taras T, Sabadakh O, Luchkevich E, Matkivskyi M. Synthesis and antimicrobial activity of nitrogen-containing anthraquinone derivatives. Iraqi J Pharm Sci 2022; 31(2): 193-201.
http://dx.doi.org/10.31351/vol31iss2pp193-201

[16]    Park ES, Moon W-S, Song M-J, Kim M-N, Chung K-H, Yoon J-S. Antimicrobial activity of phenol and benzoic acid derivatives. Int Biodeterior Biodegradation 2001; 47(4): 209-14.
http://dx.doi.org/10.1016/S0964-8305(01)00058-0

[17]    Yirdaw B, Kassa T. Preliminary phytochemical screening and antibacterial effects of root bark of *Ferula communis* (Apiaceae). Vet Med Sci 2023; 9(4): 1901-7.
http://dx.doi.org/10.1002/vms3.1170 PMID: 37392454

[18]    Zhuravleva OI, Chingizova EA, Oleinikova GK, *et al.* Anthraquinone derivatives and other aromatic compounds from marine fungus *Asteromyces cruciatus* KMM 4696 and their effects against *Staphylococcus aureus*. Mar Drugs 2023; 21(8): 431.
http://dx.doi.org/10.3390/md21080431 PMID: 37623712

[19]    Adekunle OD, Adeleke OA, Odugbemi AI, Faboro EO, Lajide L. *In vitro* and *in silico* screening and identification of potential bioactive anthraquinones of *Morinda lucida* benth against pathogenic bacterial target proteins. Discover Applied Sciences 2024; 6(6): 295.
http://dx.doi.org/10.1007/s42452-024-05832-2

[20]    Bajpai VK, Alam MB, Quan KT, *et al.* Cytotoxic properties of the anthraquinone derivatives isolated from the roots of *Rubia philippinensis*. BMC Complement Altern Med 2018; 18(1): 200.
http://dx.doi.org/10.1186/s12906-018-2253-2 PMID: 29970094

[21]    Hasanien YA, Nassrallah AA, Zaki AG, Abdelaziz G. Optimization, purification, and structure elucidation of anthraquinone pigment derivative from *Talaromyces purpureogenus* as a novel promising antioxidant, anticancer, and kidney radio-imaging agent. J Biotechnol 2022; 356: 30-41.
http://dx.doi.org/10.1016/j.jbiotec.2022.07.002 PMID: 35868432

[22]    Manhas RS, Ahmad SM, Mir KB, *et al.* Isolation and anticancer activity evaluation of rare Bisaryl anthraquinone antibiotics from novel *Streptomyces* sp. strain of NW Himalayan region. Chem Biol Interact 2022; 365: 110093.
http://dx.doi.org/10.1016/j.cbi.2022.110093 PMID: 35985519

[23]    Hwang SJ, Cho SH, Bang HJ, Hong JH, Kim KH, Lee HJ. 1,8-Dihydroxy-3-methoxy-anthraquinone inhibits tumor angiogenesis through HIF-1α downregulation. Biochem Pharmacol 2024; 220: 115972.
http://dx.doi.org/10.1016/j.bcp.2023.115972 PMID: 38072164

[24]   Chua HM, Moshawih S, Kifli N, Goh HP, Ming LC. Insights into the computer-aided drug design and discovery based on anthraquinone scaffold for cancer treatment: A systematic review. PLoS One 2024; 19(5): e0301396.
http://dx.doi.org/10.1371/journal.pone.0301396 PMID: 38776291

[25]   Hussain Y, Singh J, Raza W, *et al.* Purpurin ameliorates alcohol-induced hepatotoxicity by reducing ROS generation and promoting Nrf2 expression. Life Sci 2022; 309: 120964.
http://dx.doi.org/10.1016/j.lfs.2022.120964 PMID: 36115584

[26]   Nam W, Kim S, Nam S, Friedman M. Structure-antioxidative and anti-inflammatory activity relationships of purpurin and related anthraquinones in chemical and cell assays. Molecules 2017; 22(2): 265.
http://dx.doi.org/10.3390/molecules22020265 PMID: 28208613

[27]   Gupta RK, Thakuri GMS, Bajracharya GB, Jha RN. Synthesis of antioxidative anthraquinones as potential anticancer agents. BIBECHANA 2021; 18(2): 143-53.
http://dx.doi.org/10.3126/bibechana.v18i2.31234

[28]   Trung NQ, Thong NM, Cuong DH, *et al.* Radical scavenging activity of natural anthraquinones: A theoretical insight. ACS Omega 2021; 6(20): 13391-7.
http://dx.doi.org/10.1021/acsomega.1c01448 PMID: 34056486

[29]   Marković Z, Komolkin AV, Egorov AV, Milenković D, Jeremić S. Alizarin as a potential protector of proteins against damage caused by hydroperoxyl radical. Chem Biol Interact 2023; 373: 110395.
http://dx.doi.org/10.1016/j.cbi.2023.110395 PMID: 36758887

[30]   Liang X, Ding L, Ma J, *et al.* Enhanced mechanical strength and sustained drug release in carrier-free silver-coordinated anthraquinone natural antibacterial anti-inflammatory hydrogel for infectious wound healing. Adv Healthc Mater 2024; 13(23): 2400841.
http://dx.doi.org/10.1002/adhm.202400841 PMID: 38725393

[31]   Ferreira TS, Moreira CZ, Cária NZ, Victoriano G, Silva WF Jr, Magalhães JC. Phytotherapy: an introduction to its history, use and application. Rev Bras Plantas Med 2014; 16(2): 290-8.
http://dx.doi.org/10.1590/S1516-05722014000200019

[32]   Leite PM, Camargos LM, Castilho RO. Recent progess in phytotherapy: A Brazilian perspective. Eur J Integr Med 2021; 41: 101270.
http://dx.doi.org/10.1016/j.eujim.2020.101270

[33]   Petrovska B. Historical review of medicinal plants' usage. Pharmacogn Rev 2012; 6(11): 1-5.
http://dx.doi.org/10.4103/0973-7847.95849 PMID: 22654398

[34]   Hardwicke CJ. The world health organization and the pharmaceutical industry. Common areas of interest and differing views. Adverse Drug React Toxicol Rev 2002; 21(1-2): 51-99.
http://dx.doi.org/10.1007/BF03256183 PMID: 12140907

[35]   Espinosa A, Paz-y-Miño-C G, Santos Y, *et al.* Anti-amebic effects of Chinese rhubarb (*Rheum palmatum*) leaves' extract, the anthraquinone rhein and related compounds. Heliyon 2020; 6(4): e03693.
http://dx.doi.org/10.1016/j.heliyon.2020.e03693 PMID: 32258515

[36]   Stompor-Gorący M. The health benefits of emodin, a natural anthraquinone derived from rhubarb—A summary update. Int J Mol Sci 2021; 22(17): 9522.
       http://dx.doi.org/10.3390/ijms22179522 PMID: 34502424

[37]   Sharifi-Rad J, Herrera-Bravo J, Kamiloglu S, *et al*. Recent advances in the therapeutic potential of emodin for human health. Biomed Pharmacother 2022; 154: 113555.
       http://dx.doi.org/10.1016/j.biopha.2022.113555 PMID: 36027610

[38]   Siddamurthi S, Gutti G, Jana S, Kumar A, Singh SK. Anthraquinone: a promising scaffold for the discovery and development of therapeutic agents in cancer therapy. Future Med Chem 2020; 12(11): 1037-69.
       http://dx.doi.org/10.4155/fmc-2019-0198 PMID: 32349522

[39]   Sasidharan S, Chen Y, Saravanan D, Sundram KM, Yoga Latha L. Extraction, isolation and characterization of bioactive compounds from plants' extracts. Afr J Tradit Complement Altern Med 2011; 8(1): 1-10.
       PMID: 22238476

[40]   Tolkou AK, Mitropoulos AC, Kyzas GZ. Removal of anthraquinone dye from wastewaters by hybrid modified activated carbons. Environ Sci Pollut Res Int 2023; 30(29): 73688-701.
       http://dx.doi.org/10.1007/s11356-023-27550-9 PMID: 37195607

[41]   Su SCY, Ferguson NM. Extraction and separation of anthraquinone glycosides. J Pharm Sci 1973; 62(6): 899-901.
       http://dx.doi.org/10.1002/jps.2600620606 PMID: 4712620

[42]   Cao YC, Zhao YD. Extraction of anthraquinone derivatives from rhubarb rhizomes and their anti-bacterial tests. J Innov Opt Health Sci 2011; 4(2): 127-32.
       http://dx.doi.org/10.1142/S1793545811001290

[43]   Wu YC, Wu P, Li YB, Liu TC, Zhang L, Zhou YH. Natural deep eutectic solvents as new green solvents to extract anthraquinones from *Rheum palmatum* L. RSC Advances 2018; 8(27): 15069-77.
       http://dx.doi.org/10.1039/C7RA13581E PMID: 35541349

[44]   Khoomsab RKK. Extraction and determination of anthraquinone from herbal plant as bird repellent. Sci & Technology Asia 24(1): 14-20.

[45]   Eom T, Kim E, Kim JS. *In vitro* antioxidant, antiinflammation, and anticancer activities and anthraquinone content from Rumex crispus root extract and fractions. Antioxidants (Basel, Switzerland) 2020; 9(8). In:

[46]   Wang D, Wang XH, Yu X, *et al*. Pharmacokinetics of anthraquinones from medicinal plants. Front Pharmacol 2021; 12: 638993.
       http://dx.doi.org/10.3389/fphar.2021.638993 PMID: 33935728

[47]   Li P, Lu Q, Jiang W, *et al*. Pharmacokinetics and pharmacodynamics of rhubarb anthraquinones extract in normal and disease rats. Biomed Pharmacother 2017; 91: 425-35.
       http://dx.doi.org/10.1016/j.biopha.2017.04.109 PMID: 28475921

[48]   Campora M, Francesconi V, Schenone S, Tasso B, Tonelli M. Journey on naphthoquinone and anthraquinone derivatives: new insights in Alzheimer's disease. Pharmaceuticals (Basel) 2021; 14(1): 33.
       http://dx.doi.org/10.3390/ph14010033 PMID: 33466332

[49]    Adlakha K, Koul B, Kumar A. Value-added products of Aloe species: Panacea to several maladies. S Afr J Bot 2022; 147: 1124-35.
        http://dx.doi.org/10.1016/j.sajb.2020.12.025

[50]    Francis AL, Namasivayam SK, Kavisri M, Moovendhan M. Anti-microbial efficacy and notable biocompatibility of *Rosa damascene* and *Citrus sinensis* biomass-derived metabolites. Biomass Conv Bioref 2024; 14: 24787-807.
        https://doi.org/10.1007/s13399-023-04439-8.

[51]    Sulaiman C, George BP, Balachandran I, Abrahamse H. Photoactive herbal compounds: a green approach to photodynamic therapy. Molecules 2022; 27(16): 5084.
        http://dx.doi.org/10.3390/molecules27165084 PMID: 36014325

[52]    Bhole R, Bonde C, Kadam P, Wavwale R. A comprehensive review on photodynamic therapy (PDT) and photothermal therapy (PTT) for cancer treatment. Turk J Biol 2021; 36(1).

[53]    Le J, Ji H, Zhou X, *et al.* Pharmacology, toxicology, and metabolism of sennoside A, A medicinal plant-derived natural compound. Front Pharmacol 2021; 12: 714586.
        http://dx.doi.org/10.3389/fphar.2021.714586 PMID: 34764866

[54]    Froldi G, Baronchelli F, Marin E, Grison M. Antiglycation activity and HT-29 cellular uptake of aloe-emodin, aloin, and aloe arborescens leaf extracts. Molecules 2019; 24(11): 2128.
        http://dx.doi.org/10.3390/molecules24112128 PMID: 31195732

[55]    Lancet JE, Uy GL, Newell LF, *et al.* CPX-351 *versus* 7+3 cytarabine and daunorubicin chemotherapy in older adults with newly diagnosed high-risk or secondary acute myeloid leukaemia: 5-year results of a randomised, open-label, multicentre, phase 3 trial. Lancet Haematol 2021; 8(7): e481-91.
        http://dx.doi.org/10.1016/S2352-3026(21)00134-4 PMID: 34171279

[56]    Johnson-Arbor K, Dubey R. Doxorubicin. In: StatPearls. Treasure Island (FL): StatPearls Publishing; 2023. PMID: 29083582

[57]    Dong X, Fu J, Yin X, *et al.* Huyiligeqi, Ni J. Emodin: a review of its pharmacology, toxicity and pharmacokinetics. Phytother Res 2016; 30(8): 1207-18.
        http://dx.doi.org/10.1002/ptr.5631 PMID: 27188216

[58]    Zhou YX, Xia W, Yue W, Peng C, Rahman K, Zhang H. Rhein: a review of pharmacological activities. Evid-based Complement Altern Med 2015.

[59]    Waterfield J. Laxatives: choice, mode of action and prescribing issues. Nurse prescribing 2007; 5(10): 456-61.
        http://dx.doi.org/10.12968/npre.2007.5.10.27557

[60]    Sehgal VN, Verma P, Khurana A. Anthralin/dithranol in dermatology. Int J Dermatol 2014; 53(10): e449-60.
        http://dx.doi.org/10.1111/j.1365-4632.2012.05611.x PMID: 25208745

[61]    Alavi-Tabari SAR, Khalilzadeh MA, Karimi-Maleh H, Zareyee D. An amplified platform nanostructure sensor for the analysis of epirubicin in the presence of topotecan as two important chemotherapy drugs for breast cancer therapy. New J Chem 2018; 42(5): 3828-32.
        http://dx.doi.org/10.1039/C7NJ04430E

[62]    Carvalho C, Santos R, Cardoso S, *et al.* Doxorubicin: the good, the bad and the ugly effect. Curr Med Chem 2009; 16(25): 3267-85.

http://dx.doi.org/10.2174/092986709788803312 PMID: 19548866

[63]    Marriott JJ, Miyasaki JM, Gronseth G, O'Connor PW. Evidence report: The efficacy and safety of mitoxantrone (Novantrone) in the treatment of multiple sclerosis: Report of the therapeutics and technology assessment subcommittee of the american academy of neurology. Neurology 2010; 74(18): 1463-70.
http://dx.doi.org/10.1212/WNL.0b013e3181dc1ae0 PMID: 20439849

[64]    Borrelli F, Izzo AA. Herb-drug interactions with St John's wort (Hypericum perforatum): an update on clinical observations. AAPS J 2009; 11(4): 710-27.
http://dx.doi.org/10.1208/s12248-009-9146-8 PMID: 19859815

[65]    Pang M, Yang Z, Zhang X, Liu Z, Fan J, Zhang H. Physcion, a naturally occurring anthraquinone derivative, induces apoptosis and autophagy in human nasopharyngeal carcinoma. Acta Pharmacol Sin 2016; 37(12): 1623-40.
http://dx.doi.org/10.1038/aps.2016.98 PMID: 27694907

[66]    Tikhomirov AS, Shtil AA, Shchekotikhin AE. Advances in the discovery of anthraquinone-based anticancer agents. Recent Patents Anticancer Drug Discov 2018; 13(2): 159-83.
http://dx.doi.org/10.2174/1574892813666171206123114 PMID: 29210664

[67]    Horenstein MS, Vander Heide RS, L'Ecuyer TJ. Molecular basis of anthracycline-induced cardiotoxicity and its prevention. Mol Genet Metab 2000; 71(1-2): 436-44.
http://dx.doi.org/10.1006/mgme.2000.3043 PMID: 11001837

[68]    Fernando J, Jones R. The principles of cancer treatment by chemotherapy. Surgery 2015; 33(3): 131-5.
http://dx.doi.org/10.1016/j.mpsur.2015.01.005

[69]    Shrivastava A, Aggarwal LM, Mishra SP, Khanna HD, Shahi UP, Pradhan S. Free radicals and antioxidants in normal *versus* cancerous cells-An overview. Indian J Biochem Biophys 2019; 56(1): 7-19.

[70]    Prasanna PL, Renu K, Valsala Gopalakrishnan A. New molecular and biochemical insights of doxorubicin-induced hepatotoxicity. Life Sci 2020; 250: 117599.
http://dx.doi.org/10.1016/j.lfs.2020.117599 PMID: 32234491

[71]    Preobrazhenskaya MN, Tevyashova AN, Olsufyeva EN, Huang K, Huang H. Second generation drugs-derivatives of natural antitumor anthracycline antibiotics daunorubicin, doxorubicin and carminomycin. J Med Sci-Taipei 2006; 26(4): 119.

[72]    Tagde P, Najda A, Nagpal K, *et al.* Nanomedicine-based delivery strategies for breast cancer treatment and management. Int J Mol Sci 2022; 23(5): 2856.
http://dx.doi.org/10.3390/ijms23052856 PMID: 35269998

[73]    Hevener K, Verstak TA, Lutat KE, Riggsbee DL, Mooney JW. Recent developments in topoisomerase-targeted cancer chemotherapy. Acta Pharm Sin B 2018; 8(6): 844-61.
http://dx.doi.org/10.1016/j.apsb.2018.07.008 PMID: 30505655

[74]    Dossou AS. Enhancing the Solubility of Valrubicin *via.* Albumin and TPGS Formulations. University of North Texas Health Science Center at Fort Worth 2018.

[75]    Gangurde SA, Laddha KS, Joshi SV. A greener approach to synthesis of diacerein. Indian Drugs. 2019; 56(4).
http://dx.doi.org/10.53879/id.56.04.11784

[76]    Huang WY, Cai YZ, Zhang Y. Natural phenolic compounds from medicinal herbs and dietary plants: potential use for cancer prevention. Nutr Cancer 2009; 62(1): 1-20.
       http://dx.doi.org/10.1080/01635580903191585 PMID: 20043255

[77]    Wang P, Wei J, Hua X, *et al.* Plant anthraquinones: Classification, distribution, biosynthesis, and regulation. J Cell Physiol 2023; jcp.31063.
       http://dx.doi.org/10.1002/jcp.31063 PMID: 37393608

[78]    Foster M, Hunter D, Samman S. Evaluation of the nutritional and metabolic effects of Aloe vera. Herbal Medicine: Biomolecular and Clinical Aspects. 2nd edition. 2011. In:
       http://dx.doi.org/10.1201/b10787-4

[79]    Cheng L, Chen J, Rong X. Mechanism of emodin in the treatment of rheumatoid arthritis. Evid-based Complement Altern Med 2022.
       http://dx.doi.org/10.1155/2022/9482570

[80]    Zhou W, Bounda GA, Yu F. Pharmacological potential action of rhein and its diverse signal transduction-a systematic review. World J Pharm Res 2014; 3(3): 3599-626.

[81]    Henamayee S, Banik K, Sailo BL, *et al.* Therapeutic emergence of rhein as a potential anticancer drug: A review of its molecular targets and anticancer properties. Molecules 2020; 25(10): 2278.
       http://dx.doi.org/10.3390/molecules25102278 PMID: 32408623

[82]    Vilanova-Sanchez A, Gasior AC, Toocheck N, *et al.* Are Senna based laxatives safe when used as long term treatment for constipation in children? J Pediatr Surg 2018; 53(4): 722-7.
       http://dx.doi.org/10.1016/j.jpedsurg.2018.01.002 PMID: 29429768

[83]    Mohiuddin AK. Alternative treatments for minor GI ailments. Innov Pharm Technol 2019; 10(3)
       http://dx.doi.org/10.24926/iip.v10i3.1659

[84]    Shukla V, Asthana S, Gupta P, Dwivedi PD, Tripathi A, Das M. Toxicity of naturally occurring anthraquinones. In Adv. Advances in Molecular Toxicology 2017; 11: 1-50.
       http://dx.doi.org/10.1016/B978-0-12-812522-9.00001-4

[85]    Mitra S, Anjum J, Muni M, *et al.* Exploring the journey of emodin as a potential neuroprotective agent: Novel therapeutic insights with molecular mechanism of action. Biomed Pharmacother 2022; 149: 112877.
       http://dx.doi.org/10.1016/j.biopha.2022.112877 PMID: 35367766

[86]    Li Y, Xu Q, Shan C, Shi Y, Wang Y, Zheng G. Combined use of emodin and ginsenoside Rb1 exerts synergistic neuroprotection in cerebral ischemia/reperfusion rats. Front Pharmacol 2018; 9: 943.
       http://dx.doi.org/10.3389/fphar.2018.00943 PMID: 30233364

[87]    Zhao Q, Wang X, Chen A, *et al.* Rhein protects against cerebral ischemic-/reperfusion-induced oxidative stress and apoptosis in rats. Int J Mol Med 2018; 41(5): 2802-12.
       http://dx.doi.org/10.3892/ijmm.2018.3488 PMID: 29436613

[88]    Yang N, Ruan M, Jin S. Melanosis coli: A comprehensive review. Gastroenterol Hepatol 2020; 43(5): 266-72.

http://dx.doi.org/10.1016/j.gastrohep.2020.01.002 PMID: 32094046

[89]   Liu Y, Mapa MST, Sprando RL. Liver toxicity of anthraquinones: A combined *in vitro* cytotoxicity and *in silico* reverse dosimetry evaluation. Food Chem Toxicol 2020; 140: 111313.

http://dx.doi.org/10.1016/j.fct.2020.111313 PMID: 32240702

[90]   He Q, Liu K, Wang S, Hou H, Yuan Y, Wang X. Toxicity induced by emodin on zebrafish embryos. Drug Chem Toxicol 2012; 35(2): 149-54.

http://dx.doi.org/10.3109/01480545.2011.589447 PMID: 21834668

[91]   Oshida K, Hirakata M, Maeda A, Miyoshi T, Miyamoto Y. Toxicological effect of emodin in mouse testicular gene expression profile. J Appl Toxicol 2011; 31(8): 790-800.

http://dx.doi.org/10.1002/jat.1637 PMID: 21319176

[92]   Müller SO, Eckert I, Lutz WK, Stopper H. Genotoxicity of the laxative drug components emodin, aloe-emodin and danthron in mammalian cells: Topoisomerase II mediated? Mutat Res Genet Toxicol Test 1996; 371(3-4): 165-73.

http://dx.doi.org/10.1016/S0165-1218(96)90105-6 PMID: 9008718

[93]   Younes M, Aggett P, Aguilar F, *et al.* EFSA Panel on food additives and nutrient sources added to food (ANS), safety of hydroxyanthracene derivatives for use in food. EFSA Journal 2018; 16(1):e05090.

https://doi.org/10.2903/j.efsa.2018.5090 PMID: 32625659

CHAPTER 4

# Anthraquinone Derivatives as Potent Anti-Cancer Agents

**Vasantha Kumar[1], Prashasthi V. Rai[1], Ganavi D.[2], Vijesh A. M.[3]** and **Roopa Nayak[4],\***

[1]*Department of PG Studies and Research in Chemistry, Sri Dharmasthala Manjunatheshwara College (Autonomous), Ujire, Karnataka-574240, India*

[2]*Department of Chemistry, Sri Dharmasthala Manjunatheshwara College (Autonomous), Ujire, Karnataka-574240, India*

[3]*PG Department of Chemistry, Payyanur College, Payyanur, Kannur University, Kerala-670327, India*

[4]*Department of Radiation Biology and Toxicology, Manipal School of Life Sciences, Manipal Academy of Higher Education, Manipal, Karnataka-576104, India*

**Abstract:** Cancer is one of the high-mortality-causing diseases in the world. It causes a serious threat to mankind with an estimated 10 million deaths every year. Identifying novel drug candidates for the treatment of various types of cancer is a prime research area in medicinal chemistry. Even though many drug molecules are used in cancer treatment, they suffer from various drawbacks, such as low selectivity, toxicity, and resistance to new tumor cells. Hence, the search for novel anti-cancer agents is a continuous process in order to develop more efficient and less toxic chemotherapeutic agents. Among them, anthraquinone, a diketo derivative of anthracene, has gained much interest in the search for novel anticancer agents. Since anthraquinone is present in various natural products, it has diverse biological properties, which makes it a prominent scaffold in medicinal chemistry. Many anthraquinone classes of anticancer agents have been developed in the last decade and remain the first treatment option for cancer. The search for novel anthraquinone derivatives by the modification of the core structure or by the introduction of newer substituents to attain higher selectivity and efficacy has gained considerable interest recently. This book chapter concisely summarizes the anticancer activities of various anthraquinone derivatives reported by researchers, either derived from natural sources or synthetically prepared.

**Keywords:** Anthraquinone derivatives, Anti-cancer agents, Chemotherapy, Emodin derivatives, Heterocycles, Natural products.

---
**\* Corresponding author Roopa Nayak:** Department of Radiation Biology and Toxicology, Manipal School of Life Sciences, Manipal Academy of Higher Education, Manipal, Karnataka-576104, India; E-mail: roopa.nayak@manipal.edu

## INTRODUCTION

The group of quinones and their derivatives including benzoquinones and naphthoquinones, which constitute the large number of natural pigments are called anthraquinones. These are the aromatic organic compounds with the chemical formula $C_{14}H_8O_2$, where keto groups are located on the central ring (Fig. **1**). The compounds belonging to this class are abundantly produced from natural sources like plant parts such as roots, rhizomes, flowers, and fruits, while others are present in lichens, fungi, and animals [1].

### Cancer and the Need for New Chemotherapeutic Agents

Cancer is a highly assorted, multifactorial disease with a group of disorders characterized by abnormal cell growths in any part of the body, which is mainly caused by genetic and environmental factors [1]. The WHO highlighted that cancer is one of the leading causes of death in humans worldwide, and nearly 10 million mortalities were reported in 2020. That means one in every six deaths is due to cancer, which shows its brutal nature. Tobacco use, alcohol consumption, high body mass index, and low fruit and vegetable intake are the major causes of mortality from cancer [1]. Out of the 200 various types of cancers reported, lung, breast, stomach, colon and rectum, skin, and prostate cancers are the most common types found in humans. Most of the cancer types can be treated effectively only if they are detected in the early stages. It becomes more complicated in the final stages as the medication becomes ineffective and there is a possibility of recurrence after treatment.

Diagnosis of the correct types and stages of cancer is very much crucial for proper and effective treatment, because every cancer type requires a specific treatment protocol. Surgery, radiotherapy, and/or systemic therapy (chemotherapy, hormonal treatments, and targeted biological therapies) are commonly employed for curing the cancer based on the types of cancer and human beings treated. Most of the presently available cancer treatments are very expensive and cannot be afforded by a common man. Chemotherapeutic agents are the drugs used during chemotherapy, and it is one of the most common and effective cancer treatment options available. In general, cytotoxic agents destroy fast-growing cells, like cancer cells, and prevent them from multiplying. Non-selectivity of the chemotherapeutic agents and their side effects still remain a major source of concern. Neurotoxicity owing to present anti-cancer drugs can be resilient in the body even after the end of treatment, and it reduces the functional power and quality of life in cancer survivors. In their review article, Lustberg *et al.* wrote well about the possible side effects of cancer chemotherapeutic agents [2].

Most of the anti-cancer drugs are made up of synthetic compounds, either a single derivative or a combination of drugs. Taxoids, docetaxel (Taxotere), and paclitaxel (Taxol), among other well-known anticancer medications derived from natural products, will likely lead to the discovery of many more active molecules among the 300,000 plant species that are currently being studied [3]. Different cytotoxic agents fight cancer cells *via* different mechanisms, and many are still unknown to researchers. The development of new, efficient, cheaper, and selective anti-cancer agents with fewer side effects and different mechanisms of action is crucial to the fight against cancer and to reduce mortality. Hence, this review chapter emphasizes the progress in recent developments of anticancer drugs with respect to anthraquinone derivatives with the aim of developing new derivatives as anti-cancer agents.

**Naturally Occurring Anthraquinones in Anticancer Drug Discovery**

Anthraquinones (Fig. **1 (1)**) are naturally occurring secondary metabolites in plants [4], fungi [5], and bacteria [6]. There are 3,798 anthraquinone derivatives reported in the PubChem database. Anthraquinones have been isolated from many plants belonging to the *Rubiaceae* [7], *Fabaceae*, *Ranunculaceae*, and *Asphodelaceae* families, and anthraquinones are active ingredients used in traditional Chinese medicines [8]. Although the exact mechanism of biosynthesis of anthraquinones in plants remains unclear, polyketide and shikimate pathways are the most important and widely accepted pathways [4]. Some of the most common anthraquinones, found in plants and microbes, are emodin (Fig. **1 (2)**), aloe-emodin (Fig. **1 (3)**), rhein (Fig. **1 (4)**), chrysophanol (Fig. **1 (5)**), alizarin (Fig. **1 (6)**), quinizarin (Fig. **1 (7)**), damnacanthal (Fig. **1 (8)**), rubiadin (Fig. **1 (9)**), purpurin (Fig. **1 (10)**), physcion (Fig. **1 (11)**), and danthron (Fig. **1 (12)**). Anthraquinones have shown a wide range of pharmacological activities, including anti-inflammatory [9], antidiabetic [10], antimicrobial [11], antiviral [12], antimalarial [13], and antiplatelet [14] activities. Natural anthraquinones have been used as photosensitizers in photodynamic therapy of cancer [15].

Emodin is one of the important naturally occurring anthraquinone molecules that exhibit diverse applications in medicinal chemistry. It is isolated from various plants belonging to the *Rhamnaceae* [16], *Polygonaceae* [17], *Fabaceae* [18], and *Asteraceae* [19] families. It is also one of the fungal metabolites and is isolated from various fungal species like *Aspergillus* [20], *Cladosporium* [21], *Chaetomium* [22], and *Penicillium* [23]. It is considered one of the most promising anthraquinones (AQ) due to its diverse biological activities [24]. Several review articles have summarized the biological activities of emodin derivatives [25]. It is being extensively screened for its anticancer activity against various cancer cell lines and in-depth mechanistic studies [26,27]. Aloe-emodin is one of the major bioactive anthraquinone compounds mainly found in aloe vera (*Aloe barbadensis miller*) [28], a perennial cactus-like plant found in tropical climates worldwide. Aloe has been used as a traditional remedy in many cultures for centuries, and it continues to be extremely popular in various medicinal applications [29]. It is also one of the AQ molecules extensively screened for its anticancer activities [30-32].

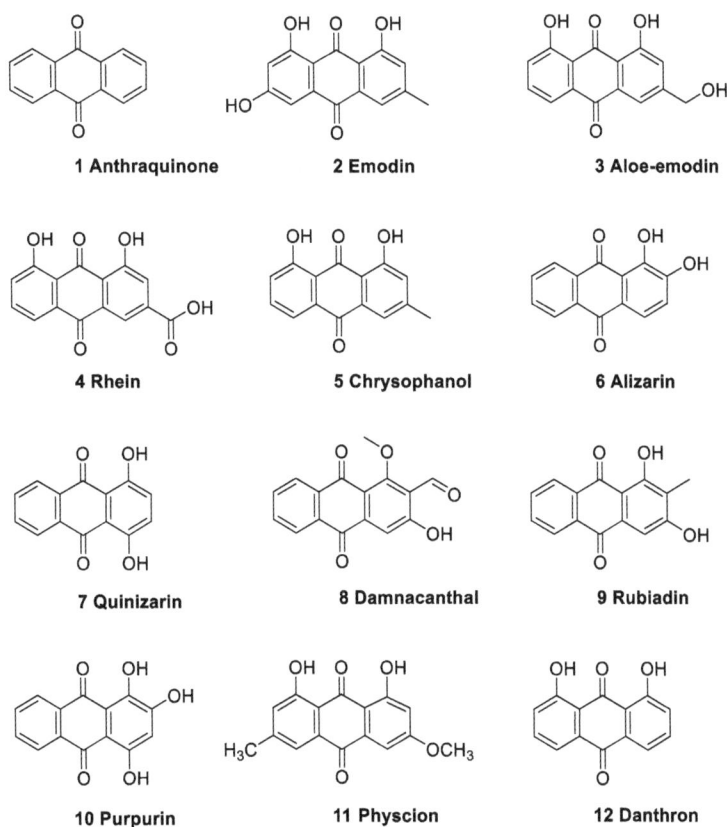

**Fig. (1).** Anthraquinone (**1**) and naturally occurring anthraquinones (**2-12**).

Rhein is isolated from medicinal herbs such as *Rheum palmatum* L., *Cassia tora* L., *Polygonum multiflorum* Thunb., and *Aloe barbadensis* Miller, which is a major traditional Chinese medicine. It is known to possess anticancer activity against various cancer cell lines, and few have noted its anticancer activity against oral [33], colorectal [34], and other cancers [35] by modulating signaling pathways such as mTOR degradation mediated by ubiquitin and suppressing mTOR signaling *in vitro* and *in vivo*. Its anticancer activities are extensively reviewed in the literature [35-37]. Chrysophanol, a constituent of anthraquinone in rhubarb, has shown cytotoxicity against breast cancer [38], lung cancer [39], colon cancer [40], and choriocarcinoma [41] by increasing the ROS production and decreasing mitochondrial membrane potential. Damnacanthal is another important natural AQ derived from the roots of *Morinda citrifolia* L., which is traditionally referred to as "noni" in several parts of the world [42]. It is used in the treatment of various diseases such as cancer [43-45], infections [46], diabetes, asthma, cough, hypertension, pain, ulcers, wounds, hemorrhoids, and rheumatoid arthritis [42]. Many other natural AQs, such as alizarin and quinizarin, have exhibited various biological activities [47,48] and are also explored for their anticancer activities [49,50]. Many compounds having anthraquinone core rings are used in clinics, including Daunorubicin (Fig. **2** (**13**)), Idarubicin (Fig. **2** (**14**)), Doxorubicin (Fig. **2** (**15**)), Epirubicin (Fig. **2** (**16**)), Valrubicin (Fig. **2** (**17**)), Mitoxantrone (Fig. **2** (**18**)), Banoxantrone (Fig. **2** (**19**)) and Losoxantrone (Fig. **2** (**20**)) in the treatment of cancer [51,52].

Keeping in view the biological importance of anthraquinone derivatives, especially their anti-cancer properties, many research groups reported the anti-cancer screening of various naturally occurring AQ derivatives and synthetic AQ derivatives.

## ANTICANCER ACTIVITY OF EMODIN DERIVATIVES

Li and his team synthesized a few derivatives of emodin and screened them for antitumor activity on eight different tumor cell lines and also on normal intestinal epithelial cells (FHC) [53]. Results showed the compound (Fig. **3** (**21**)) exhibited good activity with an $IC_{50}$ value of 108.1 μM against the HCT116 cell line and 118.9 μM against the SW480 cell line (Fig. **3**). This compound was further found to be safer as it did not show any cytotoxicity against normal FHC cell lines. Further, it caused the arrest of HCT116 cells in the G0/G1 phase (Fig. **4**) and induced apoptosis in it. Also, it increased intracellular levels of ROS in HCT116 cells.

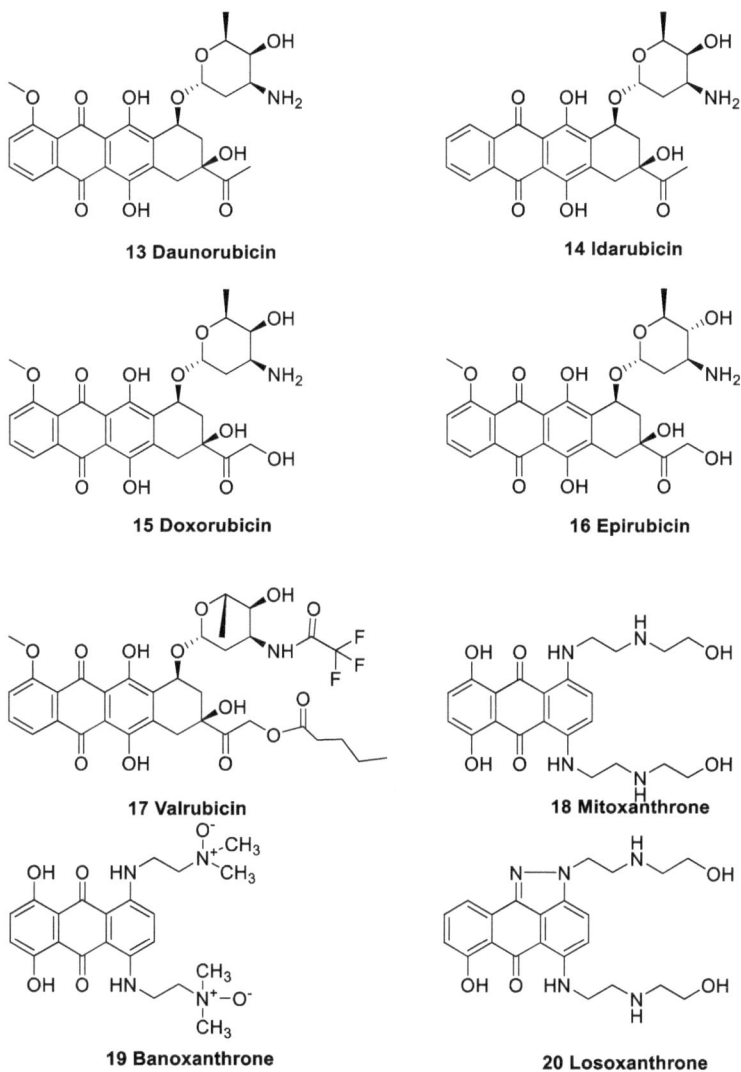

**Fig. (2).** Anthraquinone-based approved anticancer drugs.

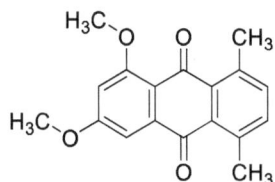

**21**

**Fig. (3).** Structure of compound **21** reported by Li *et al.* [53].

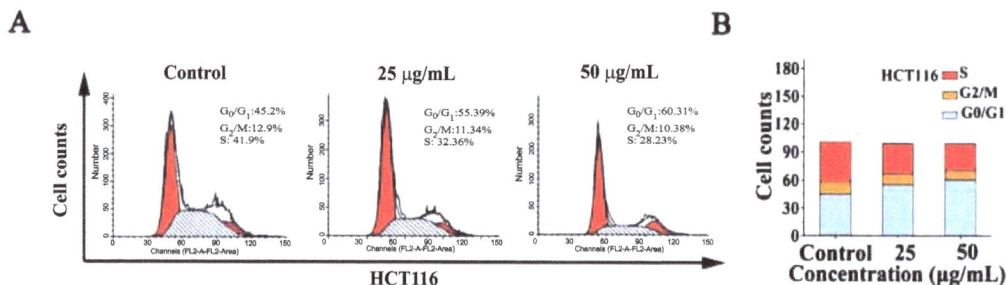

**Fig. (4).** The treatment of compound **21** leads to cell cycle arrest in the G0/G1 phase of HCT116 cells. **A)** HCT116 cell cycle arrest was detected by flow cytometry upon treatment with compound **21** (25-50 μg/mL, 24 h). **B)** Histogram analysis showing the cell distribution in different phases of the cell cycle (Reused with permission from [53].

A new series of emodin-based quaternary ammonium salt compounds were screened for cytotoxic activity against HepG2, BGC-823, AGS, and HELF cell lines by Wang and coworkers [54]. Compound mixtures **22** and **23** (Fig. **5**) exhibited excellent potency upon all tested cell lines in comparison to standard paclitaxel and emodin (Table **1**). This indicates that the presence of a quaternary ammonium group enhances the activity. Further, mechanistic studies of **22** and **23** on AGS cells indicated the inhibition of cell growth through inducing apoptosis, and it arrested the cell cycle in the G0/G1 phase. Moreover, the activation of caspase-3 and caspase-9 was observed after 24 hours of treatment. Significant *in vivo* antitumor activity of 48.6 ± 3.4% and 49.9 ± 2.5% was observed for **22** and **23,** respectively, against H22 xenograft tumor cells.

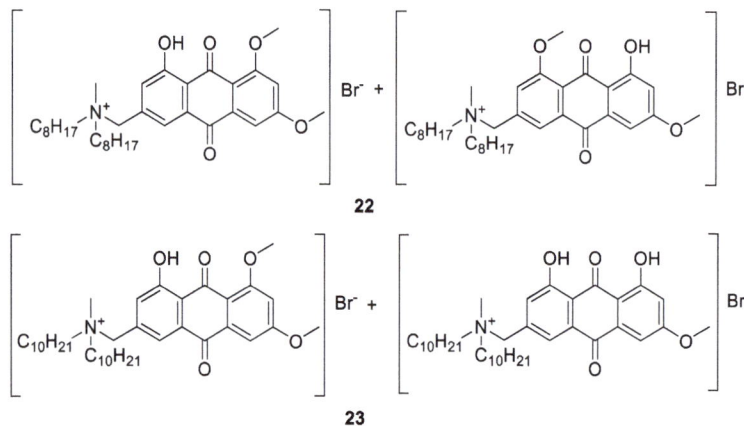

**Fig. (5).** Potent emodin-based quaternary ammonium salt compounds (**22** and **23**) reported by Wang *et al.* [54].

**Table 1. Anticancer activity of 22 and 23 against various cell lines (IC$_{50}$ in μM) [54].**

| Compound | IC$_{50}$ (μM) | | | |
|---|---|---|---|---|
| | HepG2 | BGC-823 | AGS | HELF |
| 22 | 2.11 ± 1.9 | 5.43 ± 2.7 | 4.97 ± 5.6 | 4.81 ± 2.2 |
| 23 | 3.41 ± 3.5 | 3.87 ± 1.5 | 2.91 ± 1.9 | 6.14 ± 1.3 |
| Paclitaxel | 10.32 ± 0.02** | 9.26 ± 0.14 | 15.40 ± 0.03 | 7.14 ± 0.09 |
| Emodin | >70 | >70 | >70 | >70 |

Anticancer activity of a novel series of amino acid-endowed emodin derivatives was carried out by Yang and team members on HepG2 and MCF-7 cell lines and found the most potent activity against HepG2 and MCF-7 with weaker toxicity against the normal L02 cell line [55]. A study has identified six potent compounds that are much more active than emodin and paclitaxel. In particular, compound **24** emerged to be highly potent with IC$_{50}$ 4.95 ± 0.14, 5.92 ± 0.36, and 13.96 ± 0.27 μM against HepG2, MCF-7, and L02 cell lines, respectively (Fig. **6**). Further, it induced apoptosis through the mitochondrial pathway and caused the arrest of HepG2 cells in the G0/G1 phase. Further wound healing studies depicted that it inhibited the migration of HepG2 cancer cells.

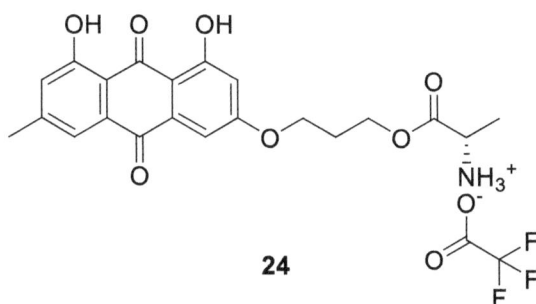

**Fig. (6).** Potent amino acid-endowed emodin compound **24** reported by Yang *et al.* [55].

A novel series of emodin derivatives having o-alkylated, amide and ester groups were synthesized by Narender *et al.* [56]. In the anticancer screening studies of these derivatives, compound **25** emerged as a potent inhibitor against HepG2, DU-145, MCF-7, PC-3, and HEK-293 cell lines with an IC$_{50}$ value of 3.5 ± 0.11, 5.6 ±

0.01, 7.5 ± 0.013, 4.5 ± 0.022, and 12 ± 0.9 µM, respectively (Fig. **7**). Compound **25** has increased annexin V and PI positive cells significantly, thereby inducing apoptosis through caspase-3 activation in HepG2 cells. Further, it was confirmed that the anticancer mechanism works through the arrest of cells in the S phase on HepG2 cells and it has shown DNA intercalation with DOX. In their other work, Narender *et al.* identified two potent anticancer emodin derivatives, **26** and **27,** out of fourteen compounds [57]. These compounds displayed remarkable activity against HepG2 ($IC_{50}$ = 5.64 and 10.44 µM) and MCF-7 ($IC_{50}$ = 13.03 and 5.027 µM) cell lines, respectively. These compounds induce apoptosis *via* caspase-3 activation in HePG2 cells. These compounds arrest the cell cycle in the G1/S phase. Further, it was observed that **27** exhibited a better DNA binding interaction than **26**.

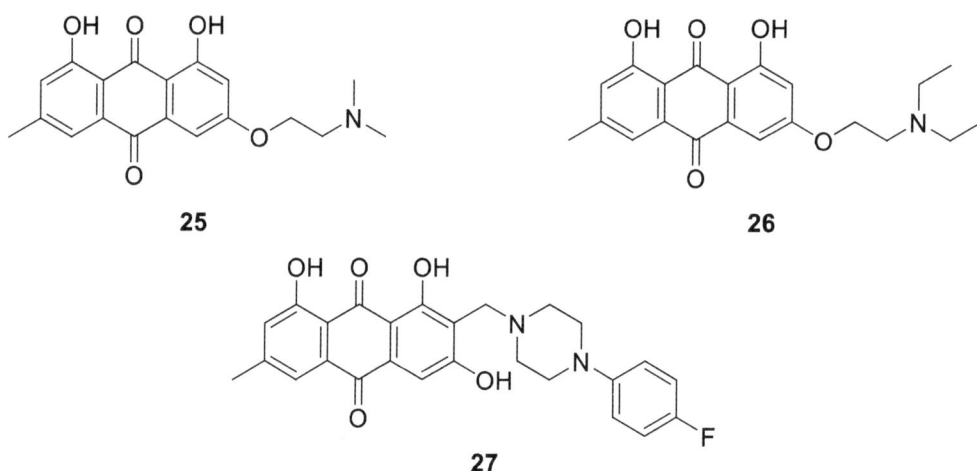

**Fig. (7).** Potent emodin derivatives (**25-27**) reported by Narender *et al.* [57].

Tan *et al.* [58] synthesized and reported the anticancer properties of eight pyrazole-endowed emodin derivatives against B16, HepG2, and LLC cell lines. Among them, compound **28** emerged as a potent agent with $IC_{50}$ values of 1.8, 2.7, and 3.8 µM against these cell lines, respectively (Fig. **8**). Moreover, all these derivatives were screened for DNA binding studies on CT-DNA (Fig. **9**), and compound **28** has shown the highest binding property with a binding constant of $7.47 \times 10^{-5}$/M, which is much more than the emodin. Further, a detailed mechanism of this derivative was studied by the same group in another work, and it was reported that it induced apoptosis and exhibited caspase-3 activation in HepG2 cells. Also, it causes the cleavage of PARP protein in HepG2 cells [59].

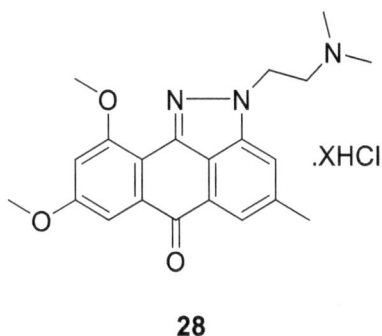

**28**

**Fig. (8).** Pyrazole endowed emodin derivative (**28**) reported by Tan *et al*. [58].

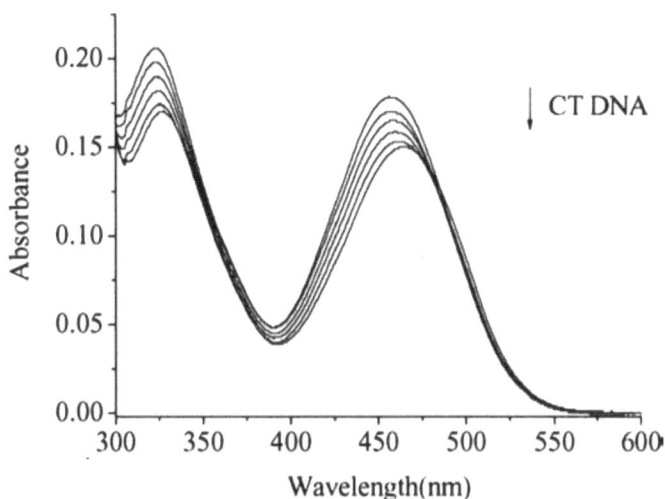

**Fig. (9).** Absorbance titration of compound **28** at an increasing CT DNA concentration (Reused with permission from [58]).

Various emodin derivatives containing amine units were synthesized and screened for their cytotoxicity on primary rat hepatocytes and HepG2 cells by Teich *et al*. [60]. Among the prepared derivatives, compound **29** exhibited excellent inhibition with an $EC_{50}$ value of $8.4 \pm 0.5$ against HePG2 cells and did not exhibit any toxicity on normal hepatocyte cells (Fig. **10**). It was shown to be bound to the second ATP-binding loop of pgp.

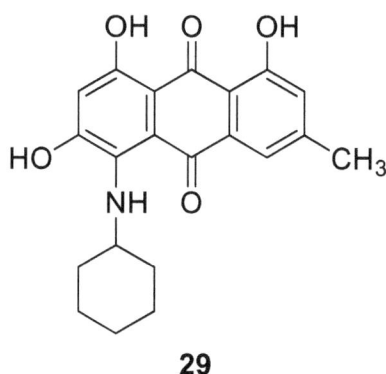

**29**

**Fig. (10).** Potent emodin amine compound **29** reported by Teich *et al*. [60].

Shao *et al*. [61] synthesized novel quaternary amine salts of emodin and screened them for their anticancer activity on BGC-823, HepG2, and AGS cancer cell lines, as well as normal HELF cell lines. Among them, compounds **30** and **31** (Fig. **11**) emerged as potent derivatives with the highest activity on all cancer cell lines and were non-toxic to normal cell lines (Table **2**). With these compounds, cell cycle progression of AGS cells was arrested in the G0/G1 phase and exerted apoptosis *via* caspase-3 and caspase-9 enzyme activation. *In vivo,* anticancer studies revealed significant inhibition at low dosages for **30** and high dosages for **31**.

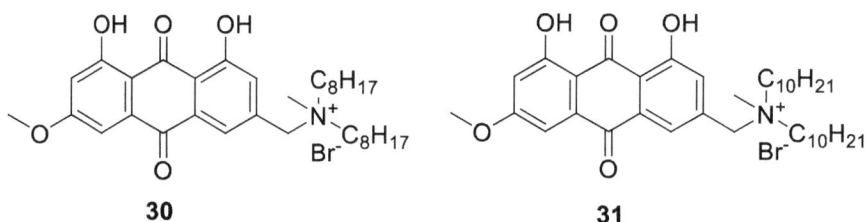

**30**          **31**

**Fig. (11).** Potent emodin quaternary amine salt derivatives (**30** and **31**).

In the studies conducted by Zheng and group [62], evaluation of anticancer activity on HepG2, A375, BGC-823, and normal HELF cell lines was carried out. Compound **32** showed the highest activity with $IC_{50}$ values of $1.39 \pm 0.02$, $2.79 \pm 0.80$, and $4.12 \pm 0.48$ µM against A375, BGC-823, and HepG2, respectively (Fig. **12**). It was non-toxic to normal HELF cell lines with an $IC_{50}$ value of $9.65 \pm 1.70$ µM. Further, it induced apoptosis in A375 cells by increasing ROS generation and by activating caspase-3 and P53.

**Table 2. Anticancer activity of compounds 30 and 31 against various cell lines (IC$_{50}$ in μM) [61].**

| Compound | IC$_{50}$ (μM) | | | |
|---|---|---|---|---|
| | HepG2 | BGC-823 | AGS | HELF |
| 30 | $6.44 \pm 3.4$ | $5.15 \pm 2.1$ | $3.76 \pm 3.5$ | $8.08 \pm 4.6$ |
| 31 | $8.03 \pm 1.9$ | $7.48 \pm 5.1$ | $3.79 \pm 3.8$ | $15.07 \pm 6.6$ |
| Paclitaxol | $10.32 \pm 0.02$ | $9.26 \pm 0.14$ | $15.40 \pm 0.03$ | $7.14 \pm 0.09$ |
| Emodin | $>70$ | $>70$ | $>70$ | $>70$ |

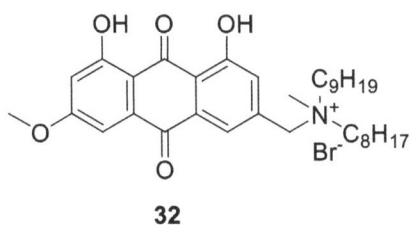

**32**

**Fig. (12).** Potent emodin quaternary amine salt derivative (**32**).

Koerner *et al.* [63] in their study revealed the ATP citrate lyase inhibition studies of some emodin derivatives with modifications at the 2$^{nd}$, 3$^{rd}$, and 5$^{th}$ positions. The study identified dibromo derivative **33** and chloro derivative **34**, which exhibited the highest ACl inhibition with IC$_{50}$ values of 2.9 and 0.283 μM (Fig. **13**).

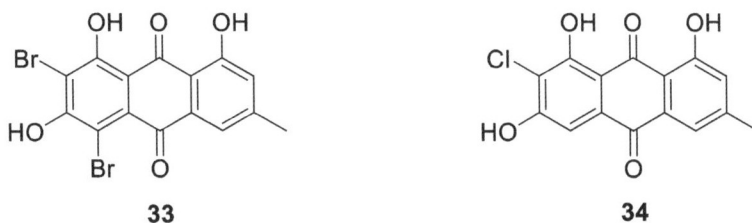

**33**                                      **34**

**Fig. (13).** Potent halogenated emodin compounds **33** and **34** reported by Koerner *et al.* [63].

## ANTICANCER ALOE-EMODIN DERIVATIVES

Cui *et al.* in their study prepared derivatives of aloe-emodin with groups at the 15 position [64]. These were tested for cytotoxic activities on HCT116 and HepG2 cell lines. Among them, compounds **35-38** (Fig. **14**) exhibited very good activities against these two cell lines (Table **3**). It was observed that compound **38** exhibited the highest activity and could be considered a new candidate for developing efficient anticancer drugs.

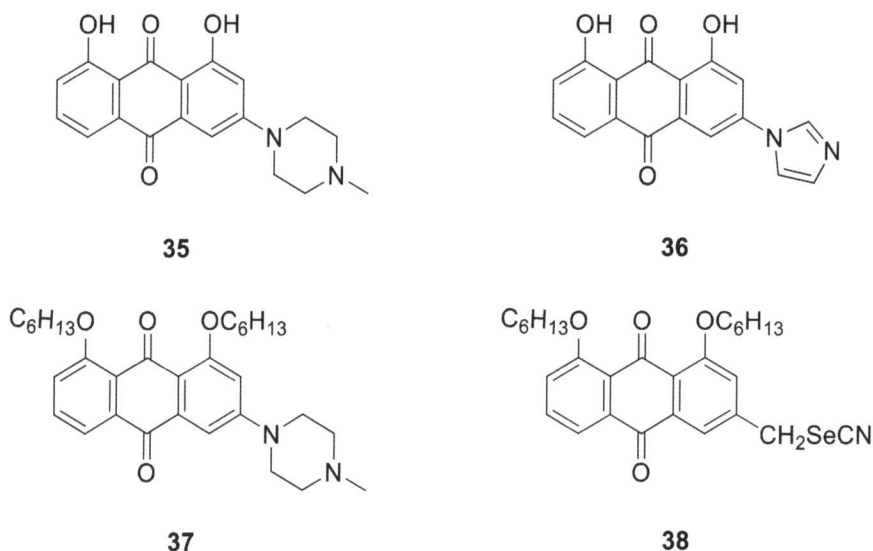

Fig. (**14**). Structures of potent aloe-emodin derivatives (**35-38**).

Table 3. Anticancer activity of compounds 35-38 [64].

| Compound | $IC_{50}$ (µM) | |
|---|---|---|
| | HCT 116 cells | Hep G2 cells |
| 35 | 2.5±0.1 | 2.3±0.4 |
| 36 | 2.5±0.3 | 2.0±0.2 |
| 37 | 2.4±0.2 | 2.2±0.3 |
| 38 | 0.2±0.08 | 0.7±0.1 |

In a study, Thimmegowda *et al.* [65] improved the water solubility and anticancer activity of aloe emodin by the combination of different substituted aromatic amines and amino acid esters. New derivatives **39** and **40** showed improved activity against HepG2 and NCI-H460 tumor cell lines (Fig. **15**). Their paper also discussed the detailed SAR with respect to its activities. Cytotoxicity of amino derivatives **39** and **40** is tabulated in Table **4**.

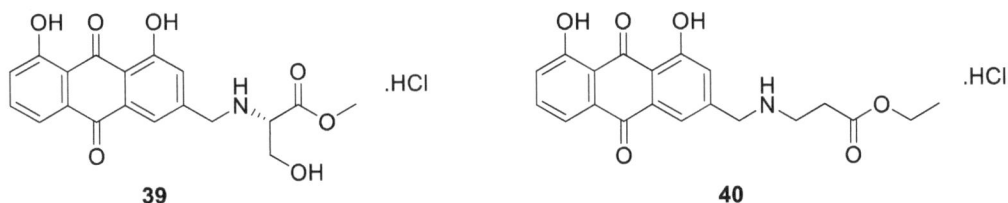

Fig. (**15**). Structures of amino derivatives of aloe-emodin (**39** and **40**).

Table 4. Cytotoxicity of amino acid containing aloe-emodin molecules 39 and 40 [65].

| Compound | $IC_{50}$ (µM) | | | |
|:---:|:---:|:---:|:---:|:---:|
| | **Hep G2** | **NCI-H460** | **PC3** | **HeLa** |
| **39** | 4.789 | 19.054 | 16.647 | >25 |
| **40** | 11.063 | 18.476 | 16.624 | 15.729 |
| **Aloe emodin** | 26.168 | 33.721 | 16.372 | 15.493 |

In a recent report, Chen and coworkers [66] synthesized new aloe-emodin derivatives through a four-step reaction, and the new compounds were evaluated for their proliferation inhibition efficiency against MDA-MB-231, A549, and HeLa cells. Studies identified compounds **41** and **42** exhibiting similar anticancer activity as that of the aloe-emodin (Fig. **16**).

Kumar and team, in their studies [67], synthesized hybrid molecules comprising emodin and pyrazole scaffolds and screened their anticancer activity. Two of the compounds, **43** and **44,** emerged as potent inhibitors of MCF-7 and MDA-MB-231cancer cell lines with $IC_{50}$ of 0.99 & 2.68 µM and 6b 1.32 & 1.6 µM, respectively (Fig. **17**). These also arrested the cell cycle in the G2/M phase and induced early and late apoptosis in MDA-MB-231 cells. Moreover, these derivatives increased the activity of caspases-2, 3, 6, 8, and 9 in MDA-MB-231 cells (Fig. **18**). In their recent work, Kumar and coworkers described the synthesis and anticancer properties of novel furan-aloe-emodin molecules [68]. Among them, compounds **45** and **46** displayed promising antitumor activity with IC$_{50}$ less than 12.5 μM, and they induced early and late apoptosis in CAL27 and SCC9 cells and arrested the cells in the G0/G1 phase in these cell lines. They also conducted docking studies to corroborate the findings of their results.

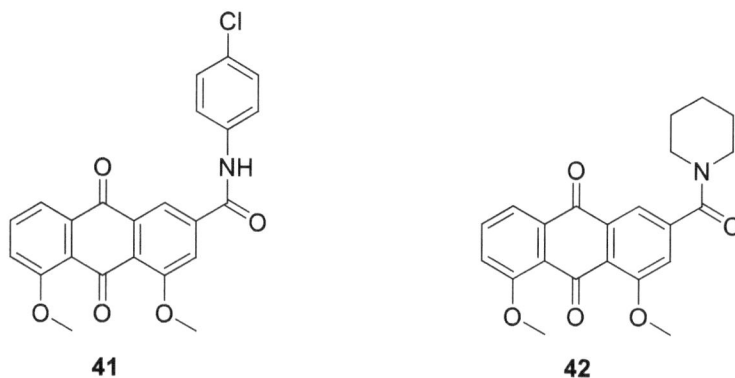

**Fig. (16).** Potent aloe-emodin carboxamide derivatives (**41** and **42**) reported by Chen and coworkers [66].

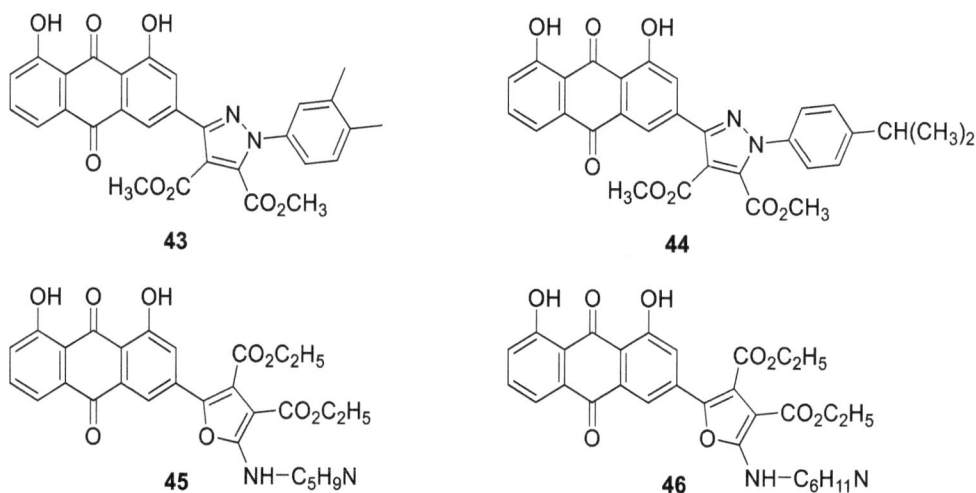

**Fig. (17).** Aloe-emodin pyrazole scaffolds (**43** and **44**) and furan-aloe-emodin compounds (**45** and **46**) reported by Kumar and team [6, 68].

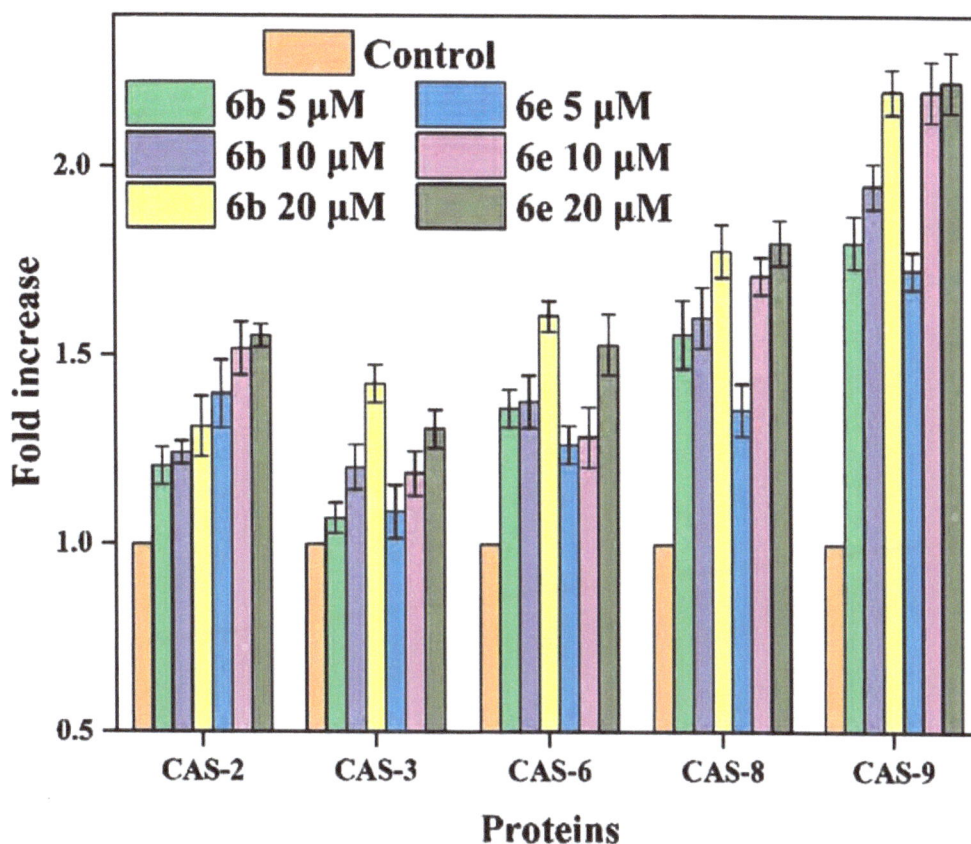

**Fig. (18)**. Measurement of caspase-2, 3, 6, 8, and 9 activities of compounds **43** and **44** in MDAMB-231 cell lines by colorimetric assay (Reused with permission from [67]).

Shang and his team members synthesized and screened the antitumor activity of aloe emodin-coumarin hybrids against various cancer cell lines [69]. Out of 18 derivatives, four compounds **47-50** (Fig. **19**) exhibited activity less than 5 μM/L, which is comparable to the etoposide drug, and the activity data is depicted in Table **5**. Among them, compound **48** was the most promising inhibitor with $IC_{50}$ values of 1.12, 0.90, 0.99, 1.59, and 0.48 μM/L against A549, SGC-7901, HepG2, MCF-7, and HCT-8 cell lines respectively. SAR studies suggested that the attachment of the triazole-emodin unit at the $3^{rd}$, $4^{th}$, and $7^{th}$ positions of coumarin has increased its activity more than at the $6^{th}$ position.

Even though aloe-emodin was studied for its anti-tumor activities by many researchers, its poor cell membrane permeability and low bioavailability limit its application as an anti-cancer agent. In order to overcome these shortcomings of aloe-emodin, derivatives comprising introduced α-amino phosphate units were reported by Zhang and his coworkers [70]. Aloe emodin α-amino-phosphate derivatives **51** and **52** (Fig. **20**) exhibited promising anticancer activity and the values are tabulated in Table **6**. These compounds were studied for their intracellular distribution in MDA-MB-231 cells, and the study results indicated that these derivatives partially entered the nucleus and surrounded the chromatin. Moreover, the DNA-binding property confirmed the interaction of compound **51** with DNA *via* intercalative interactions, whereas no confirmatory information was observed for compound **52**. Molecular docking experiments were also performed in order to get the other information to prove the findings.

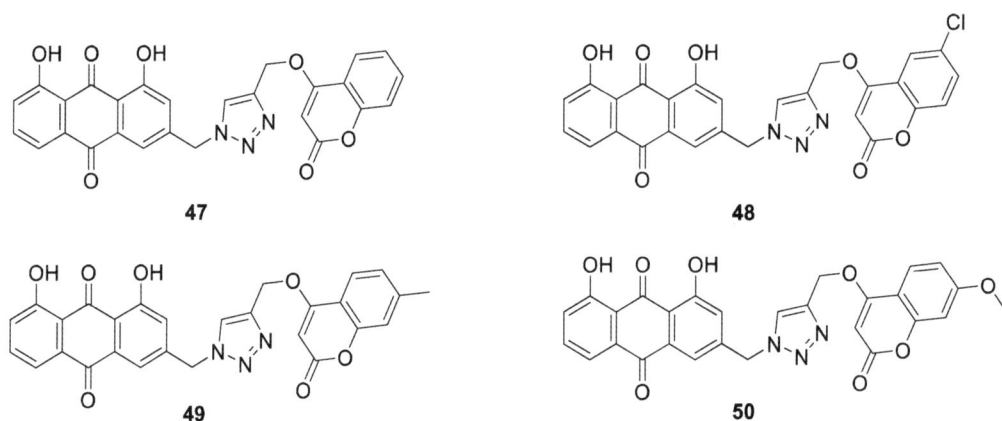

**Fig. (19).** Structures of potent anticancer aloe-emodin-coumarin hybrid derivatives (**47-50**).

**Table 5. Anticancer activities of aloe-emodin-coumarin hybrid derivatives (47-50) [69].**

| Compound | $IC_{50}$ ($\mu$mol/L) | | | | | |
|---|---|---|---|---|---|---|
| | A549 | SGC-7901 | HepG2 | MCF-7 | HCT-8 | Hk-2 |
| **47** | 2.57±0.48 | 3.28±0.67 | 7.06±0.27 | 2.34±0.48 | 1.56±0.45 | <1.0 |
| **48** | 1.12±0.09 | 0.90 ±0.12 | 0.99±0.12 | 1.59±0.58 | 0.48 ±0.06 | <1.0 |
| **49** | 3.82±0.31 | 2.17±0.24 | 1.02±0.40 | 1.68±0.21 | 4.49±0.24 | <10 |
| **50** | 4.05±1.99 | 9.04 ±1.17 | 5.01±1.27 | 3.60±0.92 | 5.79±1.24 | <10 |
| **Aloe-emodin** | 16.38±2.82 | 10.78±3.54 | 3.61±0.81 | 10.8±0.41 | 16.33±0.67 | - |
| **Etoposide** | 3.18 ±1.36 | 8.30±1.09 | 5.77±0.99 | 5.45±1.14 | 3.40±0.42 | - |

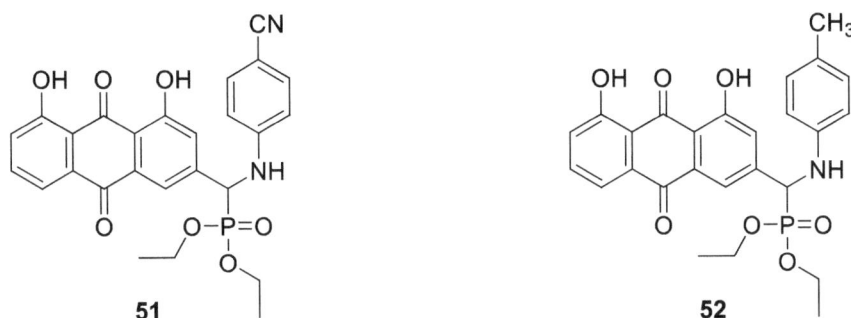

**Fig. (20).** Structures of aloe-emodin- α-amino-phosphate derivatives (**51** and **52**).

**Table 6. Anticancer activities ofaloe-emodin-α-amino-phosphate derivatives (51 and 52).**

| Comp. | $IC_{50}$ (µM) | | |
|---|---|---|---|
| | A549 | MDA-MB-231 | HepG2 |
| **51** | $10.89 \pm 1.27$ | $14.50 \pm 3.87$ | $6.47 \pm 0.25$ |
| **52** | $7.54 \pm 1.13$ | $26.89 \pm 4.21$ | $25.81 \pm 3.80$ |
| **Aloe-emodin** | $41.11 \pm 3.80$ | $50.15 \pm 2.76$ | $36.67 \pm 2.27$ |

## ANTICANCER RHEIN DERIVATIVES

Rhein is another kind of anthraquinone derivative widely used for the treatment of several diseases like cancer, diabetic nephropathy, and arthritis with a unique mechanism of action. Significant efforts have been made by researchers in the past decade to get more efficient anti-cancer derivatives by the execution of structural modification of rhein. In an article, Yang and coworkers reported the preparation of a series of rhein analogs by the modifications at the third position of the rhein moiety, and the final derivatives were evaluated for their cytotoxicity against two different cell lines (HeLa and MOLT4) [71]. Few compounds from **53** to **55** exhibited promising activities. In particular, 2-(chloro-N-(4, 5-dihydroxy-9, 10-dioxo-9, 10-dihydroanthracen-2-yl)) acetamide (**53**), a structurally modified rhein analog, showed improved potency towards the HeLa and MOLT4 with $IC_{50}$ of 2.7 and 0.6 µM, respectively, which was similar to doxorubicin against the HeLa cell line with $IC_{50}$ of 0.98 µM. The structures of the compounds **53-55** are depicted in Fig. (**21**), and their anti-cancer efficiencies are tabulated in Table **7**.

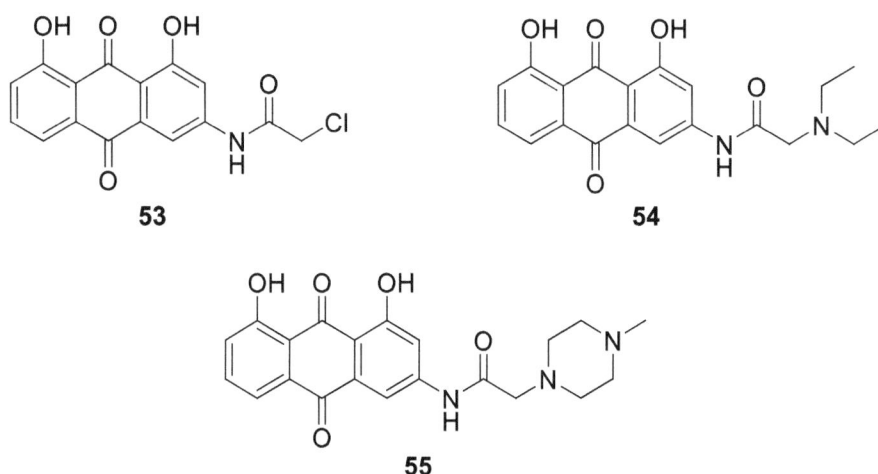

**53**

**54**

**55**

**Fig. (21).** Structures of the rhein-amide derivatives (**53-55**).

**Table 7. Anticancer activities of rhein-amide derivatives (53-55).**

| Compound | $IC_{50}[\mu M]$ | |
|---|---|---|
| | HeLa | MOLT4 |
| **53** | 2.7 | 0.6 |
| **54** | 6.1 | 3.0 |
| **55** | 5.8 | 3.1 |
| **Doxorubicin** | 0.98 | 0.04 |

Wang *et al.* [72] synthesized the new derivative of tetramethylpyrazine-rhein from two main ingredients of Huazhenghuisheng Pian, a Chinese traditional medicinal recipe. Compound **56** was screened for antiproliferative activity against the Bel-7402 cancer cell line and emerged as a promising agent with $IC_{50} = 26.4$ μM (Fig. **22**). Compound (**56**) tested for *in vivo* toxicity revealed that it did not affect the oral administration. Further, stability was tested by using artificial gastric juice and artificial intestinal juice , which showed that the solution of artificial digestion did not have any effect on structure or hydrolysis in 24 h.

**56**

**Fig. (22).** Potent tetramethylpyrazine-rhein compound **56** reported by Wang *et al.* [72].

In the study conducted by Ye and his team [73], rhein-phosphonate conjugates were synthesized and tested for their antitumor efficacy on CNE, Spca-2, Hela, HepG2, and Hct-116 cell lines. Compound **57** (Fig. **23**) was reported as a potential inhibitor with $IC_{50}$ values of 8.82 and 9.01 μM against HepG2 and Spca-2 cell lines, respectively. Further, this compound induced 70.4% apoptosis in HepG2 cells, and cell cycle analysis indicated the disruption of cells in different phases. DNA binding studies were carried out for compound **57** by various methods, and it strongly correlated with rhein and showed strong binding interaction with DNA. The same group in another study described the synthesis and antiproliferative activity of a new series of rhein α-amino phosphonate analogs against various human cancer cell lines [74]. The most promising compound (**58**) emerged as a potent inhibitor (Fig. **23**) with an $IC_{50}$ of 5.32 μM, and it induced cell apoptosis against the HCT116 cell line *via* the membrane death receptor pathways. Further, it also influenced the arresting of the cell cycle at the G1 stage in the HCT116 cell line. Further, it showed a moderate interaction with ct-DNA.

**57**                    **58**

**Fig. (23).** Structures of rhein-phosphonate conjugates compound **57** and **58**.

Liu and co-workers [75] analyzed the rhein lysinate (**59**) for its growth inhibition effect on human glioma U87 xenograft in BALB/c nude mice with 31.9% inhibition and 40% inhibition in human glioma U87 cells. It also induced apoptosis and ROS production in U87 cells. Upon treatment with this compound, there was a substantial decrease in the Bcl-2 and cyclin D expressions and an increase in the BAX and Bim expressions in glioma U87 cells. The structure of the compound is shown in Fig. (24), and its antitumor activities are depicted in Fig. (25).

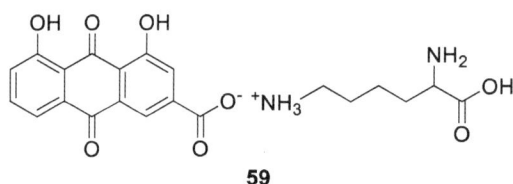

**Fig. (24).** Structure of rhein lysinate (59).

Huang *et al.* [76] synthesized a new series of rhein esters containing amines as the terminal group, and the synthesized compounds were screened for their cytotoxic effect against various cell lines. Among the synthesized compounds, derivative **60** emerged as a potential inhibitor against the proliferation of cells. Further, it induced cell apoptosis and trapped the cell cycle at the G2/M phase in HCT116 cells. The structure of **60** is given in Fig. (26), and its cytotoxicity efficiency is tabulated in Table **8**.

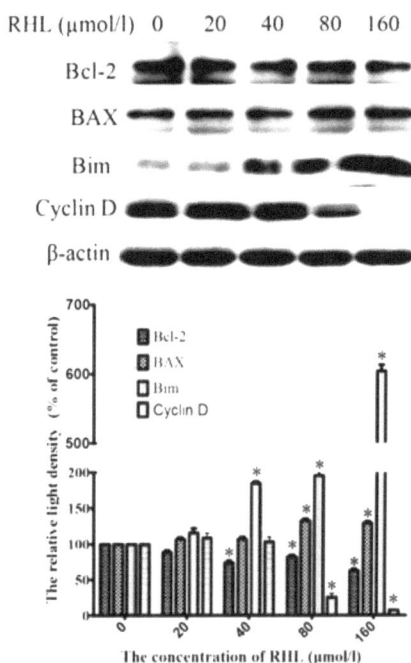

**Fig. (25).** Effect of rhein lysinate (59) on the expression of proteins associated with apoptosis and the cell cycle. Cells were treated with compound **59** at different concentrations for 48 h and then lysed and subjected to western blotting (Reused with permission from [75]).

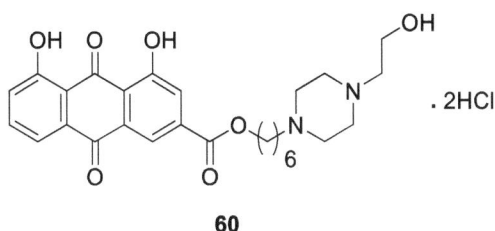

**60**

**Fig. (26).** Structure of potent anticancer rhein ester derivative (**60**).

**Table 8. Anticancer activities of rhein esters 60.**

| Compd | Cell Lines and IC$_{50}$ (µM) | | | | | |
|---|---|---|---|---|---|---|
| | HepG2 | HCT116 | A549 | MCF-7 | Bel-7402 | Bel-7402/5-FT |
| **60** | 0.85±0.02 | 0.31±0.01 | 5.28±0.31 | 7.63±0.42 | 2.36±0.08 | 4.48±0.29 |
| **Rhein** | 100.8±7.89 | 115.7±7.93 | 120.4±9.87 | 105.7±8.92 | 120.27±9.26 | 136.19±10.6 |
| **5-FU** | 5.82±0.32 | 4.30±0.19 | 10.32±0.87 | 4.32±0.23 | 7.57±0.43 | 214.33±11.7 |

A novel compound derived from Rhein, **61** (Fig. **27**), was synthesized by attaching a benzyl and piperazine unit [77]. Anticancer screening studies revealed strong inhibition against MDA-MB-231 and MCF-7 with IC$_{50}$ of 12.80 ± 0.83 and 7.54 ± 1.25 µg/mL, respectively. The docking study showed the strong binding of compound **61** with Rac-1, which is further evidenced by the inhibition of Rac1 promoter activity and downregulation of Rac1 protein expression. Another set of similar structural derivatives with a slight structural modification in the amide unit was reported by the same group in their recent studies [78]. Several compounds showed good antitumor activity, and in particular, compound **62** demonstrated excellent anticancer activities with IC$_{50}$ values of 4.7, 2.1, 3.2, and 1.5 µM against CNE-1, CNE-2, HepG2, and SMMC-7721 cell lines, respectively (Fig. **27**). It kills the tumor cells by paraptosis rather than apoptosis. SAR and in-depth studies indicated that its bis-benzyloxy unit is responsible for inducing ER stress and increases the ratio of LC3II/I and cytoplasmic vacuolization, which was caused by its N-(2-hydroxyethyl) formamide unit.

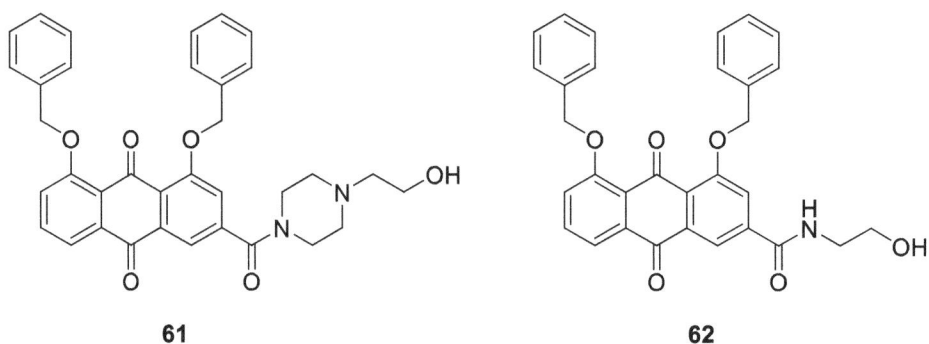

**Fig. (27).** Structures of rhein amide compounds **63** and **64**.

In an effort focused on the development of radiosensitizers, Su and co-workers [79] applied in-silico studies to design piperazine-endowed rhein derivatives. Among them, compound **63** was found to be a strong binder of Rac1 protein (Fig. **28**). It inhibited the growth of CNE1 and CNE2 cells, as evidenced by the results in Table **9**. Further, it has shown significant apoptosis ability at its non-toxic concentration when combined with 2 Gy irradiation. By activating the Rac1/NADPH signaling pathway and its downstream JNK/AP-1 pathway, compound **63** successfully modified the radiosensitivity of both CNE1 cells and CNE2 cells. Another work from the same research lab [80] describes the synthesis of a new series of benzyloxy-endowed modified rhein-quinazoline molecules and their anticancer activities against various human cancer cell lines. Compound **64** emerged as the best hit among them with $IC_{50}$ values of 23.3, 2.1, 3.2, 34.9, and 12.5 µM against A549, CNE-1, MDA-MB-231, HepG2, and SKOV3 cell lines, respectively. This compound was found to be a potent EGFR and p-EGFR inhibitor with inhibition much better than the gefitinib drug. Further, it generated a notable apoptotic impact, aided in the disintegration of the cytoskeleton, and the reorganization of F-actin filaments. Moreover, it has increased the radiosensitivity of the A549 cells, making it a promising radiosensitizer in anticancer treatment, and it has high DNA damage activity when combined with irradiation.

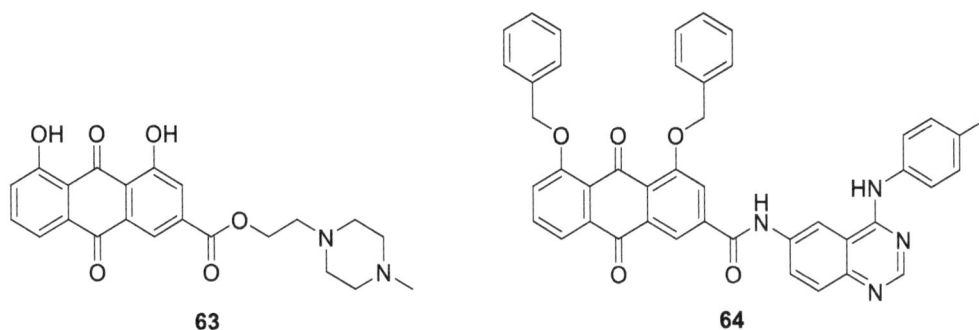

**Fig. (28).** Potent rhein amide derivatives (**63** and **64**).

**Table 9. The inhibitory effect of 63 on NPC CNE1 and CNE2 cells.**

| Concentration (µg mL$^{-1}$) | Inhibition Rate (%) | |
|---|---|---|
| | CNE1 | CNE2 |
| 6.25 | 6.13±1.52 | 15.74±1.37 |
| 12.5 | 21.02±1.67 | 30.77±1.22 |
| 25 | 26.24±1.33 | 37.32±1.86 |
| 50 | 36.08±1.55 | 55.74±1.55 |
| 100 | 65.86±1.43 | 59.83±1.14 |

Quaglio *et al.* [81] developed and synthesized hybrid-modified rhein-amide derivatives and tested them for the antiproliferation of cells against Med1-MB cells and SHH-MB cells. Compounds **65** and **66** emerged as promising agents for HH inhibitory growth of cells by the action on primary target SMO through the HH pathway. Compound **65** contains 4-chloro cyclohexyl, and compound **66** contains a 4-methoxy cyclohexyl group in their final structures, as shown in Fig. (**29**).

## ANTICANCER CHRYSOPHANOL, DAMNACANTHAL, ALIZARIN, QUINIZARIN DERIVATIVES

Koyama *et al.* developed a few derivatives of chrysophanol combined with nitrogen mustard groups and screened their anticancer activity against the 1210 murine leukemic cell line [82]. Among them, compounds **67** and **68** (Fig. **30**) emerged as potent anticancer agents with $ID_{50}$ values of 0.13 and 0.023 μM, respectively. Further in their continued study [83], they determined the cell cycle analysis of compound **67** on 1210 murine leukemic cell lines. Results clearly showed the arrest of the cells at the G2 phase and in the early S phase as the dose increased.

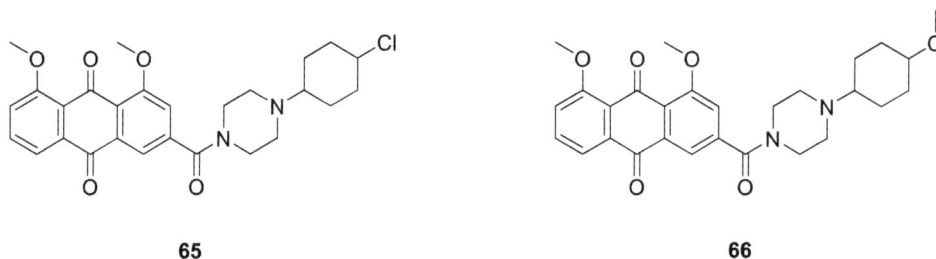

**Fig. (29).** Structures of anthraquinone-amide derivatives (**65** and **66**).

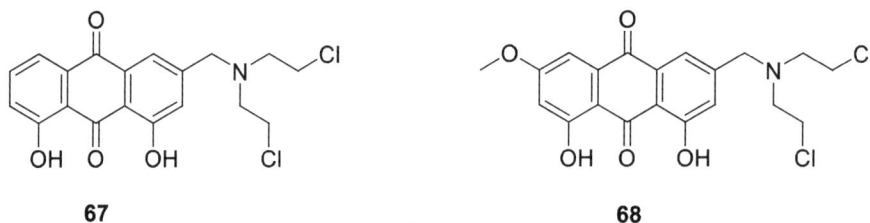

**Fig. (30).** Structures of chrysophanol derivatives (**67** and **68**).

New analogs of damnacanthal and nor-damnacanthal were synthesized and investigated for their cancer cell antiproliferation by Saha *et al.* [84]. Among the synthesized derivatives, 2-bromomethyl-1,3-dimethoxyanthraquinone (**69**) displayed potent activity with $IC_{50}$ values of 8, 2, 2, 4, and 5 μM against MCF7, MES-SA, MES-SA/DX5, DU145, and H460 cell lines, respectively (Fig. **31**). The cytotoxic activity of these compounds was discovered to be significantly influenced by the bromomethyl group at C-2 and hydroxyl or methoxyl groups at C-3 and C-1 positions of the anthraquinone.

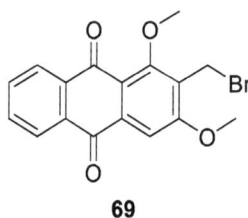

**Fig. (31).** Structure of potent anti-cancer damnacanthal analogue (**69**).

A new set of alizarin derivatives was prepared by Yao *et al.* [85] and screened for their anticancer activity against Spca-2, HepG2, MGC-803, CNE, and Hct-116 cell lines with reference drug 5-fluorouracil. Among them, compound **70** (Fig. **32**) showed good activity against these cell lines (Table **10**), and in particular, compound **70** emerged with a potent cytotoxic effect with an $IC_{50}$ of 9.08 µM against CNE cell lines and induced cell apoptosis in it. The cell cycle of CNE cells got arrested in the G1 stage, and it showed moderate interaction with ct-DNA.

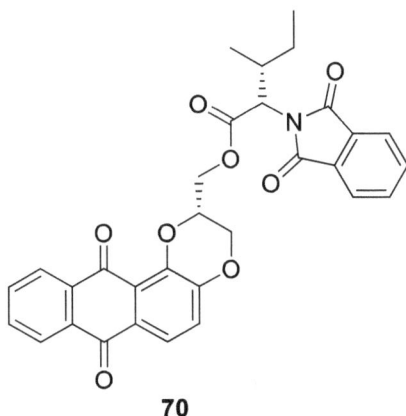

**70**

**Fig. (32).** Structure of potent alizarin derivative (**70**).

**Table 10. Anticancer activities of alizarin derivative (70), alizarin, and 5-fluorouracil (5-FU).**

| Compd | $IC_{50}$ [µM] | | | | |
|---|---|---|---|---|---|
| | **CNE** | **Spca-2** | **Hct-116** | **Hep G-2** | **MGC-803** |
| **70** | $9.08 \pm 0.87$ | $10.35 \pm 1.32$ | $22.12 \pm 1.45$ | $27.74 \pm 1.20$ | $15.04 \pm 2.00$ |
| **Alizarin** | >100 | >100 | $88.93 \pm 1.22$ | $96.21 \pm 5.98$ | $98.03 \pm 8.87$ |
| **5-FU** | $45.10 \pm 2.33$ | $58.92 \pm 3.02$ | $10.05 \pm 6.34$ | $38.34 \pm 3.24$ | $46.93 \pm 2.09$ |

Hu and his group designed and synthesized a series of quinizarin derivatives and tested them for antitumor activity against various leukemia cells [86]. Among them, four derivatives (**71-73**) exhibited good activity against K562 and HL60 cancer cell lines. Structures of compounds **71-73** are depicted in Fig. (**33**) and the results of anti-cancer screening of the tested compounds are tabulated in Table **11**. It is to be noted that compound **71** emerged as highly potent with $IC_{50}$ values of $2.31\pm0.37$ and $1.40\pm0.81$ µM against K562 and HL60 cell lines, respectively. This was further studied in detail for its mechanistic studies. Upon exposure of compound **71** in different doses to the Molt-4 and Jurkat cells, cells get arrested at the G0/G1 phase of the cell cycle. This compound induced apoptosis with Bcl-2 and c-myc protein degradation.

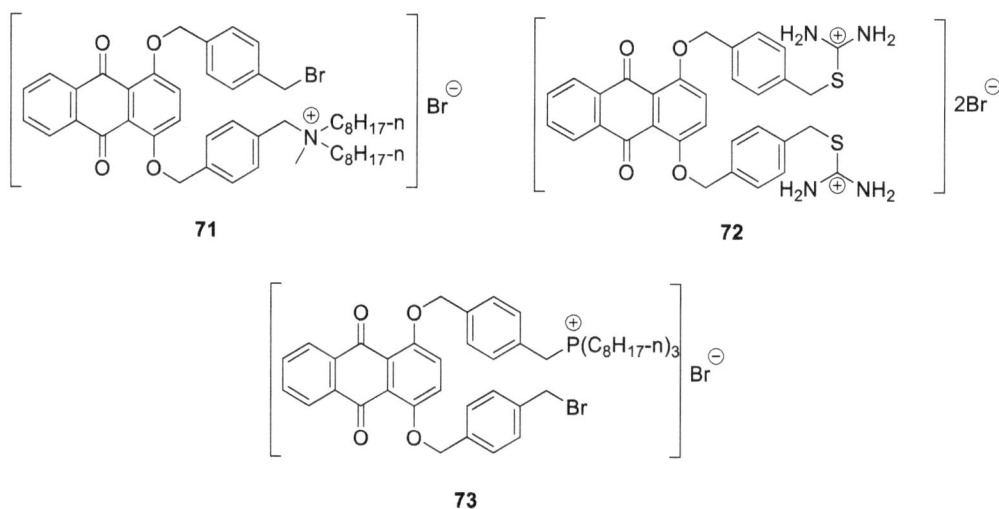

**71**          **72**

**73**

**Fig. (33).** Potent quinizarin derivatives (**71-73**).

**Table 11. Anticancer activities of quinizarin derivatives (71-73).**

| Compound | $IC_{50}[\mu M]$ | | |
|---|---|---|---|
| | **K562** | **HL60** | **HEK-293** |
| 71 | $2.31\pm0.37$ | $1.40\pm0.81$ | $2.44\pm0.38$ |
| 72 | $1.81\pm1.60$ | $3.35\pm0.84$ | $3.19\pm0.92$ |
| 73 | $1.24\pm2.38$ | $0.90\pm2.55$ | $2.17\pm1.91$ |
| ADM | $0.63\pm0.18$ | $0.14\pm0.01$ | $0.37\pm0.14$ |

A novel set of anthraquinone derivatives was prepared by derivatizing quinizarin and screened for anti-cancer activity by Zhao and co-workers [87]. These analogs were screened for cytotoxicity against HeLa, MDA-MB-231, MIA PaCa-2, and MCF-7 cell lines. The compound mixture (**74**) having a bromo group at the 7th and 8th positions (Fig. **34**) exhibited potent inhibition against these cell lines with $IC_{50}$ values of 6.22, 6.25, 6.55, and 6.57 µM, respectively.

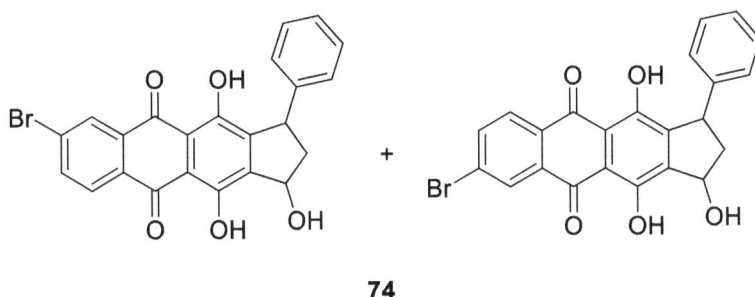

**74**

**Fig. (34).** Quinizarin derived potent AQ (**74**).

A new series of AQ derivatives containing an aryl/thiophene ring and the nitrogen-mustard group were synthesized starting from quinizarin by Liu and his research group [88], and they were designed to bind covalently with topoisomerase II. They were screened for selective toxicity against various human cancer cell lines. Against HepG2, compound **75** showed excellent anti-proliferation with an $IC_{50}$ of 12.5 µM (Fig. **35**). It also induced apoptosis in HepG2 cells in a dose-dependent manner, and docking studies with the human DNA-Topo II complex (1ZXM) protein showed that it could interact with topoisomerase II of the catalytic active site.

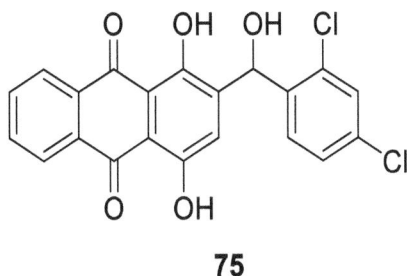

**75**

**Fig. (35).** Structure of quinizarin derivative (**75**).

# ANTICANCER ACTIVITY OF SYNTHETIC ANTHRAQUINONE DERIVATIVES

Jin and his group synthesized a new class of anthraquinone derivatives by a multistep synthetic route with the introduction of *N,N*-dimethylamino groups at the $1^{st}$ and $4^{th}$ positions and aryl units at the $2^{nd}$ position with the aim of identifying newer anticancer agents for leukemic cancer (p388 mouse tumor cell line) [89]. Among them, compounds **76-79** (Fig. **36**) exhibited excellent activity toward tested tumor cells (Table **12**). The electron-releasing amino group has the highest activity with $ED_{50} = 0.7$ µg/mL.

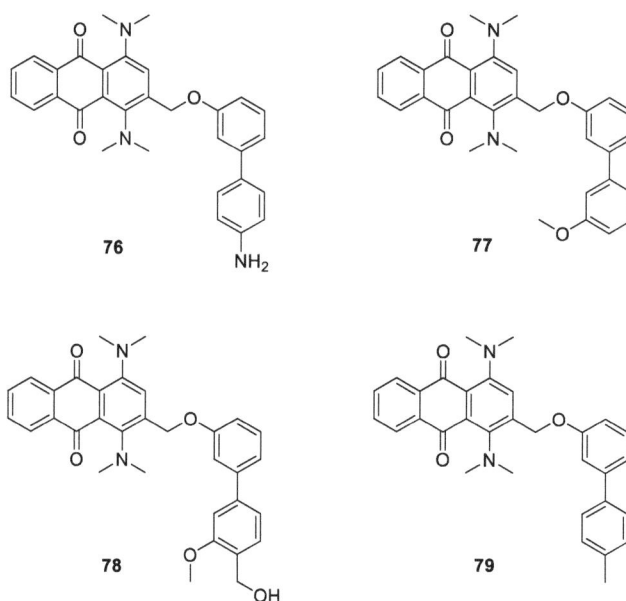

**Fig. (36).** Anthraquinone derivatives (**76-79**) reported by Jin and coworkers [89].

**Table 12. Anticancer activity of compounds 76-79 against various cell lines ($ED_{50}$ in µg/mL).**

| Compound | $ED_{50}$ (µg/mL) |
|----------|-------------------|
| 76 | 10.7 |
| 77 | 1.3 |
| 78 | 1.5 |
| 79 | 1.8 |

In the study conducted by Tu *et al.*, the synthesis and evaluation of the cytotoxic properties of a few 2-alkoxy-substituted amine-1-hydroxy anthraquinone derivatives were reported [90]. Study results revealed that compounds **80-83** displayed excellent potency against PC3 and NTUB1 cells. Additionally, compound **80** showed apoptosis and arrested cells at the G2/M phase and exerted the upregulation of p21 and cyclin B1 expressions in the NTUB1 cell line. Structures of the compounds and their anticancer screening are given in Fig. (**37**) and Table **13**, respectively.

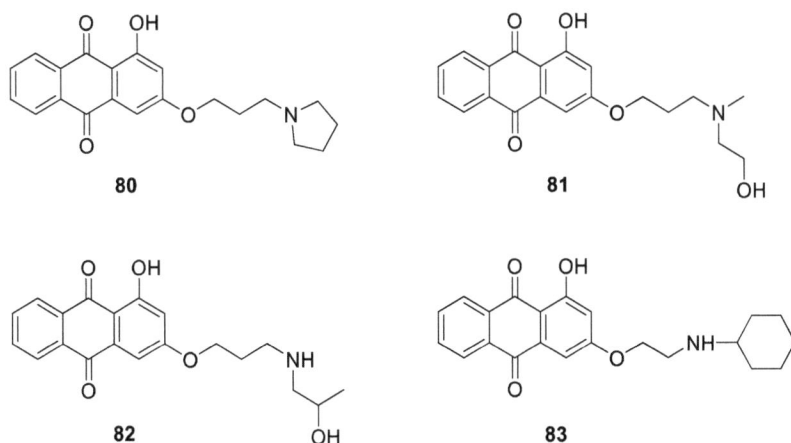

**Fig. (37).** Potent 2-alkoxy-substituted amine-1-hydroxy anthraquinone derivatives (**80-83**) reported by Tu and coworkers [90].

**Table 13. Anticancer activity of compounds 80-83 against various cell lines (ED$_{50}$ values in µM).**

| Compound | ED$_{50}$ (µM) | |
|---|---|---|
| | **NTUB1** | **PC3** |
| **80** | $9.72 \pm 1.21$ | $7.64 \pm 0.68$ |
| **81** | $12.30 \pm 3.05$ | $8.89 \pm 0.09$ |
| **82** | $12.00 \pm 1.19$ | $9.07 \pm 0.00$ |
| **83** | $8.37 \pm 0.30$ | $9.12 \pm 0.32$ |
| **Cisplatin** | $3.27 \pm 0.10$ | $4.56 \pm 0.76$ |

In their study, Lin and his group [91] identified novel anticancer agents containing the sulfonamide group at the 1st position of AQ as STAT3 phosphorylation inhibitors with potential antiproliferative activity. Compound **84** exhibited cell proliferation with $IC_{50}$ = 0.97, 0.29, 0.16, 0.21, and 0.81 µM against MDA-MB-231, PANC-1, HPAC, U87, and U373 cell lines, respectively. Docking studies suggest the occupancy of this molecule in the pTyr705 binding pocket of STAT3 with three important hydrogen bonds. This compound induced apoptosis and inhibited STAT3 phosphorylation in these cell lines. Inhibition of STAT3 DNA binding rather than STAT1 indicates the selectivity of this compound (Fig. **38**). Further, in their continued work, a new series of sulfonamide-AQ with substitutions at the 1st and 2nd positions were synthesized and screened for their anticancer activity against HT-29 and DU-145 cell lines [92]. Compound **85** (Fig. **38**) displayed excellent potency in the series with the highest activity against the HT-29 cell line ($IC_{50}$ = 6.5 µM) and the DU-145 cell line ($IC_{50}$ = 8.7 µM)

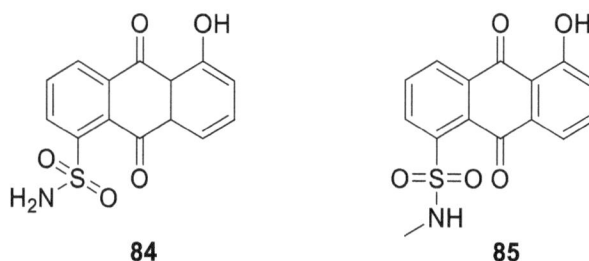

**Fig. (38).** Potent anthraquinone derivatives (**84** and **85**).

Abu *et al.* evaluated the anti-breast cancer potential of 3-bromo-1-hydroxy-9,10-anthraquinone (**86**, Fig. **39**) against MDA-MB-231 and MCF-7 cell lines [93]. It exhibited good $IC_{50}$ values of 24.40 and 29.40 µM, comparable to tamoxifen. Cell cycle arrest at the G1 phase in MCF-7 cells induced apoptosis. Moreover, its wound-healing assay confirms the inhibition of cell migration in MDA-MB-231.

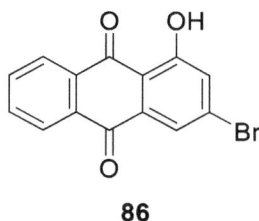

**Fig. (39).** Potent anthraquinone derivative (**86**) reported by Abu and coworkers [93].

Thiosemicarbazone-attached AQ derivatives were prepared and screened for antiproliferative activity against HeLa, A549, K562, MDA-MB-453, and MDA-MB-361 cell lines by Markovic *et al.* [94]. Most of the compounds showed variable potency against the tested cell lines, with compounds **87** and **88** emerging as the most potent among the series (Fig. **40**, Table **14**) against tested cells. These compounds arrested the cell cycle progression at the sub-G1 phase with the induction of apoptosis by the activation of caspase-3, 8, and 9 in the HeLa cell line.

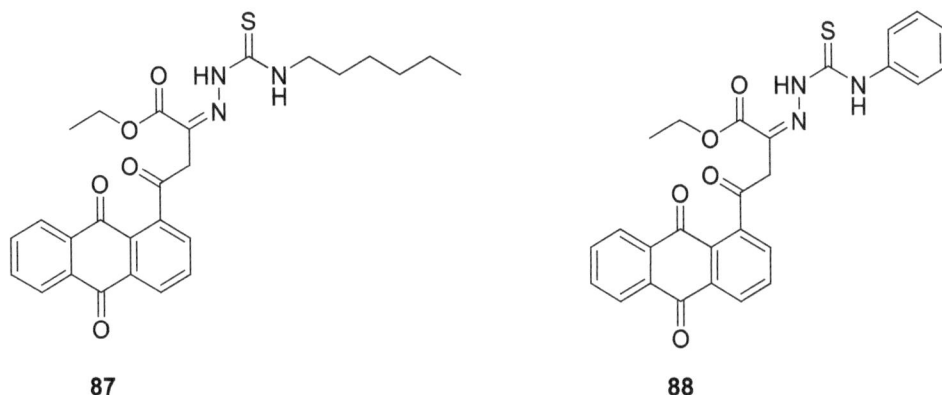

**Fig. (40).** Potent thiosemicarbazone attached AQ derivatives (**87** and **88**) reported by Markovi´c and coworkers [94].

**Table 14. Anticancer activity of compounds 87 and 88 against various cell lines (IC$_{50}$ values in µM).**

| Comp. | IC$_{50}$ (µM) | | | | | |
|---|---|---|---|---|---|---|
| | HeLa | MDA-MB-361 | MDA-MB-453 | K562 | A549 | MRC-5 |
| 87 | 7.66±2.44 | 4.45±0.68 | 19.55±3.92 | 5.86 ±3.05 | 43.62±2.08 | 42.13±4.92 |
| 88 | 8.19±1.93 | 6.61±1.51 | 6.91±0.56 | 3.59 ±1.67 | 9.89 ±1.94 | 11.79±0.36 |
| Cisplatin | 2.1±0.20 | 14.74±0.36 | 3.75±0.12 | 5.54±1.03 | 11.92±2.19 | 14.44±1.90 |

In the synthesis reported by Castro *et al.* [95], three series of derivatives comprising annulated AQ and simple substituted AQ were prepared (Fig. **41**). Anticancer properties of these were tested on J82, SK-MES-1, AGS, and HL-60 cell lines. In the first series, compound **89** showed good activity against MRC-5 and AGS cell lines with $IC_{50}$ values of 10.2 and 4.5 µM, respectively. Also, in the $1^{st}$ series, compound **90** showed the highest activity with $IC_{50}$ 3.3 and 2.4 µM against AGS and J82 cell lines, respectively. In the second series, compound **91** displayed $IC_{50}$ = 1.5 and 0.86 µM against AGS and J82 cell lines, respectively. In the $3^{rd}$ series, compound **92** showed the highest activity with $IC_{50}$ 4.1 against the HL-60 cell line (Table **15**).

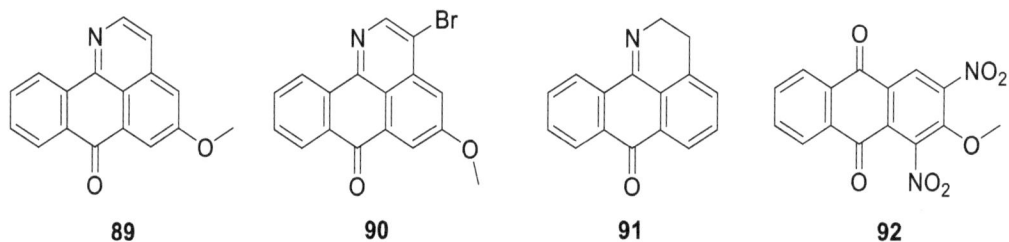

**Fig. (41).** Potent annulated AQ (**89-91**) and simple substituted AQ (**92**) reported by Castro *et al.* [95].

**Table 15. Anticancer activity of annulated AQ (89-91) and simple substituted AQ (92).**

| Comp. | $IC_{50}$ (µM) | | | | |
|---|---|---|---|---|---|
| | MRC-5 | AGS | SK-MES-1 | J82 | HL-60 |
| **89** | 10.2±0.5 | 4.5±0.3 | 17.1±1.0 | 25.4±1.3 | - |
| **90** | 28.9±2.0 | 3.3±0.2 | 9.2±0.5 | 2.4±0.1 | - |
| **91** | 28.0±1.7 | 1.5±0.1 | 28.9±2.0 | 0.86±0.04 | - |
| **92** | 11.4±0.7 | 11.7±0.8 | 15.4±1.0 | 18.1±1.5 | 4.1±0.3 |
| **Etoposide** | 3.9±0.2 | 0.36±0.02 | 2.5±0.2 | 2.8±0.2 | 1.8±0.1 |

In the detailed anticancer studies conducted by Lee with his team [96], they synthesized and screened numerous amido-endowed AQ derivatives. All the synthesized compounds were initially screened for cytotoxicity on the PC-3 cell line by SRB assay, and among them, compound **93** emerged as a potent inhibitor with $IC_{50}$ = 0.95 µM. Further, five compounds were selected for NCI 60 cell line screening studies. Among them, compounds **94** and **95** (Fig. **42**) showed potent activity with $GI_{50}$ ranging from 0.78 to 13.2 µM against the melanoma LOX IMVI cell line and 2.09 to > 100 µM against non-small cell lung cancer HOP-92 cell line, respectively (Table **16**). Moreover, it is noted that compound **94** exhibited stronger cancer-inhibitory effects than compound **95**, with $LC_{50}$ and TGI values of 2.83 and 1.01 µM in tested cell lines, respectively.

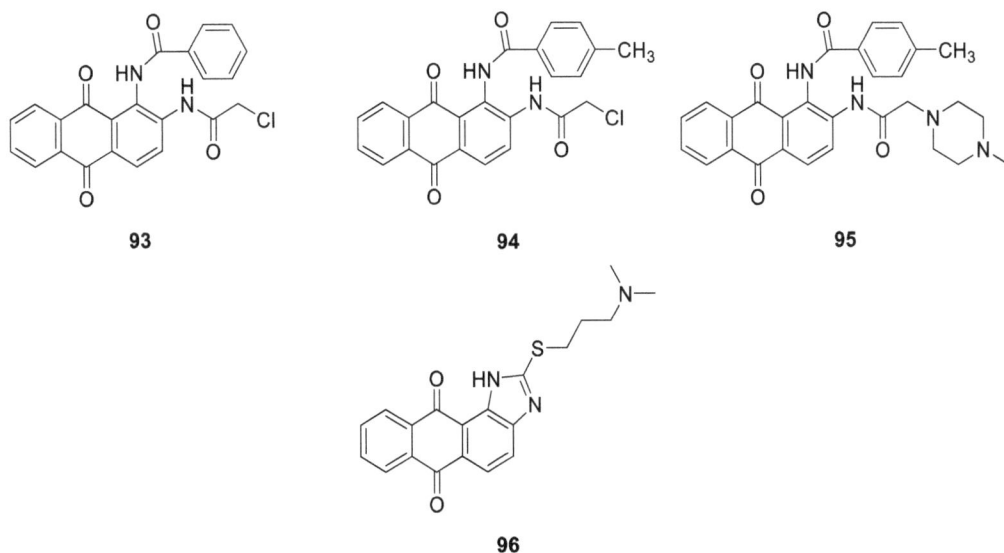

**Fig. (42).** Structures of potent AQ-amides (**93-95**) and AQ-imidazole derivative (**96**).

**Table 16. Anticancer activity of AQ-amides (94 and 95).**

| Comp | $GI_{50}$ | | | | | | | | |
|---|---|---|---|---|---|---|---|---|---|
| | RPMI-8226 | NCI-H522 | HCT-116 | U251 | LOX IMVI | OVCAR-3 | SN12C | DU145 | MCF7 |
| 94 | 1.35 | 0.88 | 1.05 | 1.80 | 0.78 | 1.35 | 1.22 | 2.69 | 1.12 |
| 95 | 3.25 | - | 3.08 | 3.64 | - | 2.43 | - | >100 | 3.13 |

Furthermore, these derivatives were screened for telomerase inhibition studies, and none of the compounds showed activity against it. In another work, they screened AQ-imidazole and amide derivatives for PARP-1 inhibition activity [97]. Two of these compounds, **94** and **96**, showed good cytotoxicity against A3 and A549 cell lines and were safer with less toxicity against the normal HEL299 cell line (Table 17). Among them, compounds **96** and **94** showed excellent PARP enzyme inhibition with 65% and 52% inhibition at 10 μM concentration.

**Table 17. PARP inhibitory activity of compounds 96 and 94.**

| Comp | IC$_{50}$ (μM) | | | PARP Inhibitory Activity (%) | | |
|------|------|------|------|------|------|------|
|  | A3 | A549 | HEL299 | 0.1 (μM) | 1 (μM) | 10 (μM) |
| 96 | 5.88 | 12.20 | 27.80 | 30 | 57 | 65 |
| 94 | 4.85 | 6.62 | 27.10 | 42 | 51 | 52 |

c-Met kinase inhibition studies of newer amino-AQs and di-AQs were reported by Liang *et al.* [98]. Two di-AQ derivatives (**97** and **98**, Fig. **43**) emerged as promising inhibitors with IC$_{50}$ values of 1.2 and 4.0 μM, respectively. Also, they suppressed the HGF-stimulated c-Met phosphorylation in A549 cells. It was further found that the inhibitory effect of compound **97** on HGF-mediated c-Met phosphorylation is associated with direct engagement with HGF, and it has a very good binding affinity with a K$_d$ value of 1.95 (Fig. **44**).

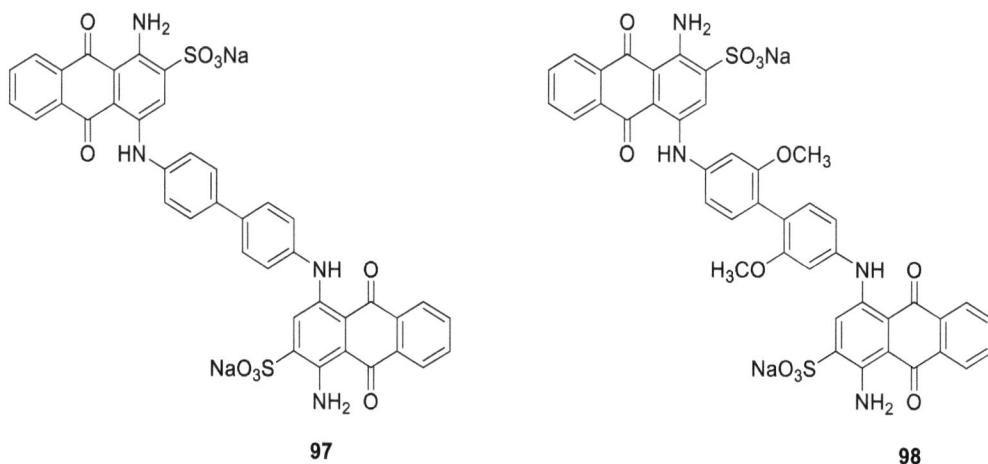

**Fig. (43).** Potent anthraquinone derivatives (**97** and **98**) reported by Liang with his team [98].

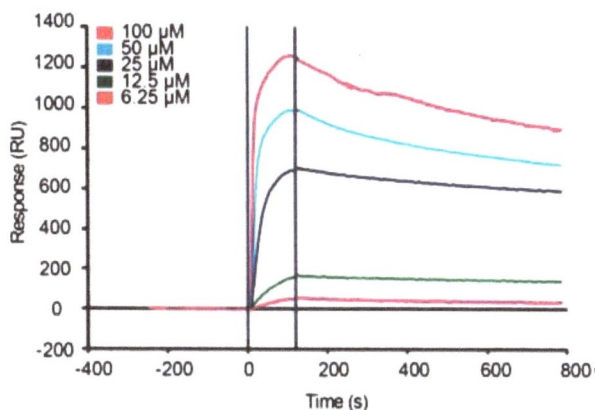

**Fig. (44).** Sensorgrams of compound **97** with immobilized HGF. Serial concentrations of compound **97** were injected over the HGF-immobilized GLH sensor chip surface. The SPR measurement yielded a $K_d$ value of 1.95 μM for compound **97** (Reused with permission from [98]).

In the studies conducted by Cogoi *et al*. [99], a new series of anthrathiophenediones having guanidino-alkyl side chains with different lengths were synthesized with the aim of studying their interaction with DNA and RNA G-quadruplexes, their uptake in malignant and non-malignant cells, and their capacity to modulate gene expression and inhibit cell growth. According to flow cytometry, these derivatives infiltrate malignant T24 bladder cells more effectively than they do on non-malignant embryonic kidney 293 or fibroblast NIH 3T3 cells. Among them, compound **99** is taken up by endocytosis in T24 malignant cells, but compounds **100** and **101** are transported by passive diffusion (Fig. **45**).

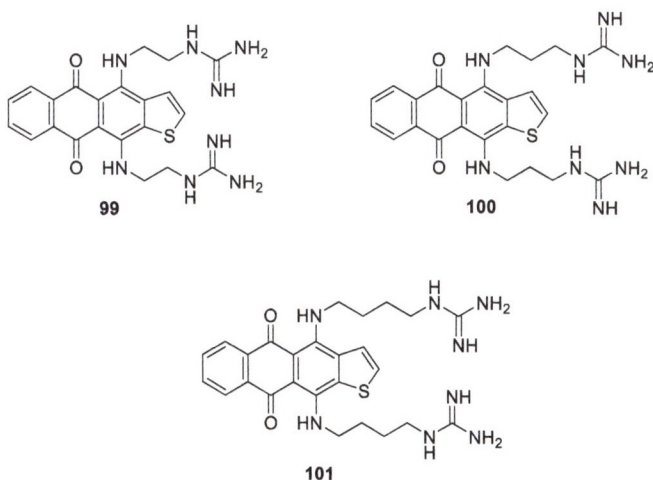

**Fig. (45).** Potent anthrathiophenediones (**99-101**).

A series of novel anthraquinone derivatives were synthesized and screened for anticancer activity by Sangthong with his team on a human cervical cancer cell line (CaSki), with compound **102** (Fig. **46**) showing promising inhibition with $IC_{50} =$ 0.3 µM and was found to be safer towards normal WI-38 cell line with $IC_{50} => 10$ µM [100]. It was observed that compound **102** completely blocked the formation of colonies in CaSki cells at 5 µM compared to the standard cisplatin drug and induced apoptosis in it (Fig. **47**). This compound has upregulated p53 expression while downregulating the Bcl-2 gene and has stopped cell growth in the G2/M phase of the cell cycle in CaSki cells.

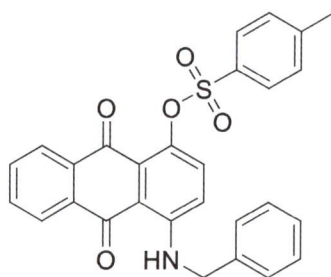

**102**

**Fig. (46).** Potent anthraquinone derivative (**102**) reported by Sangthong with his team [100].

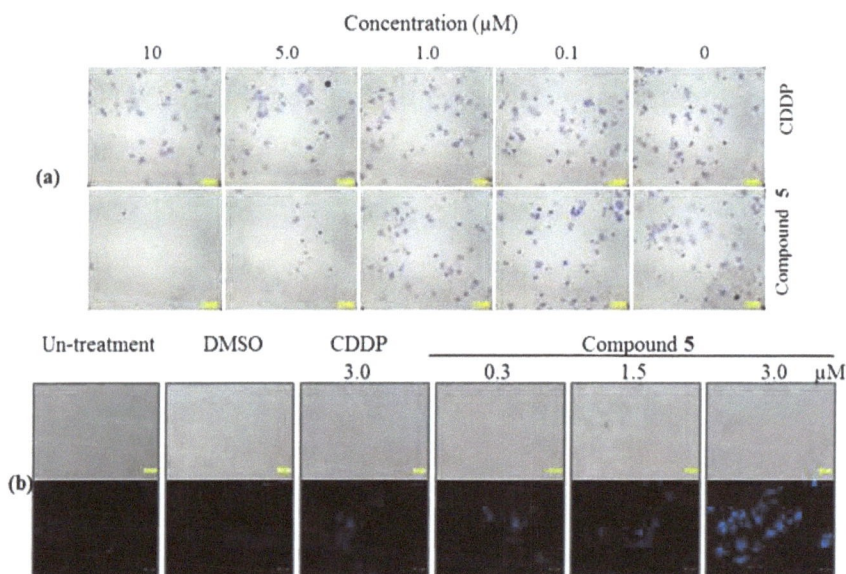

**Fig. (47).** (**a**) Colony formation in CaSki treated with CDDP or compound **102** at 10, 5, 1, 0.1, and 0.01 mM (**b**) apoptosis induction with compound **102** in a dose-dependent manner (Reused with permission from [100]).

In the studies conducted by Almutairi *et al.*, novel compounds were prepared by combining anti-inflammatory drugs [101]. These compounds were screened for HepG2 cell proliferation. Three synthesized molecules (**104-106**) displayed promising inhibition with $IC_{50}$ values of 3.74 and 3.92 µg/mL, respectively (Fig. **48** and Table **18**). In addition, all the derivatives were studied for their biochemical effects on the enzymes alanine and aspartate aminotransferases (ALT and AST) and alkaline phosphatase (ALP) and also screened for their total lipids, cholesterol, triglycerides, bilirubin, albumin, globulin, and creatinine in the serum of mice. The study concluded that these derivatives have a moderate safety level with regard to these parameters.

**Fig. (48).** Structures of anthraquinone derivatives (**104**, **105**, and **106**) reported by Almutairi with his team [101].

**Table 18. Anticancer activity of compounds 104-106 against HepG2 cell lines.**

| Compound | $IC_{50}$ (µg/mL) |
|---|---|
| 104 | 3.74 |
| 105 | 4.31 |
| 106 | 3.92 |
| 5-Fluorouracil | 5.00 |
| Doxorubicin | 3.56 |

The synthesis of a series of imine unit-containing chalcones of AQs as anticancer agents was carried out against A549, HeLa, MRC5, and LS174 cell lines [102]. Some of the molecules, **107-110**, prevented the growth of HeLa cells with $IC_{50}$ of 1.45 and 1.82 µM, respectively (Fig. **49**). Moreover, caspase-3 and 8 were activated upon treatment of these in HeLa cells. The structures of the compounds are depicted in Fig. (**49**), and their anticancer screening results are given in Table **19**.

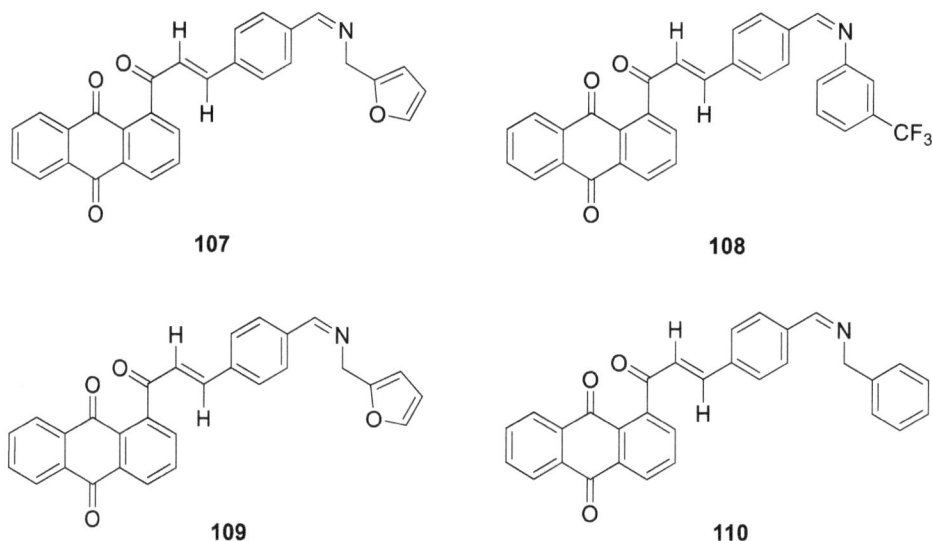

Fig. (49). Potent chalcones of AQs (**107-110**) reported by Kolundžija with his team [102].

Table 19. Anticancer activity of compounds 107-110 against various cell lines ($IC_{50}$ values in µM).

| Compound | $IC_{50}$ (µM) | | | |
|---|---|---|---|---|
| | HeLa | LS174 | A549 | MRC5 |
| 107 | $1.45 \pm 0.34$ | $6.61 \pm 0.49$ | $7.78 \pm 0.26$ | $14.55 \pm 0.72$ |
| 108 | $1.82 \pm 0.68$ | $1.76 \pm 0.78$ | $6.11 \pm 0.45$ | $15.11 \pm 3.56$ |
| 109 | $1.93 \pm 0.89$ | $17.18 \pm 2.24$ | $9.99 \pm 0.46$ | $16.69 \pm 0.85$ |
| 110 | $5.34 \pm 1.13$ | $5.30 \pm 0.45$ | $6.01 \pm 0.09$ | $11.87 \pm 0.65$ |
| Cisplatin | $2.10 \pm 0.20$ | $5.54 \pm 1.03$ | $11.92 \pm 2.19$ | $14.21 \pm 1.54$ |

The synthesis of anthraquinone-endowed chalcone molecules was reported by Markovic *et al.* with the aim of screening them for cytotoxicity activity against HeLa, LS174, A549, and MRC-5 cell lines [103]. Compounds (**111-113**) are potent inhibitors of HeLa and LS174 cell lines and moderate towards A549 and MRC-5 cell lines, having $IC_{50}$ values ranging from 2.36 to 2.73 μM (Fig. **50**, Table **20**). These compounds exert cell cycle arrest at the S and G2/M phases of the cell cycle and further exert apoptosis through caspase-3 activation in HeLa cells.

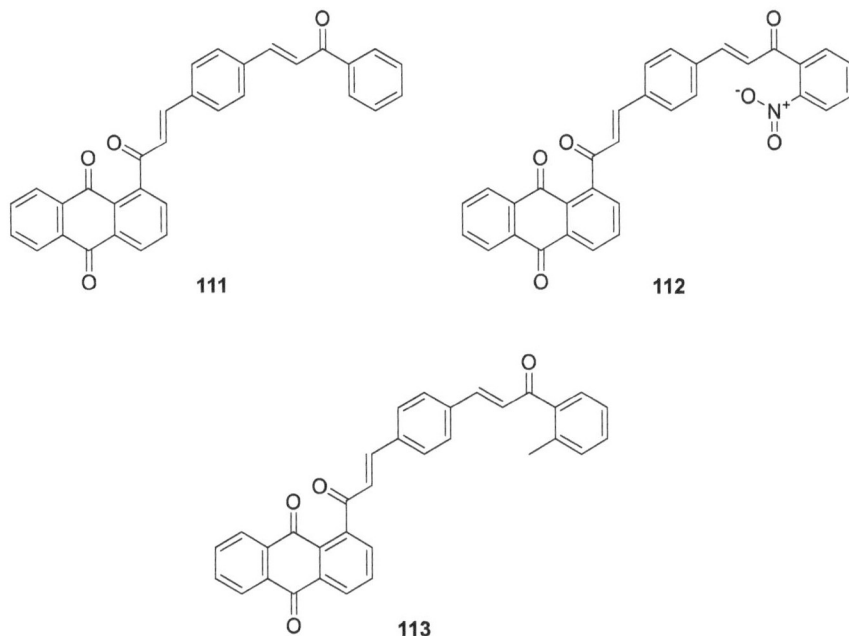

**Fig. (50).** Potent anthraquinone-endowed chalcone molecules **111-113**.

**Table 20. Anticancer activities of AQ-chalcone molecules 111-113.**

| Compound | IC$_{50}$ (μM) | | | |
|---|---|---|---|---|
|  | **HeLa** | **LS174** | **A549** | **MRC-5** |
| **111** | 2.41 ± 0.10 | 4.56 ± 0.49 | 26.20 ± 1.47 | 33.57 ± 0.17 |
| **112** | 2.36 ± 0.10 | 3.13 ± 0.42 | 29.05 ± 0.25 | 41.87 ± 0.16 |
| **113** | 2.73 ± 0.08 | 6.44 ± 0.99 | 28.84 ± 1.75 | 48.76 ± 1.24 |
| **Cisplatin** | 2.10 ± 0.20 | 5.54 ± 1.03 | 11.92 ± 2.19 | 14.21 ± 1.54 |

A new set of compounds comprising N-mustard units attached to hydroxyanthraquinone was screened for anticancer activity against MDA-MB-231, HeLa, MCF-7, and A549 cell lines by Zhao and his team [104]. Three compounds with fluoro substitutions, **114-116** (Fig. **51**, Table **21**), emerged as highly potent, and **116** showed activity with $IC_{50}$ = 0.263 nM, which is higher than the standard drug doxorubicin $IC_{50}$ = 0.294 nM.

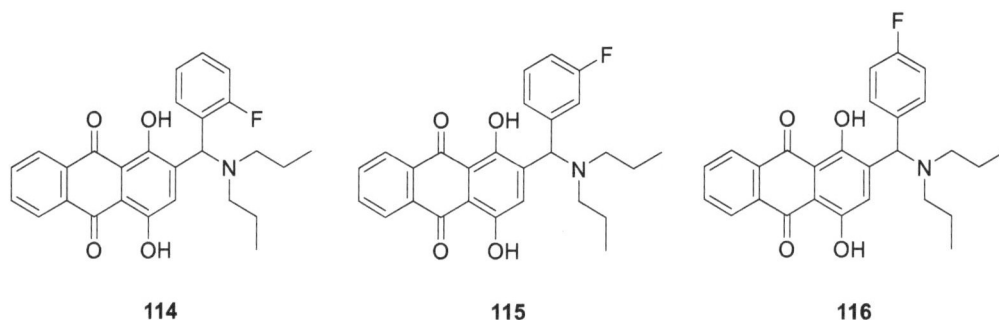

Fig. (**51**). N-mustard derivatives of anthraquinones (**114-116**).

Table 21. Anticancer activities of anthraquinones (114-116).

| Compound | $IC_{50}$ (nM) | | | |
|---|---|---|---|---|
| | MDA-MB-231 | HeLa | MCF-7 | A549 |
| **114** | 0.59 | 615 | 1.57 | 462 |
| **115** | 59.6 | 166 | 0.295 | 495 |
| **116** | 80.7 | 249 | 0.263 | 421 |
| **Chlorambucil** | 520 | 943 | 26.4 | 85.7 |
| **Doxorubicin** | 0.414 | 4.16 | 0.294 | 5.58 |

A novel anthraquinone comprising a tetrahydropyran moiety **117** (Fig. **52**) was prepared by Yang *et al.* and evaluated for its DNA-binding property by fluorescence assay study [105]. Results have shown that it binds the DNA by intercalation of base pairs. The molecular docking studies further confirm its interaction with DNA along with hydrogen bonding with key residues. This compound also inhibited the growth of HepG2 cells at higher concentrations. In another work of the same group [106], a novel anthraquinone derivative **118** was developed and screened for cytotoxicity against CCK-8 cells (Fig. **52**). The compound was treated in a dose-dependent manner to the cell line. In this study, as the concentration of the compound was increased, the viability of the cells decreased, and the concentration of $1.60 \times 10^{-7}$ M showed the viability of cells at about 99.51%. In molecular docking studies, the compound interacted with HSA and ctDNA.

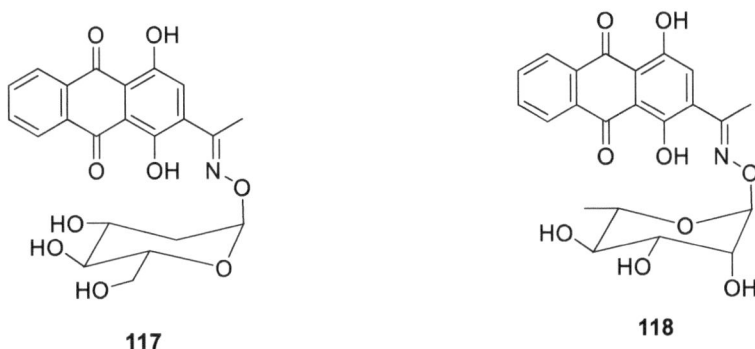

**Fig. (52).** Structures of anthraquinone derivatives comprising tetrahydropyran moiety (**117** and **118**).

A new series of [1,2,5]-thiadiazole-connected AQs having various thio-substitutions at the $3^{rd}$ position were synthesized by Lee *et al.* [107] and screened for anticancer activity on NCI-60 cell lines. Many of these derivatives showed good activity against these cell lines, and in particular, compounds **119** and **120** (Fig. **53**) inhibit the proliferation of PC-3 (IC$_{50}$ = 3.69 and 5.39 µM) and DU-145 cell lines (IC$_{50}$ = 4.53 and 2.55 µM), respectively, and are more potent than standard doxorubicin. It was also found to be non-toxic to SV-HUC-1, WMPY-1, and RWPE-1 cells, whereas doxorubicin was found to be toxic. It inhibited the ERK1/2 and p-38 signaling pathways with the induction of cell apoptosis through the morphological changes in DU-145 cells. Fig. (**54**) explains the morphological changes by **119** in DU-145 cells.

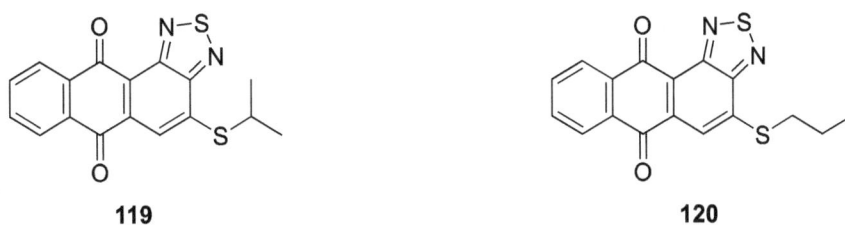

**Fig. (53).** Potent [1,2,5]-thiadiazole connected AQs (**119** and **120**).

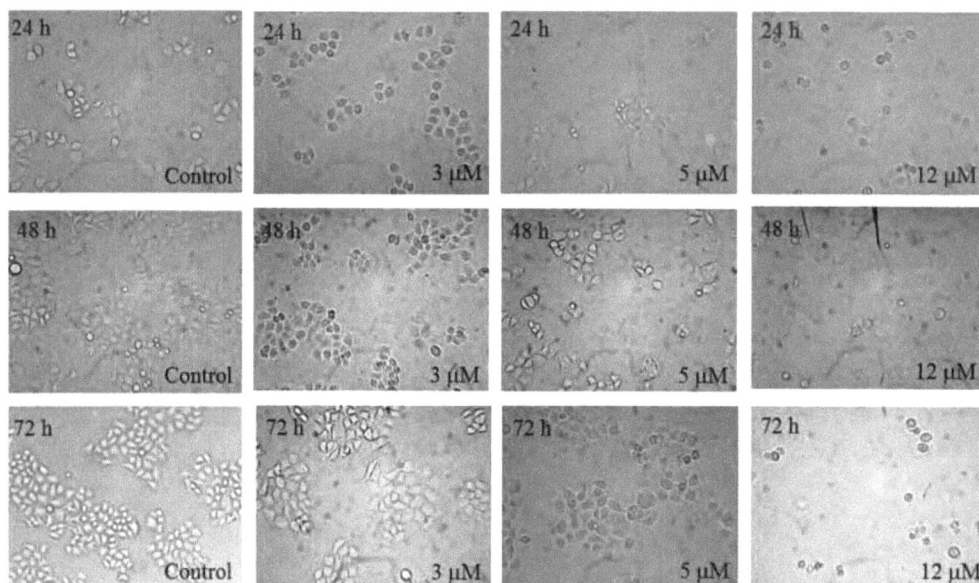

**Fig. (54).** Morphological changes and apoptosis induced by compound **119** in DU-145 cells (Reused with permission from [107]).

Yeap and his research team [108] synthesized 1,3-dihydroxy-9,10-anthraquinone-2-carboxylic acid (**121**) by the oxidation of 2-hydroxymethyl-1,3-dimethoxy-9,10-anthraquinone and tested it for its cytotoxic potency on the MCF-7 cell line. This compound exhibited a good cytotoxic effect with the $IC_{50}$ = 12.80 µg/mL and was also reported with good selectivity against MCF-7 cells (Fig. **55**). MCF-7 cells were arrested at the G2/M phase by the exposure of this compound with the inhibition of PLK1 gene expression. Further, apoptosis is induced through the upregulation of BAX, p53, and cytochrome c levels in MCF-7 cells (Fig. **56**).

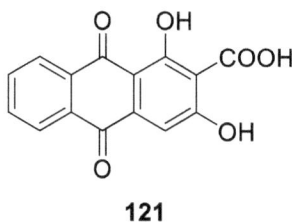

**121**

**Fig. (55).** Structure of 1,3-dihydroxy-9,10-anthraquinone-2-carboxylic acid (**121**).

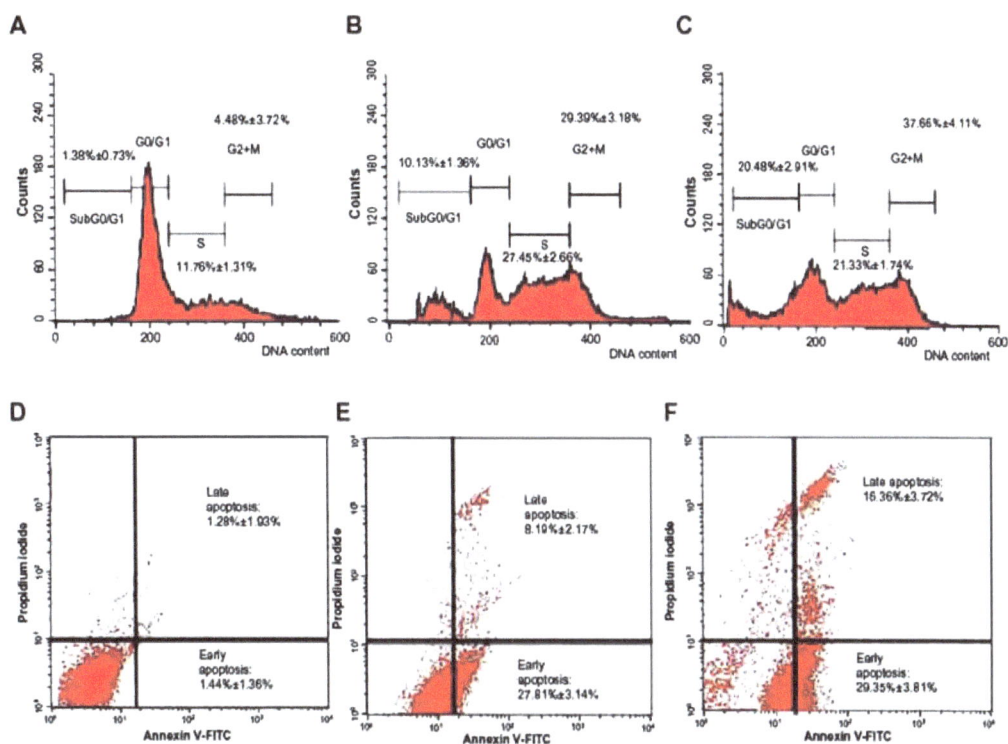

**Fig. (56).** Cell cycle progression (**A–C**) and apoptosis (**D–F**) of compound **121** treated ($IC_{25}$ and $IC_{50}$ values) and untreated MCF-7 cells after 48 hours of incubation (Reused with permission from [108]).

The study of cytotoxicity for synthesized amino-anthraquinone derivatives was analyzed by Tikhomirov and his research developers [109]. The compounds were screened against L1210, CEM, HeLa, and HCT116 cancer cell lines. The analog **122** (Fig. **57**) emerged as a potential active agent against the proliferation of cells with $IC_{50}$ values of $0.10 \pm 0.02$ μM, $0.19 \pm 0.06$ μM, $0.17 \pm 0.01$ μM, and $0.10 \pm 0.09$ μM, respectively, with reference to Dox. The compounds inhibited the ratio of $NAD^+/NADH$ by the suppression of tumor-associated NADH oxidase (tNOX) and also inhibited Sirtuin 1 (SIRT1) deacetylase activity. Further, through cell apoptosis, it activated caspase.

Cogoi and co-workers [110] synthesized anthrathiophenediones containing two chloroacetamide with side chains as G-quadruplex binders and screened them for anticancer activity. Compounds **123** and **124** exhibited potent inhibition against T24 cancer cell lines with $IC_{50}$ values of $2.7 \pm 0.2$ μM and $2.9 \pm 0.01$ μM, respectively (Fig. **58**). These compounds induced apoptosis, and cells get trapped at the G2 phase of the cell cycle.

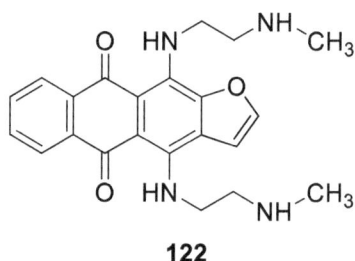

**122**

**Fig. (57).** Potent amino-anthraquinone compound **122**.

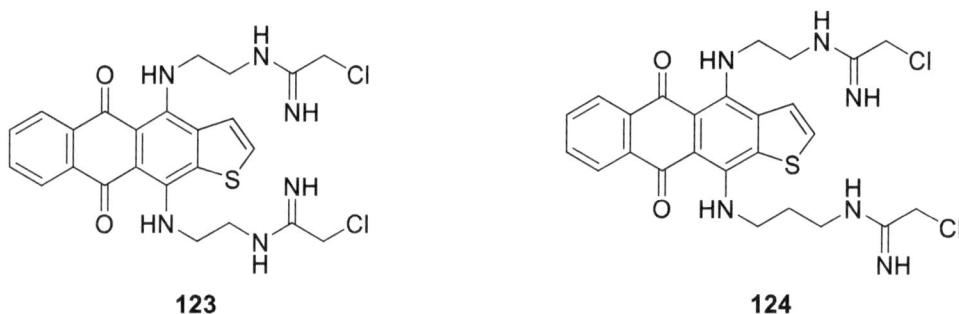

**123**     **124**

**Fig. (58).** Potent amino-anthraquinone compounds **123** and **124**.

Juhan *et al.* [111] prepared a novel set of amino-anthraquinone derivatives and screened them for anticancer activity. The compounds were tested for cytotoxicity, and among them, the compound (**125** and **126**) (Fig. **59**) exhibited a strong cytotoxic effect on MCF-7 and Hep-G2 cell lines (Table **22**). Compound **126** has emerged as a highly potent with $IC_{50}$ values of 2.0 µg/mL and 1.1 µg/mL against these cell lines, respectively.

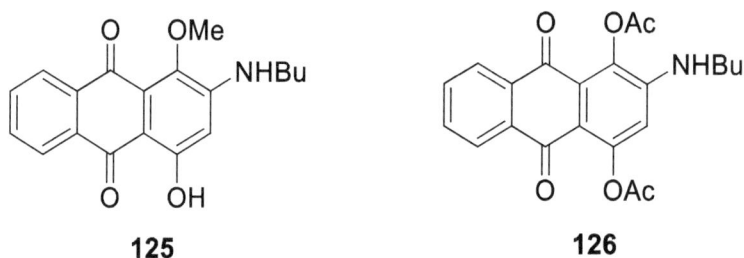

**125**     **126**

**Fig. (59).** Potent amino-anthraquinone derivatives (**125** and **126**).

**Table 22. Anticancer activities of amino-anthraquinone derivatives (125 and 126).**

| Compound | IC$_{50}$ ($\mu$g/mL) | |
|---|---|---|
| | MCF-7 | Hep-G2 |
| 125 | 7.0 | 14.0 |
| 126 | 2.0 | 1.1 |

Choi and co-workers [112] in their study reported the synthesis of a new sulfonamide-endowed anthraquinone and evaluated its anticancer activity against various cell lines. Compound **127** emerged as an effective cytotoxic agent against HCT116, HCT116 p53-/-, HT29, and A549 cell lines (Fig. **60**). This has arrested the process of the cell cycle at the S and G2/M phases by DNA damage and induced apoptosis. In wild-type p53, p53 knockout HCT116, and HT29 cells, this substance boosted caspase-3 cleavage. It also significantly elevated the phosphorylation of histone H2AX and Chk1 in both cell types. Moreover, it enhanced the chemosensitivity of HCT116 and HCT116 p53$^{-/-}$ cancer cells and the radiosensitivity of HT29 cancer cells.

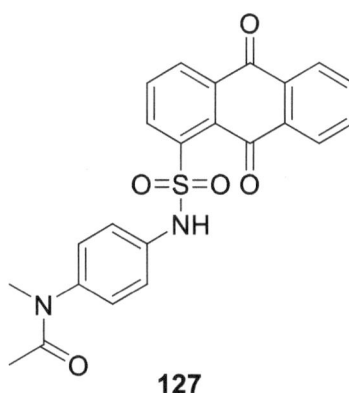

**127**

**Fig. (60).** Potent sulphonamide endowed anthraquinone (**127**).

Antitumor activity of 23 new anthraquinone-chalcone hybrids comprising an amide functional group was synthesized and screened for cytotoxic studies by Stanojkovic and co-workers [113]. All the compounds exhibited good cytotoxic effects, and in particular, compounds **128-131** emerged as highly potent against K562, Jurkat, and HL-60 leukemia cell lines. Subsequently, it showed low toxicity against the normal

MRC-5 cell line (Fig. **61** and Table **23**). Compounds **128**, **129**, and **130** induced the extrinsic and intrinsic apoptotic pathways and downregulated the expression of MMP2, MMP9, and VEGF. Compound **131** exhibited the extrinsic apoptosis pathway in K562 cells and upregulated the expression of miR-155 in K562 cells.

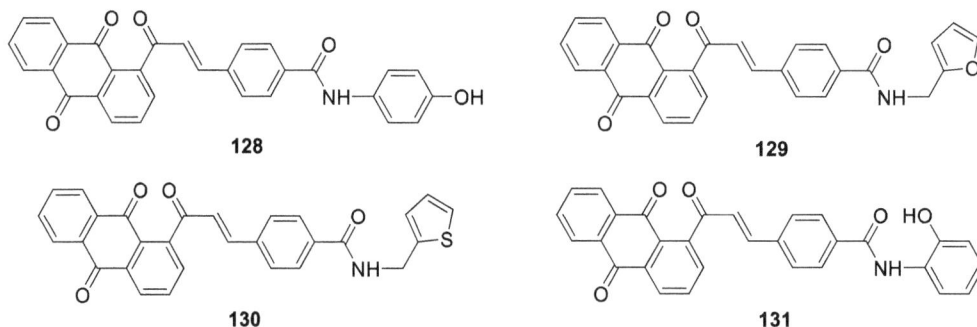

**Fig. (61).** Potent anthraquinone-chalcone hybrid molecules **128-131**.

**Table 23. Anticancer activity of anthraquinone-chalcone hybrid molecules 128-131.**

| Compound | $IC_{50} \pm SD$ ($\mu M$) | | | |
|---|---|---|---|---|
| | **K562** | **Jurkat** | **HL-60** | **MRC-5** |
| **128** | $3.87 \pm 0.17$ | $3.23 \pm 0.04$ | $1.89 \pm 0.22$ | $60.36 \pm 4.17$ |
| **129** | $5.99 \pm 0.24$ | $3.66 \pm 0.41$ | $3.50 \pm 0.67$ | $190.74 \pm 7.73$ |
| **130** | $4.45 \pm 0.73$ | $4.25 \pm 0.67$ | $3.16 \pm 1.12$ | $161.01 \pm 6.57$ |
| **131** | $4.35 \pm 0.47$ | $4.25 \pm 0.78$ | $4.43 \pm 0.78$ | $188.28 \pm 11.20$ |
| **DOX** | $0.14 \pm 0.03$ | $0.11 \pm 0.03$ | $0.05 \pm 0.01$ | $0.29 \pm 0.05$ |

As a part of identifying potent anticancer agents, a novel series of anthraquinone hydrazones comprising different structural units was synthesized by Lozynskyi and co-workers [114]. Among them, compound **132** having a thioxothiazolidine ring showed good anti-cancer activity on NCI-H460 (GP = 13.25%) and HCT-116 (GP = 13.69%) cell lines of lung and colon cancer, respectively (Fig. **62**).

Cytotoxic properties of new amino group AQ derivatives having different groups, including amines, were reported in the article of Sweidan and his group [115].

Inhibition results of compounds (**133** and **134**) showed more potent activity against K562 (IC$_{50}$ 2.1 and 1.1 μM) than MCF-7 and fibroblast cells (Fig. **63**, Table **24**).

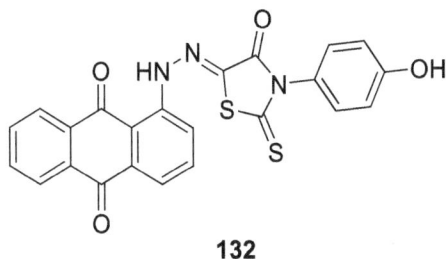

**132**

**Fig. (62).** Structure of the potent thioxothiazolidine-AQ derivative (**132**).

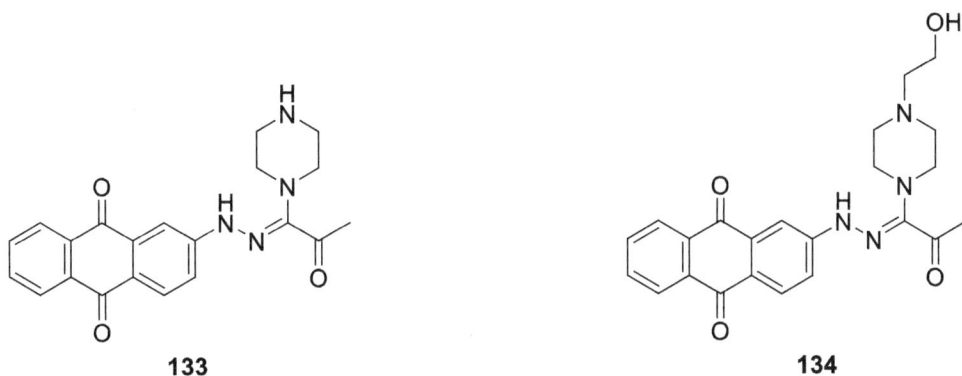

**133**

**134**

**Fig. (63).** Potent amino-AQ derivative (**133** and **134**).

**Table 24. Anticancer activities of compounds 133 and 134 (IC$_{50}$ in μM).**

| Compound | MCF-7 | K562 | Fibroblast |
|---|---|---|---|
| 133 | 20 | 2.1 | 2.2 |
| 134 | 4.5 | 1.1 | 15 |
| Doxorubicin | 0.31 | 1.4 | - |

Volodina *et al.* [116] synthesized a novel series of anthra[2,3-b]furan-2-carboxamides derivatives and studied their potency towards antiproliferative activity. The compounds were screened for their cytotoxic effect against K562,

K562/4, HCT116, and HCT116p53KO cell lines. Three compounds (**135-137**, Fig. **64**) displayed excellent activity, with compounds **135** and **136** emerging as potent against K562 and K562/4 cell lines with $IC_{50}$ = 0.9 and 1 µM and $IC_{50}$ = 0.5 and 0.6 µM, respectively. These compounds were inactive towards HCT116 and HCT116p53KO cell lines. Further, compound **137** showed potent inhibition against all cell lines with $IC_{50}$ values of 0.5 ± 0.1 µM and 0.5 ± 0.1 µM against K562 and K562/4 cell lines, respectively. Also, 1.5 ± 0.2 µM and 1.2 ± 0.2 µM against HCT116 and HCT116p53KO cell lines, respectively. Further, an in-depth study was carried out on compounds **135** and **136**. These compounds arrested cells of K562 at the G1 phase and further induced apoptosis by inhibiting topoisomerase 1 through unwinding of DNA.

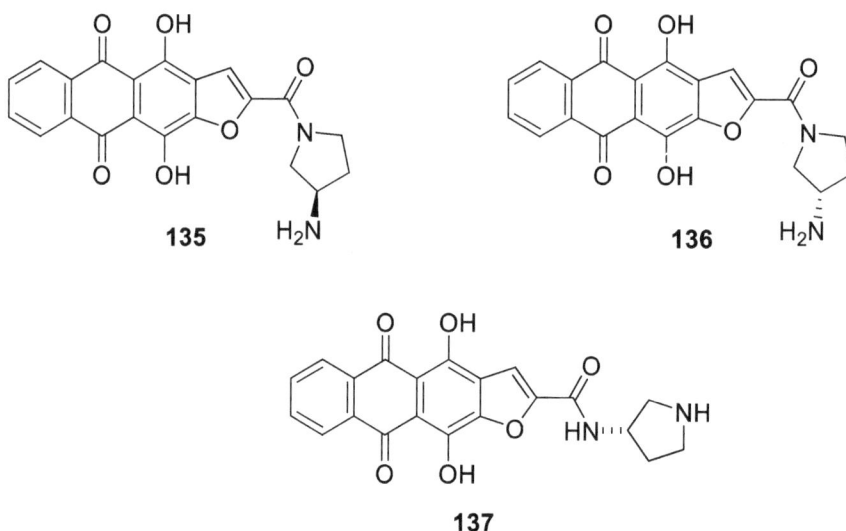

**Fig. (64).** Structures of anthra[2,3-b]furan-2-carboxamides compounds **135-137**.

Phosphoglycerate mutase 1 inhibition activity of sulfonamide units containing 1,2-dihydroxyanthraquinone derivatives was reported by Huang and co-workers [117]. Compound **138** (Fig. **65**) was co-crystallized with PGAM1, and the compound was found to occupy a novel allosteric site PGAM1. Moreover, against H1299, A549, and PC9 cells, it has displayed promising anticancer activity with $IC_{50}$ values of 6.4±1.3, 10.3±0.8, and 4.8±0.05 µM, respectively. Further, it exhibited $IC_{50}$ = 0.25 µM against PGAM1 enzyme and hence was further studied *in vivo* for its pharmacokinetic properties. Cancer cells treated with compound **138** showed decreased rates of glycolysis and oxygen consumption, which led to the depletion of adenosine 5'-triphosphate (ATP) synthesis and the consequent activation of 5' adenosine monophosphate-activated protein kinase (AMPK). Further, it showed remarkable inhibition of tumor growth in the H1299 xenograft model.

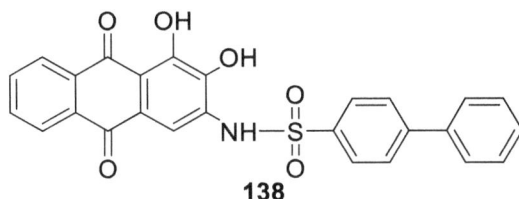

**Fig. (65).** Potent sulfonamide endowed HAQ **139**.

Shahab *et al.* [118] designed and synthesized four anthraquinone-imidazole molecules having hydroxy aryl units and screened for antitumor activities against different cell lines (Fig. **66**). Compound **139** exhibited excellent growth inhibition against renal cancer cell lines CAKI-1 and ACHN with $GI_{50}$ of 0.36 and 0.43 µM. Compound **140** was found to be highly potent against the CNS cancer SNB-75 cell line with $GI_{50}$ = 0.81 µM, **141** against leukemia cancer K-562 cell line with $GI_{50}$ = 0.43 µM, and **142** against melanoma cancer LOX IMVI ($GI_{50}$ = 0.55 µM) and UACC-257 ($GI_{50}$ = 0.86 µM) cell lines.

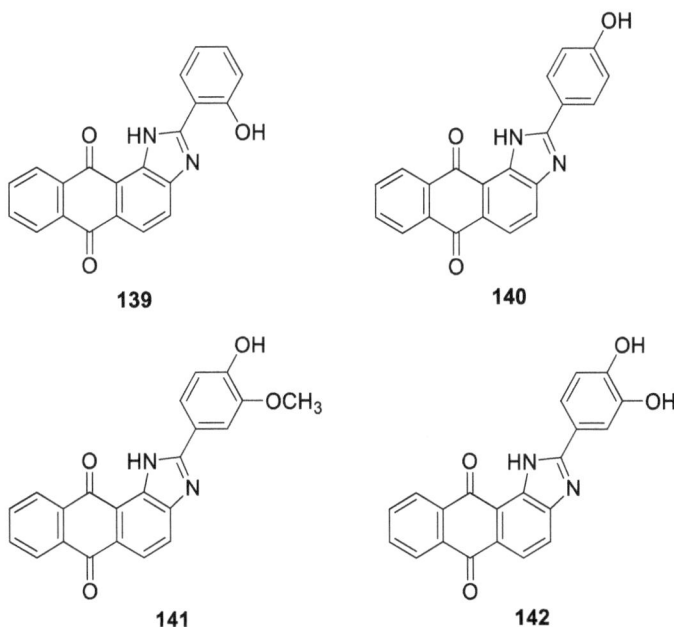

**Fig. (66).** Potent anthraquinone-imidazole molecules **139-142**.

In their study, Niedziałkowski *et al.* [119] synthesized anthraquinone derivatives with one or two piperidine units. These compounds were screened for *in vitro* antiproliferative activity against leukemia cancer HL-60, mitoxantrone-resistant HL-60/MX2, colon cancer LoVo, and doxorubicin-resistant LoVo/Dx cell lines, and also tested against normal BALB/3T3 mouse fibroblast cell lines. In their studied compounds, 1-(piperidin-1-yl)-9,10-anthraquinone (**143**, Fig. **67**) emerged as the most promising anti-proliferating agent against these cell lines with $IC_{50}$ values of 3.67, 2.50, 3.29, and 2.31 µg/mL, respectively. Also, it was found to be a safer candidate with $IC_{50}$ = 16.04 µg/mL against the normal cell line.

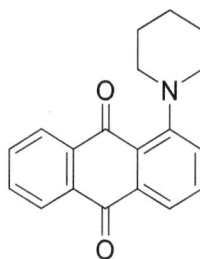

**143**

**Fig. (67).** Potent anticancer 1-(piperidin-1-yl)-9,10-anthraquinone (**143**).

Awasthi and her research developers reported novel anthraquinone derivatives having sulfonamide substituents and tested for anticancer activity [120]. The cytotoxicity effect was evaluated with reference to MTX against PC-3, MCF-7, and Hep2C cell lines. Compound **144** (Fig. **68**) exhibited excellent cytotoxicity against Hep2C cell lines with $IC_{50}$ = 1.008 µg/ml. This compound arrested the cell cycle at the G0/G1 phase in HeLa cells and also showed good binding with ct-DNA through the interaction mode.

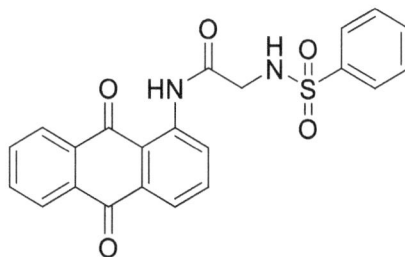

**144**

**Fig. (68).** Potent AQ-sulfonamide compound **144**.

The nature of redox properties and cytotoxicity of the synthesized anthraquinone derivatives was studied with the influence of attached hydroxyl and methoxy functional groups by Okumura and co-workers [121]. With the help of cyclic voltammetry, the initial behavior of analogs was analyzed, and the nature was reported by using molecular orbital calculations. Among the synthesized compounds, molecule **145** showed potent cytotoxic activity against the human leukemia HL-60 cell line and $H_2O_2$-resistant HP100 cells (Fig. **69**).

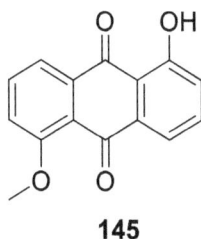

**145**

**Fig. (69).** Structure of potent hydroxy AQ (**145**) reported by Okumura and co-workers [121].

Sirazhetdinova and colleagues [122] synthesized a new library of disubstituted and trisubstituted hydroxy-anthraquinone derivatives. MDA-MB-231, prostate cancer DU-145, glioblastoma SNB-19, breast cancer, and human telomerase (h-TERT) immortalized lung fibroblast cells were used to test the anticancer efficacy of each drug. The majority of the compounds exhibited very good anticancer activity against different cell lines tested. The findings revealed that compounds **146-149** (Fig. **70**) exhibited significant activity comparable to that of the standard drug, doxorubicin. Compound **146** was potent against the DU-145 cell line ($IC_{50}$ = 1.1 μM), **147** against MDA-MB-231 ($IC_{50}$ = 6.8 μM), **148** against U-87MG ($IC_{50}$ = 8.2 μM), and **149** against SNB-19 ($IC_{50}$ = 5.77 μM). All these compounds were found to be safe for human telomerase (h-TERT) immortalized lung fibroblast cells. Further, cell cycle arrest was achieved at the sub-G1 phase in DU-145 cells upon treatment with compound **146** at its $IC_{50}$ concentration and subsequently induced apoptosis in it. Also, it enhanced the DNA synthesis in SNB-19 cells.

In the work described by Tikhomirov *et al.* [123], the synthesis of a new family of naphtho[2,3-*f*]indole-3-carboxamides and anthra[2,3-*b*]thiophene-3-carboxamides was reported. Utilising an MTT assay, these substances were examined against five tumor cell lines). Capan-1 pancreatic adenocarcinoma, HCT116, NCIeH460, HL60, and K562. Among the series, compound **150** displayed potent activity compared to the standard drug doxorubicin (Dox) and anthrafuran. (Fig. **71**, Table **25**).

**146**    **147**

**148**    **149**

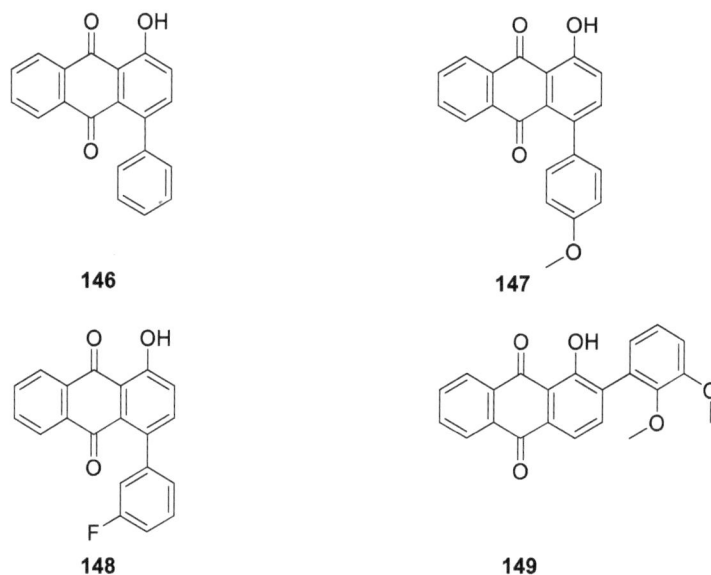

**Fig. (70).** Potent disubstituted and trisubstituted hydroxy-anthraquinone derivatives (**146-149**).

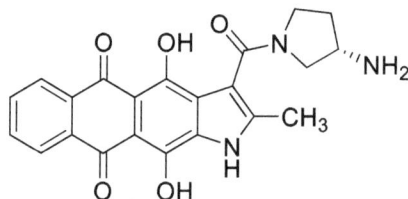

**150**

**Fig. (71).** Potent anthraquinone derivative (**150**) reported by Tikhomirov and colleagues [123].

**Table 25. Anticancer activities of compound 150 and doxorubicin.**

| Comp. | IC$_{50}$ (µM) | | | | | |
|---|---|---|---|---|---|---|
| | **Capan-1** | **HCT116** | **NCI-H460** | **HL60** | **K562** | **hTERT RPE-1** |
| **150** | $0.5 \pm 0.2$ 0. | $0.9 \pm 0.1$ | $0.9 \pm 0.2$ | $0.9 \pm 0.1$ | $0.8 \pm 0.1$ | $15.8 \pm 0.8$ |
| **Dox** | $1.7 \pm 0.2$ | $0.2 \pm 0.1$ | $0.4 \pm 0.1$ | $0.2 \pm 0.1$ | $0.1 \pm 0.0$ | $1.0 \pm 0.2$ |

Oliveira *et al.* synthesized two series of *O-* and *N*-alkylated anthraquinones (Fig. **72**) and screened them against three cancer cell lines, MCF-7, HeLa, and M059J, as well as a normal cell line GM-07492A [124]. The majority of the compounds showed very good activity, and among them, the *N*-alkylated compound **151** with hexyl substitution exhibited the maximum cytotoxicity against the three cancer cells. With compound **152**, which had IC$_{50}$ values that were lower than those of MTX, had strong cytotoxic activity. The excellent activity was shown by the O-alkylated compound **153** with a pentyl substitution (Table **26**).

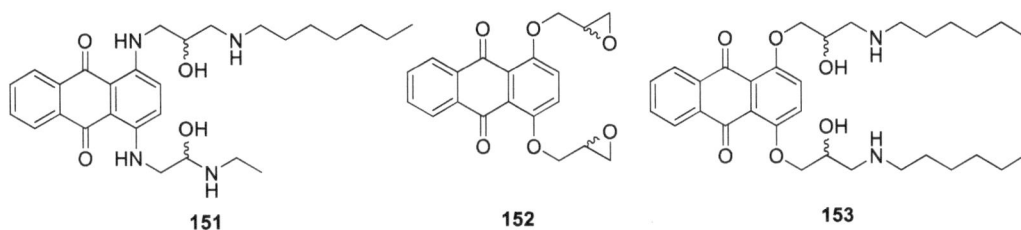

**151**            **152**            **153**

**Fig. (72).** Potent anthraquinone derivatives (**151-153**) reported by Oliveira and colleagues [124].

**Table 26.** *O-* and *N*-alkylated anthraquinone derivatives (151-153).

| Compound | logPa | IC$_{50}$ (µM) | | | |
|---|---|---|---|---|---|
| | | GM07492A | MCF-7 | HeLa | M059J |
| 151 | 5.76 | 10.3 ± 0.6 | 13.6 ± 0.1 | 14.1 ± 0.4 | 14.8 ± 1.5 |
| 152 | 3.35 | 164.8 ± 10.8 | 64.0 ± 0.9 | 93.9 ± 3.3 | 72.9 ± 7.5 |
| 153 | 6.16 | 53.6 ± 6.2 | 28.6 ± 1.1 | 32.3 ± 1.7 | 53.4 ± 0.3 |
| MTX | 0.36 | 141.1 ± 7.2 | 146.3 ± 3.5 | 146.3 ± 3.5 | 150.0 ± 8.3 |

Li with his group synthesized a new class of 1-nitro-2-acyl anthraquinone-amino acids and assessed their antitumor effects on two normal human cell lines (liver HL7702 and colorectal FHC) as well as eight human cancer cell lines (HCT116, QBC939, MCF-7, SGC-7901, EC9706, HepG2, HeLa, and SW480) [125]. The majority of the substances showed encouraging anti-proliferative effects against cancer cell lines in testing while exhibiting no harm towards normal cells. Compound **154** from the series showed the greatest suppression of HCT116 cell activity with an IC$_{50}$ of 17.80 µg/mL (Fig. **73**). Additionally, compound **154** efficiently destroyed cancer cells by phosphorylating JNK, inducing mitochondrial stress, and activating JNK and reactive oxygen species (ROS), all of which increased ROS levels. Using the comparative molecular field analysis (CoMFA) and comparative molecular similarity index analysis (CoMSIA) methodologies, the structure-activity relationship of the compounds was studied. The results of the studies showed that the nitro group at the C-1 position had a major role in determining the activity of the compounds; the higher the capacity for electron withdrawal, the higher the inhibitory activity of the compound.

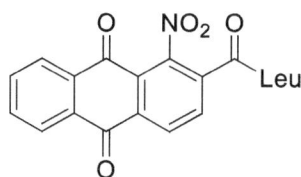

**154**

**Fig. (73).** Potent anthraquinone derivative (**154**) reported by Li with his group [125].

Anifowose *et al.* synthesized 2-chloro-*N*-(9,10-dihydro-4,5-dihydroxy-9,10-dioxo-2-anthracenyl)acetamide analogs to assess the compounds' structural activity relationship in acute lymphoblastic leukemia (ALL) cells [126]. The WST-8 assay was used to evaluate each compound's biological activity. With an $IC_{50}$ value of 0.74 μM, compound **155** (Fig. **74**) from the series showed potential cytotoxic action against the leukemia cell line. It became clear that, in contrast to its reference drug, the active component **155** demonstrated cytotoxicity in a distinct way. In contrast to the reference, the 2-chloro-*N*-(9,10-dihydro-4,5-dihydroxy-9,10-dioxo-2-anthracenyl)acetamide increased the expression of p53 but did not cause MDM2 to degrade. The structural analysis showed that substituting a methylene group in place of the -NH in the chloroacetamide group decreased the cytotoxic activity but did not completely stop it.

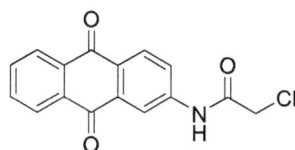

**155**

**Fig. (74).** Potent anthraquinone derivative (**12**) reported by Anifowose with his group [126].

A new library of polyfunctionalized anthraquinone hydrazone derivatives was synthesized by Lozynskyi *et al.* [127]. Some derivatives were evaluated for their *in vitro* anticancer activity. The compound **156** displayed excellent antimitotic activity when tested towards tumor cells with mean GI50/TGI values of 4.06/78.52 µM. All compounds were screened for antimicrobial and antifungal activities (Fig. **75**).

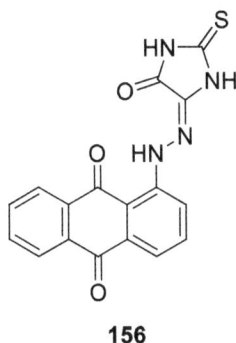

**156**

**Fig. (75).** Potent anthraquinone derivative (**156**) reported by Lozynskyi with his group [127].

Awasthi *et al.* [128] synthesized two different compounds **(157 and 158)** with the 1- or 2-position of the 9,10-anthraquinone ring substituted with 1-oxo-3-phenyl-2-(benzosulfonamide)-propylamide (Fig. **76**). These synthesized compounds had anticancer activity similar to those of the standard drug mitoxantrone. Two compounds were tested for cytotoxicity against the MCF-7, PC-3, and Hep2C cancer cell lines. It was found that molecule **157** with position 1 substituted was more effective against the Hep2C and MCF-7 cell lines, with IC$_{50}$ values of 9 µg/ml and <10 µg/ml, respectively.

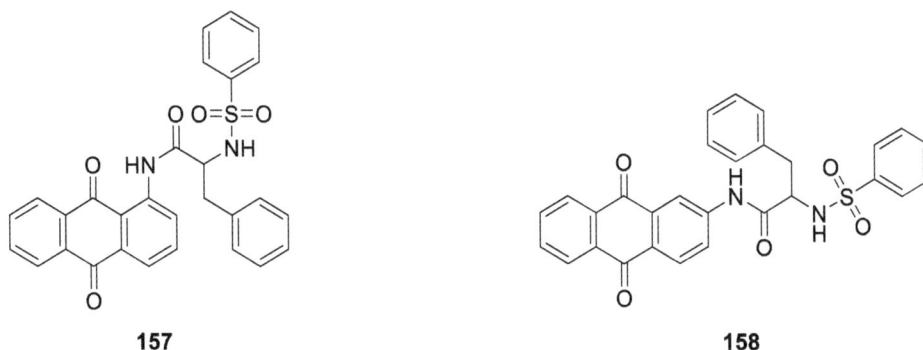

**157**                                 **158**

**Fig. (76).** Potent anthraquinone derivatives (**157** and **158**) reported by Awasthi with his group [128].

Two novel anthraquinone-based aspirin derivatives were reported by Lin *et al*. and tested for effectiveness in preventing the growth of gastric cancer [129]. Two compounds (159 and 160) have $IC_{50}$ values ten times lower than aspirin that prevented the development of gastric cancer cells (SGC 7901) (Fig. 77). With $IC_{50}$ values that were around 2 times greater than the similar $IC_{50}$ values discovered with SGC7901 cells, these drugs were less hazardous to stomach mucosal cells. Compound 160 did not significantly reduce COX-1 production in gastric mucosal cells. However, compound 159 and aspirin both drastically lowered COX-1 production when measured by Western blot, and both compounds also caused SGC7901 cells to undergo apoptosis. However, the two compounds had the same impact on the COX-2 level in stomach cancer cells, causing up to 90% and 95% reduction in COX-2 production.

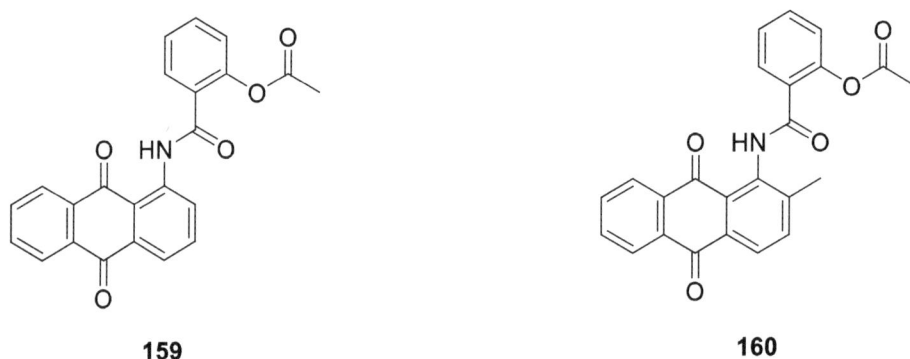

**Fig. (77).** Structures of anthraquinone derivatives (159 and 160) reported by Lin with his group.

Lin *et al*. [130] synthesized a new library of 13 anthraquinone derivatives and tested them for anticancer efficacy with the reference drug cisplatin. When tested in NTUB1 and PC3 cells, compound 161 showed promising cytotoxicity with $IC_{50}$ values of 1.51 µM and 12.78 µM, respectively (Fig. 78). They used MTT tests and autophagy to further evaluate the drugs' effectiveness. In NTUB1 cells, compound 161 caused DNA damage and triggered apoptosis at doses of 1 and 3 µM. According to the structural analysis of 161, the hydroxy group at position C-1 improved the antiproliferative action. The cytotoxicity was simultaneously decreased by swapping out the bromo atom in the side chain of C-3.

A new series of thiophene-2-carboxamide anthraquinone derivatives was synthesized and screened for antitumor activity by Volodina *et al*. [131]. Five synthesized compounds, 162-166, showed excellent activity against different cell lines (Table 27), and among them, compound 164 was the most effective in the series against drug-resistant tumor cell lines (Fig. **79**). Compound **164** (2.5 M, 3-6 h) induced apoptosis in cells and was associated with caspase 3 and 9 activations, cleavage of poly(ADP ribose) polymerase, an increase in annexin V/propidium iodide double-stained cells, DNA fragmentation (sub G1 fraction), and a decrease in mitochondrial membrane potential.

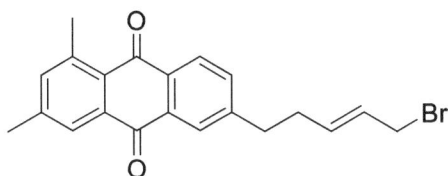

**161**

**Fig. (78).** Potent anthraquinone derivative (**161**) reported by Lin with his group [130].

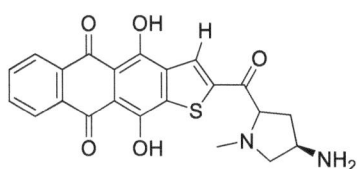

**162**

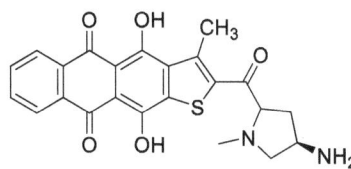

**163**

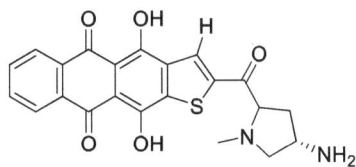

**164**

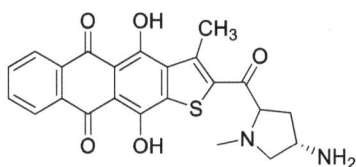

**165**

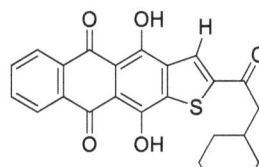

**166**

**Fig. (79).** Structures of anthraquinone derivative (**162-166**).

**Table 27.** Anticancer activity of thiophene-2-carboxamide anthraquinone derivatives (**162-166**).

| Compound | Values of $IC_{50}$ in µM | | | |
|---|---|---|---|---|
| | **Capan-1** | **NCI-H460** | **DND-41** | **HL60** |
| **162** | $1.5 \pm 0.3$ | $0.4 \pm 0.1$ | $0.4 \pm 0.0$ | $1.1 \pm 0.2$ |
| **163** | $2.0 \pm 0.2$ | $1.2 \pm 0.2$ | $1.4 \pm 0.3$ | $1.8 \pm 0.5$ |

*(Table 27) cont.....*

| 164 | 0.7 ± 0. | 0.4 ± 0.1 | 0.4 ± 0.1 | 0.3 ± 0.0 |
|---|---|---|---|---|
| 165 | 2.2 ± 0.4 | 5.0 ± 1.0 | 1.6 ± 0.3 | 0.2 ± 0.0 |
| 166 | 0.6 ± 0.1 | 1.4 ± 0.1 | 1.6 ± 0.4 | 3.2 ± 0.2 |
| Dox | 1.7 ± 0.2 | 0.4 ± 0.1 | 0.7 ± 0.1 | 0.2 ± 0.0 |

Mohamadzadeh *et al.* [132] synthesized a novel series of 2-azetidinones comprising anthraquinone derivatives and tested their cytotoxicity against various human cancer and healthy cell lines. Towards PC3, MCF7, SKNMC, and HCT116 cell lines, some of the compounds showed moderate to considerable cytotoxicity. From the series, compound **167** (Fig. **80**) was chosen for research on fibroblast (Hu02) cells and compared with doxorubicin. Results showed reduced cytotoxicity against fibroblast (Hu02) cells than doxorubicin and comparable cytotoxicity against cancer cell lines compared to doxorubicin. Through hydrogen bonds and hydrophobic interactions, molecular docking experiments revealed that compound **167** significantly fit the active site of topoisomerase II (PDB 4G0V).

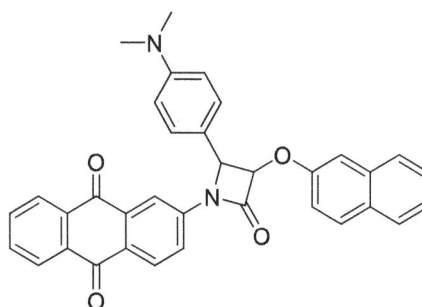

**167**

**Fig. (80).** Potent anthraquinone derivative **167** reported by Mohamadzadeh with his group [132].

Sirazhetdinova *et al.* [133] synthesized a new class of anthraquinone derivatives containing an alkyne unit as potential anticancer agents. The most promising cytotoxic action against glioblastoma cancer cells can be found in several anthraquinone-propargylamine derivatives (**168-171**, Fig. **81**, Table **28**). In comparison to the standard medicine doxorubicin, two compounds (**168** and **169**) showed excellent efficacy towards the DU-145 cell line (IC$_{50}$ = 6.55 and 6.16 µM) and compounds (**170** and **171**) on the MCF-7 cell line (IC$_{50}$ = 5.45 and 6.02 µM).

Further docking studies were performed with G-quadruplex DNA motifs to understand their binding interactions.

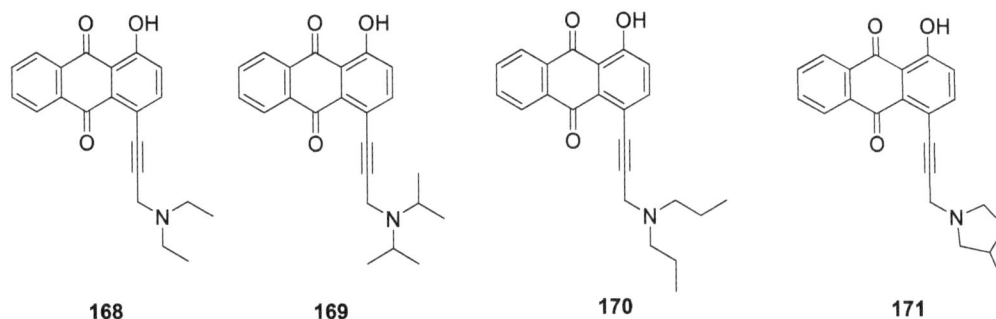

**Fig. (81).** Potent anthraquinone derivatives **168, 169, 170,** and **171** reported by Sirazhetdinova with his group [133].

**Table 28. Anticancer activity of anthraquinone derivatives containing alkyne unit (168-171).**

| Compound | Growth Inhibition of Cells (GI$_{50}$ in μM) | | | | |
|---|---|---|---|---|---|
| | MCF-7 | DU-145 | SNB-19 | U-87 MG | hTERT lung fibroblasts |
| **168** | 15.66 ± 2.14 | 6.55 ± 0.77 | 21.45 ± 1.33 | 17.74 ± 0.84 | 18.24 ± 0.88 |
| **169** | 12.33 ± 2.74 | 6.16 ± 0.67 | 8.64 ± 1.07 | 9.66 ± 1.02 | 15.22 ± 2.34 |
| **170** | 5.45 ± 0.87 | 8.06 ± 0.45 | 4.23 ± 0.85 | 9.24 ± 0.64 | 8.23 ± 0.54 |
| **171** | 6.02 ± 1.11 | 9.48 ± 1.55 | 7.08 ± 1.31 | 8.15 ± 0.72 | 15.46 ± 2.07 |
| **DOX** | 5.11 ± 0.54 | 6.61 ± 0.34 | 7.62 ± 0.69 | 6.11 ± 0.15 | 3.18 ± 0.21 |

A series of anthraquinone-heteroarene-fused compounds were reported by Singh *et al.* [134] and investigated through virtual screening to find possible AurB inhibitors. Comparing their binding profiles to their parent analog, anthrafuran, two derivatives showed improvement in binding. Compound **172** from the derivatives had the best *in vitro* AurB inhibition (7.3 μM). Compounds **172** and **173** displayed good antiproliferative activity against AurB overexpressing MDA-MB-453 and Saos-2 cell lines as well as noncancerous HEK 293T cell line (Fig. **82**, Table **29**).

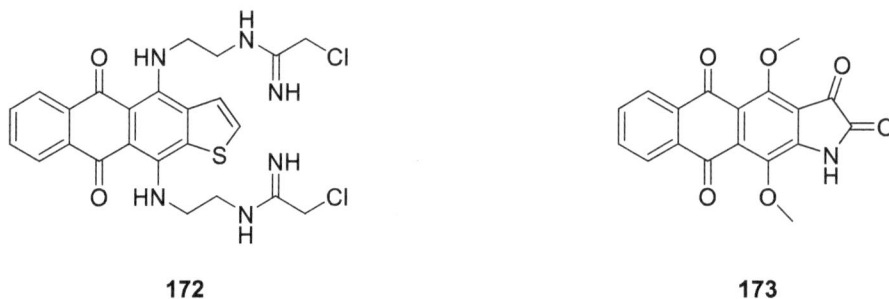

Fig. (82). Potent anthraquinone derivatives (172 and 173) reported by Singh with his group [134].

**Table 29. Anticancer activity of anthraquinone-heteroarene compounds 172 and 173.**

| Compound | IC$_{50}$ in µM | | |
|---|---|---|---|
| | **MDA-MB-453** | **Saos-2** | **HEK 293T** |
| 172 | 4.0 | 7.0 | 15 |
| 173 | 2.7 | 79.5 | >100 |

A new library of diketo-anthraquinone derivatives was synthesized by Zou *et al.* [135] and screened for HDAC inhibition. Among the series, compounds (174-177, Fig. 83) exhibited excellent inhibition in enzymatic activity, with compound 175 emerging as a potent inhibitor with IC$_{50}$ = 8.6 nM. Further, these four compounds were screened for HDAC1, 2, 3, and 6 enzyme inhibition, and the results are depicted in Table 30. Compound 175 emerged as a potent inhibitor of HDAC1 and 6 enzymes, whereas compound 177 of the HDAC6 enzyme. Further, these derivatives showed very good antiproliferative activity against K562, MCF-7, HeLa, and H9c2 cell lines (Table 31). Among them, compound 177 was highly potent and comparable to the standard drug vorinostat.

Tikhomirov *et al.* [136] prepared a new series of anthraquinones fused with heterocycles comprising amide derivatives and studied their anticancer activity. Compound 178 exhibited as a potential agent for antiproliferation of cells against MCF-7 and K562/4 cells with IC$_{50}$ values of 2.3 ± 0.3 and 1.0 ± 0.1 µM, respectively. Compound 179 showed potent activity at a concentration of 3.8 ± 0.5 µM against the fibroblast cell line. Also, compound 180 emerged as a potential anti-cancer molecule against HCT116, HCT116p53KO, and K562 cells with 1.0 ± 0.1,

$1.8 \pm 0.2$, and $0.4 \pm 0.1$ μM (Fig. **84**, Table **32**). At very low concentrations, compound **178** induced cell apoptosis and emerged as a potential therapeutic agent for anticancer action with the presence of fused pyridine with anthraquinone derivatives. From the studies of the Pgp model, we conclude that compound **178** interacted strongly with this model.

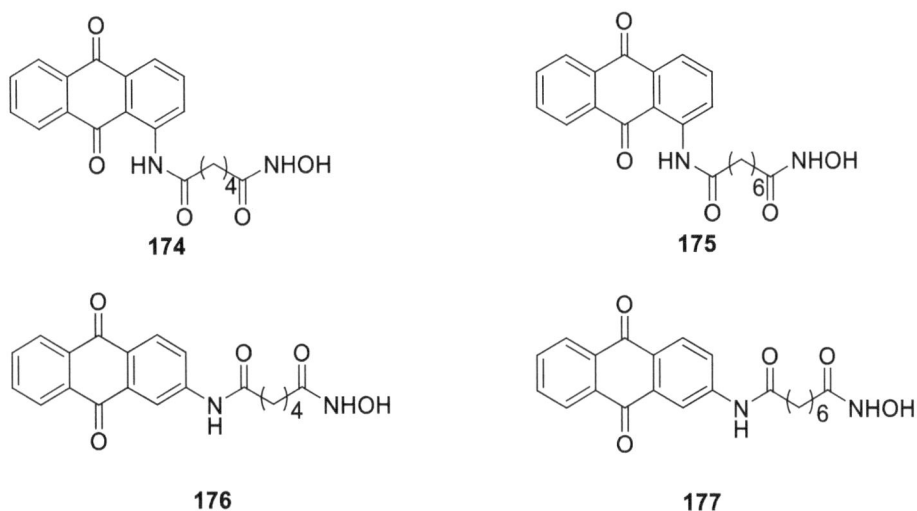

**174**

**175**

**176**

**177**

**Fig. (83).** Potent diketo-anthraquinone derivatives (**174-177**).

**Table 30. HDAC inhibition activity of diketo-anthraquinone derivatives (174-177).**

| Comp. | IC$_{50}$ in nM | | | |
|---|---|---|---|---|
| | HDAC1 | HDAC2 | HDAC3 | HDAC6 |
| 174 | $70.1 \pm 1.3$ | - | - | $11.5 \pm 0.6$ |
| 175 | $7.7 \pm 2.1$ | $41.4 \pm 0.4$ | $12.6 \pm 1.3$ | $5.9 \pm 0.2$ |
| 176 | $29.7 \pm 0.4$ | $104.2 \pm 3.4$ | $41.6 \pm 1.0$ | $11.4 \pm 1.0$ |
| 177 | $51.1 \pm 0.2$ | $96.4 \pm 0.6$ | $39.0 \pm 0.2$ | $5.1 \pm 0.4$ |
| Vorinostat | $93.6 \pm 1.7$ | >200 | $89.7 \pm 0.6$ | $20.2 \pm 0.9$ |

**Table 31. Anticancer activities of diketo-anthraquinone derivatives (174-177).**

| Comp. | IC$_{50}$ in µM | | | |
|---|---|---|---|---|
| | K562 | MCF-7 | HeLa | H9c2 |
| 174 | 9.7 ± 1.1 | 10.6 ± 1.8 | 12.0 ± 1.0 | 31.2 ± 1.5 |
| 175 | 5.3 ± 1.2 | 5.3 ± 0.2 | 7.5 ± 1.7 | 10.0 ± 1.7 |
| 176 | 6.0 ± 2.6 | 10.4 ± 0.4 | 5.0 ± 1.1 | 5.0 ± 0.5 |
| 177 | 2.5 ± 1.3 | 3.4 ± 0.4 | 3.0 ± 1.3 | 2.0 ± 0.2 |
| Vorinostat | 1.9 ± 0.2 | 2.6 ± 0.6 | 2.7 ± 0.2 | 4.6 ± 0.4 |

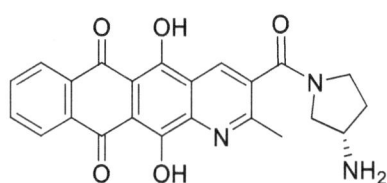

**178**

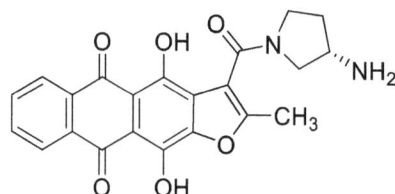

**179**

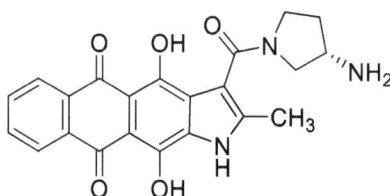

**180**

**Fig. (84).** Potent anthraquinone fused with heterocycles comprising amide derivatives (**178-180**).

Morgan *et al.* [137] synthesized an anthraquinone derivative and evaluated it for anticancer activity. Compound **181** showed potent activity against the proliferation of PC3 cells with an IC$_{50}$ value of 4.65 µM (Fig. **85**). By the activated caspases, it induced cell apoptosis, and the cells got arrested at the G1/M phase of the cell cycle of PC3 cells.

**Table 32. Anticancer activities of anthraquinone fused with heterocycles comprising amide derivatives (178-180).**

| Compd. | IC$_{50}$ ($\mu$M) | | | | | |
|---|---|---|---|---|---|---|
| | MCF-7 | HCT116 | HCT116p53KO | K562 | K562/4 | Fibroblasts |
| 178 | 2.3 ± 0.3 | 2.5 ± 0.3 | 2.2 ± 0.4 | 0.5 ±0.1 | 1.0 ±0.1 | 4.7 ± 0.3 |
| 179 | 3.0 ± 0.4 | 1.9 ± 0.2 | 2.0 ± 0.2 | 0.9 ±0.1 | 1.4 ±0.2 | 3.8 ± 0.5 |
| 180 | 3.2 ± 0.5 | 1.0 ± 0.1 | 1.8 ± 0.2 | 0.4 ±0.1 | 50.0±7.0 | 5.0 ± 0.7 |
| DOX | 0.40±0.05 | 0.30±0.04 | 1.0 ± 0.1 | 0.20±0.02 | 14.0±1.7 | 0.30 ± 0.04 |

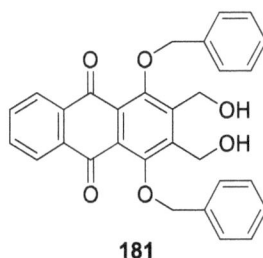

**181**

**Fig. (85).** Structure of dibenzyloxy-AQ derivative (**181**).

The recent work of Tikhomirov *et al.* described the synthesis of novel series of amides attached to the cyclopentane ring of the anthraquinone-fused cyclopentane moiety and evaluated their anticancer activity against Capan-1, HCT-116, LN-229, NCI–H1975, Z138, DND-41, HL60, K562, K562/4, and fibroblast cell lines [138]. Among them, compound **182** with the amino-cyclopentane derivative emerged as highly promising against all the cell lines (Fig. **86**). The compound was also studied for topoisomerase inhibition activity and found to be an effective inhibitor that affected topoisomerase 1-mediated unwinding of supercoiled DNA at a 5 $\mu$M concentration.

Olszewski and group in their studies highlighted that amide derivatives of anthraquinone (**183-185**, Fig. **87**) effectively induced apoptosis, inhibited tyrosine kinase proteins (PTKs), and caused DNA damage in cancer cells, making them potent anticancer agents [139] (Table **33**). These three compounds showed selective cytotoxicity against NSCLC cells while sparing healthy human kidney cells, positioning them as promising candidates for lung cancer treatment.

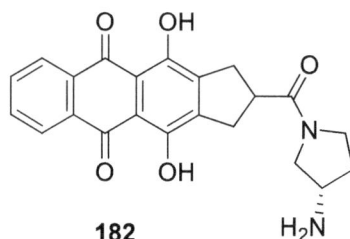

Fig. (86). Structure of cyclopentane-AQ amide derivative (182).

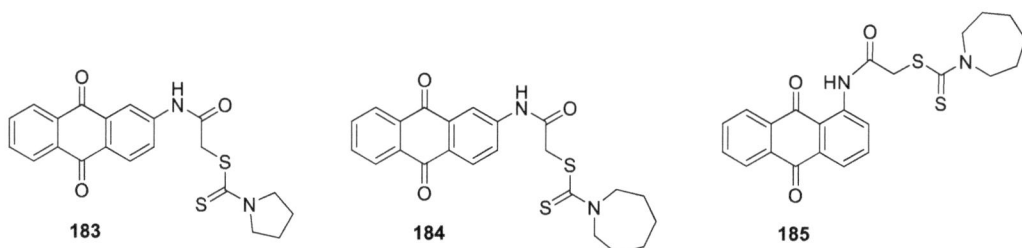

Fig. (87). Potent anthraquinone-amide compounds (183-185).

Table 33. Anti-proliferative activity of anthraquinone comprising amide derivatives (183-185).

| Comp. | IC$_{50}$ values (± SD, μM) | | | | |
|---|---|---|---|---|---|
| | A549 | H226 | H460 | HEK293 | NHBE2594 |
| 183 | 2.49 ± 0.04 | 7.88 ± 1.17 | 11.02 ± 0.56 | 33.75 ± 1.24 | 3.05 ± 0.22 |
| 184 | 1.35 ± 0.21 | 1.54 ± 0.28 | 4.21 ± 0.81 | 0.89 ± 0.12 | 1.02 ± 0.14 |
| 185 | 0.81 ± 0.12 | 3.87 ± 0.54 | 5.02 ± 0.81 | 37.2 ± 1.63 | 1.62 ± 0.38 |
| Etoposide | 0.54 ± 0.21 | 0.39 ± 0.01 | 0.83 ± 0.15 | 1.91 ± 0.97 | 4.21 ± 0.23 |
| Cisplatin | 29.01 ± 0.12 | 17.47 ± 2.12 | 21.49 ± 1.87 | 28.45 ± 1.97 | ND |

Zhou *et al*. [140] designed and synthesized two novel series of anthraquinone-based benzenesulfonamide derivatives. Among them, two compounds, **186** and **187** (Fig.

**88**), emerged as potent carbonic anhydrase inhibitors (CAIs), showing strong inhibition against both off-target human carbonic anhydrase II (hCA II) and tumor-associated human carbonic anhydrase IX (hCA IX). The $IC_{50}$ values for compound **186** were 41.58 nM for hCA II and 40.58 nM for hCA IX, while compound **187** exhibited activity with $IC_{50}$ values of 21.19 nM and 46.19 nM for hCA II and hCA IX, respectively. These results indicate their potential as effective dual-target inhibitors for cancer therapy. These novel compounds demonstrated significant inhibitory activity against hCA II and IX isoforms, with the most potent compounds, **186** and **187**, exhibiting promising antitumor effects in MDA-MB-231, MCF-7, and HepG2 cell lines under both normoxic and hypoxic conditions. Further assays revealed enhanced apoptosis and reduced cell viability in these compounds. The study was also supported by molecular docking studies on hCA II and hCA IX proteins. Additionally, ADME predictions confirmed the compounds' favorable pharmacokinetic and physicochemical properties, indicating their potential as effective therapeutic agents.

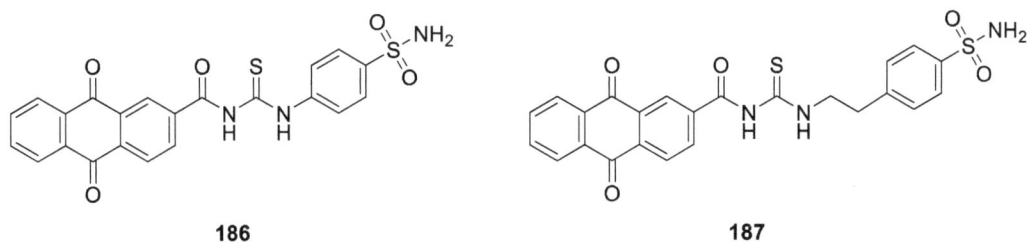

**186**            **187**

**Fig. (88).** Potent anthraquinone-sulphonamides derivatives (**186** and **187**).

Andreeva *et al.* [141] described the synthesis and antiproliferative activity of hetero-fused anthraquinones-carboxamides. Among them, thiazole-fused compounds, **188** and **189**, exhibited potent activity against K562, HCT116, and hFB-hTERT6 human cancer cells (Fig. **89**). Compound **188** showed an $IC_{50}$ value of $5.5 \pm 0.9$ μM for K562, while the $IC_{50}$ values for both HCT116 and hFB-hTERT6 were >50.0 μM. Similarly, compound **189** demonstrated an $IC_{50}$ of $11.7 \pm 1.1$ μM for K562 but showed $IC_{50}$ values >50.0 μM for HCT116 and hFB-hTERT6 cell lines. These results indicate selective anticancer activity, particularly against the K562 cell line. However, their activity was lower compared to anthra[2,3-*d*]thiophene analogs, highlighting the significance of the thiazole core in the antitumor efficacy of heteroarene anthraquinones. The same group, in their extended studies, synthesized azole-fused anthraquinones, focusing on the development of G4 ligands with amino- or guanidinoalkylamino side chains [142].

In the series, two compounds, **190** and **191**, exhibited excellent anticancer activity against K562, HCT116, MDA-MB-231, and hFB-hTERT6A cell lines (Table **34**). Furthermore, in particular, compound **191**, having a selenadiazole core, demonstrated a strong affinity for the c-MYC G4 structure *in vitro* and effectively downregulated the expression of the c-MYC oncogene in cellular environments. Further analysis showed that in a dose- and time-dependent manner, it induced cell cycle arrest and apoptosis, inhibiting the growth of K562 cells. These findings highlight the critical impact of structural variations in heterocycles on the biological properties of G4 ligands. Compound **190** preferentially stabilized telomeric quadruplex structures (22AG and 22CTA), whereas selenadiazole **191** exhibited similar potency across all G4 structures (Fig. **89**). Notably, substituting the sulfur atom in **190** with selenium in **191** improved the selectivity of **191** for G4 structures over the DNA duplex, though this modification resulted in a reduction of all $\Delta T_m$ values.

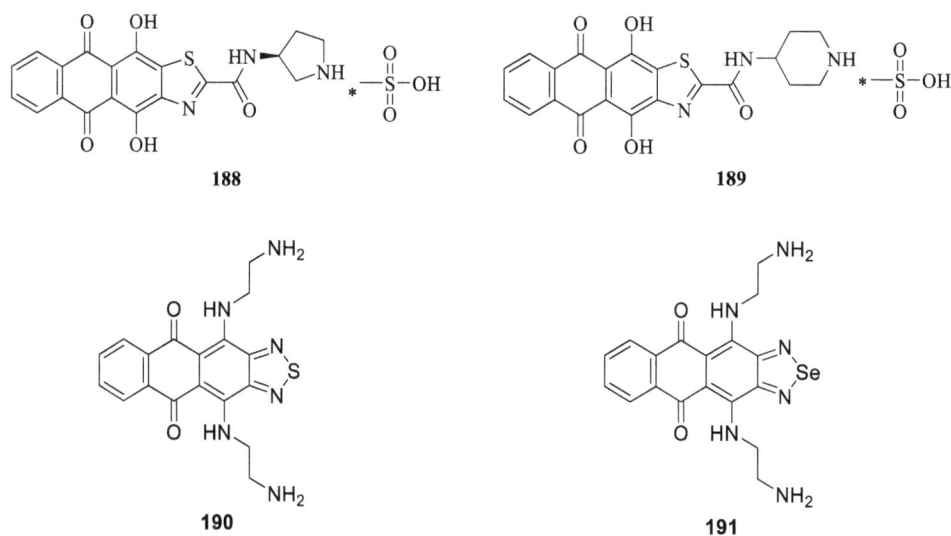

**Fig. (89).** Potent anthraquinone-thiazole-carboxamide (**188** and **189**), anthraquinone-thiadiazole (**190**), and anthraquinone-selenadiazole (**191**) compounds.

**Table 34. The antiproliferative activity of azole-fused anthraquinones (190 and 191).**

| Comp. | IC$_{50}$ (μM) | | |
|---|---|---|---|
| | **K562** | **HCT116** | **MDA-MB-231** |
| **190** | $0.8 \pm 0.2$ | $0.6 \pm 0.1$ | $0.6 \pm 0.1$ |

*(Table 34) cont.....*

| 191 | $4.0 \pm 0.4$ | $0.8 \pm 0.1$ | $1.0 \pm 0.1$ |
|---|---|---|---|
| **DOX** | **$0.37 \pm 0.03$** | **$0.55 \pm 0.04$** | **0.790.06** |

## Mechanism of Action of Anthraquinone Derivatives

It has been suggested that anthraquinone derivatives primarily exert their anticancer effects through apoptosis, cycle arrest, and DNA damage [143, 144]. However, the exact mechanism of action is not fully studied, but it is found that apoptosis is one of the main mechanisms through which they exhibit activity. On the other hand, few new research studies have demonstrated that novel anthraquinone compounds can also suppress cancer through paraptosis and autophagy [78, 145-148]. Moreover, many of these compounds work by inhibiting the different pathways that lead to the growth of cancer cells. Anthraquinone derivatives were reported as topoisomerase inhibitors, telomerase inhibitors, matrix metalloproteinase (MMP) inhibitors, and kinase inhibitors [149-151].

Topoisomerases are promising targets for the development of new chemotherapeutic agents. These enzymes play a vital role in DNA replication and transcription within cells. They are responsible for relaxing or introducing supercoils in DNA. These processes are crucial for maintaining cell survival. Topoisomerases are classified into 3 types). topoisomerase I, topoisomerase II, and topoisomerase III. Type I topoisomerases alter DNA topology by creating a temporary break in a single strand of DNA. Type I topoisomerases function without ATP and rely on the strain energy within supercoiled DNA. Type II topoisomerases generate double-strand breaks in DNA. Type II topoisomerases require ATP to function and have DNA-binding and ATP-binding domains [152]. Most of the anthraquinone derivatives are type II inhibitors, and their general pathway is depicted in Fig. (**90**) and Table **35**.

Telomerases are ribonucleoprotein complexes that maintain and extend the telomere regions at the ends of chromosomes. They play a role in regulating cellular aging and the process of cellular senescence. Telomerase is a reverse transcriptase enzyme that plays a critical role in maintaining telomere length. This enzyme allows cancer cells to proliferate indefinitely, which is a hallmark of malignancy. The components of basal human telomerase enzymes are the catalytic subunit hTERT and the human telomerase RNA template. Moreover, several proteins are required for telomerase activity. The telomers and G-tetraplexes are also potential targets where drugs directly bind telomerase and cause telomere structural disruption. Many anthraquinone derivatives are known to exhibit potential inhibition of telomerase. Table **36** summarizes it, and the telomerase pathway mechanism is depicted in Fig. (**91**).

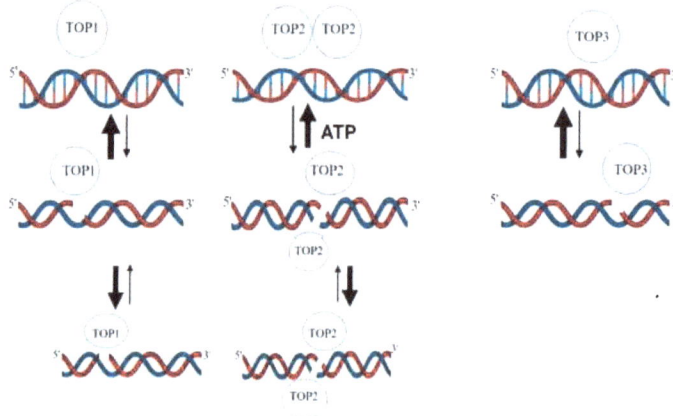

**Fig. (90).** Mechanism of topoisomerase pathway and inhibition by anthraquinone drugs.

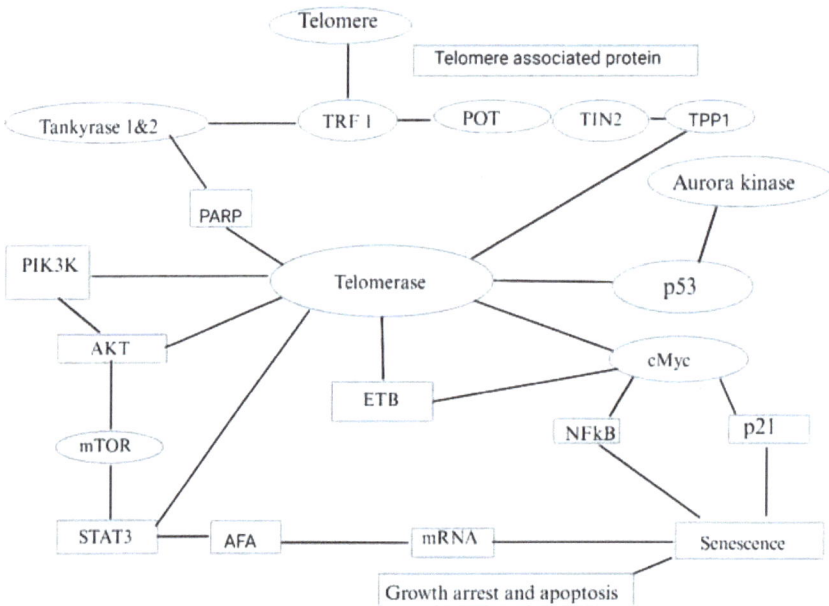

**Fig. (91).** Telomerase pathway mechanism.

**Table 35. Anthraquinone derivatives acting as topoisomerase inhibitors.**

| Compound | Therapeutic Potential and Mechanism of Action | Refs. |
|---|---|---|
| **192** <br> **193** <br> **194** <br> **195** | •DNA topoisomerase II-dependent cytotoxicity, bio-reductive agent. | [153-155] |

*(Table 35) cont.....*

| | | |
|---|---|---|
| <br>**196** | •Specific topoisomerase IIα DNA covalent complexes inhibitor, effective anticancer activity against doxorubicin and cisplatin-resistant ovarian cancer cell lines. | [156] |
| <br>**197**<br><br>**198** | •Potent topoisomerase I inhibition. | [157] |
| <br>**199** | •Anticancer activity against MCF-7 cell lines and topoisomerase I and II at 30 and 60 µM | [158] |

**Table 36. Anthraquinone derivatives acting as telomerase inhibitors.**

| Compound | Therapeutic Potential And Mechanism of Action | Refs. |
|---|---|---|
| **200** | •Telomerase inhibition with $IC_{50}$ = 0.1 μM and anticancer activity against cancer cell lines. | [159] |
| **201** | •Potent telomerase inhibitor targeting the potassium form of telomeric G-quadruplex DNA at micromolar levels. | [160] |
| **202** | •Telomerase inhibitory activity through the activation of hTERT expression. | [161] |

*(Table 36) cont.....*

| | | |
|---|---|---|
| **203** <br> **204** | •Compound **203** is a potent telomerase inhibitor with IC$_{50}$ = 5 μM and **204** causes a selective repression of hTERT expression. | [162] |
| **205** | •DNA intercalation and 43.3% Telomerase inhibition at 10 μM concentration. | [163] |
| **206** <br> **207** | •Potent telomerase inhibition and activated hTERT expression. | [164] |

*(Table 36) cont.....*

| | | |
|---|---|---|
| **208** | | |
| **209** <br> **210** <br> **211** | •Compound **209** and **210** are equipotent to doxorubicin, telomerase enzyme, and Taq polymerase inhibitors. Compound **211** is an excellent telomerase inhibitor (IC$_{50}$ = 4.5 μM). | [165] |

Protein kinases phosphorylate proteins by transferring γ-phosphate groups to them, whereas phosphatases remove phosphate groups from proteins. Phosphorylation is the most prevalent reversible post-translational modification of proteins [166]. Approximately 50% of all proteins are phosphorylated, and this process is closely regulated by various kinases and phosphatases. Nearly 538 kinases have been identified in humans. These kinases keep cells running by changing the protein activity of most protein kinases involved in signaling networks that use phosphorylation to control target protein activities. Kinases play crucial roles in nearly all cellular processes that promote cell survival, proliferation, metabolism, and migration. Kinase aberrant expression causes cancer and other illnesses [167]. Tyrosine kinases catalyze the transfer of a γ-phosphate group of tyrosine residues to hydroxyl groups on target proteins. Tyrosine phosphorylation of signal transduction molecules is a key process that regulates and closely controls the majority of fundamental cell activities, including the cell cycle, proliferation, differentiation, motility, and survival or death of the cell. Tyrosine kinase receptors have come into greater focus due to recent and rapid advances in cellular signaling of tyrosine kinase receptors in both normal and malignant cells [168]. Some of the anthraquinone derivatives have been reported to exhibit tyrosine kinase inhibition and are given in Table **37**.

**Table 37. Anthraquinone derivatives acting as tyrosine kinase inhibitors.**

| Compound | Therapeutic Potential and Mechanism of Action | Refs. |
|---|---|---|
| **212** | •Tyrosine kinase inhibitor<br><br>•Significant *in vitro* and *in vivo* anticancer activities.<br><br>•Marked downregulation of Her2/neu protein expression in dose- and time-dependent manners in both MDA-MB-453 and Calu-3 cells.<br><br>•Inhibited the downstream .MAPK and PI3K-Akt signaling pathway | [169] |

| **213** | •More potent than emodin<br>•It stops the tyrosine phosphorylation of p185neu by decreasing the cell proliferation.<br>•It transforms the overexpression of HER-2/neu in cells of human breast cancer. | [170] |
|---|---|---|
| **214** | •Selective inhibitor of p56lck tyrosine Kinase.<br>•$IC_{50} = 17$ nM for inhibition of p56lck autophosphorylation.<br>•$IC_{50} = 620$ nM for phosphorylation of an exogenous peptide by p56lck.<br>•Used to treat T-cell malignancies and autoimmune diseases. | [171] |
| **215** | •By inhibiting receptor Src tyrosine kinase leads to apoptosis in malignant cells.<br>•$IC_{50} = 33$ μM against GST-v-src protein tyrosine kinase<br>•$IC_{50} = 67$ μM against SPC-A-1 total protein tyrosine kinase. | [172] |

Metallo-ecto enzymes called ectonucleotidases hydrolyze extracellular nucleotides into nucleosides. NTPDases, NPPs, GPI-anchored ecto-5'-nucleotidase, and GPI-anchored alkaline phosphatases are the four main groups into which they are separated. P1 (nucleoside) and P2 (nucleotide) receptor-mediated signaling is regulated by ecto-nucleotidases [173]. Changes in the amount of adenosine and extracellular nucleotides can either increase or decrease P1 and P2 receptor activation. One potentially helpful and unique approach to cancer therapy is the inhibition of adenosine generation in the tumor cell environment by reducing the activity of the enzyme in cases of melanomas, gliomas, and breast cancers [174]. AQ derivatives showing ecto-nucleotidase inhibition are given in Table **38**.

**Table 38. Anthraquinone derivatives acting as Ecto-5'-nucleotidase inhibitors.**

| Compound | Therapeutic Potential and Mechanism of Action | Refs. |
|---|---|---|
| **216** **217** | • Ecto-5'-nucleotidase inhibitors with $K_i$ values of 260 nM for 216, and 150 nM for 217. | [175] |

## Toxicity of Anthraquinones

Toxicity is one of the main parameters that have to be evaluated for all drug candidates before their use in clinical trials. Any drug candidate that exerts toxicity on normal cells will pose a greater threat to life. Hence, even though having superior pharmacological activity, it must be non-toxic to healthy cells. Many studies have also identified the toxic nature of anthraquinones since these molecules resemble the toxic analog, anthracene. In many review articles, toxicity studies of anthraquinone drugs have been discussed [176,177]. A notable anticancer drug, doxorubicin has cumulative dose-dependent cardiotoxicity [178]. However, the exact mechanism of the toxicity of anthraquinones is still unclear, though it is thought to be due to redox cycling and the generation of free radicals by these compounds. Moreover, *in vitro* studies of natural laxative anthraquinones were found to be mutagenic in bacteria and also mutagenic and genotoxic in mammalian cells [179-184]. In the research study by Liu *et al.*, rhein has shown potential liver toxicity against HuH-7 cells with $EC_{50}$ = 93.9 µM [185]. Emodin (2) was found to exhibit toxicity against mouse and rat fetuses, where it exerted adverse effects against mice and rats at 17 and 60 mg/kg or higher doses. Also, it was found to be toxic with $LC_{50}$ = 0.19 µM against brine shrimp [186]. Moreover, it also exerts hepatotoxicity as studied by *in vitro* and *in vivo* studies by various mechanisms, including increased ROS production, apoptosis, and disrupting mitochondria membrane potential, leading to disease conditions such as fatty acid β oxidation disorder [187]. Further, in the studies of He *et al.*, emodin was found to exhibit toxicity effects on zebrafish embryos at a concentration of 0.25 µg/mL and higher with multiple abnormalities such as edema, a crooked trunk, and aberrant morphogenesis, thereby impacting embryo survival [188].

A number of diseases, including renal failure, rhabdomyolysis, nephrotoxicity (regular intake), dehydration, anorexia (long-term exposure), genotoxicity in mammalian cells, mutation, tumor promotion, chromosomal aberration, and liver enlargement, have been linked to natural anthraquinones such as aloe-emodin, physcion, rhein, chrysophanol, and emodin. Furthermore, it is reported that prolonged use of anthracoid laxatives can cause symptoms similar to melanosis coli, which is characterized by dark pigmentation of the colonic mucosa. In a small number of cases, this is also thought to cause morphological changes in the colonic myenteric system. This is due to the presence of chromophores in anthraquinone-based medications, which give colonic epithelial cells a vivid yellow color and are typically reversible. They may also result in long-term bodily harm, though. Hence, in-depth toxicity experiments and their mechanism of action need to be studied before using them for various purposes.

## CONCLUSION AND FUTURE PROSPECTS

Cancer has been identified as one of the major causes of human hazards in the world in recent times. Even though a large number of anti-cancer agents are available in the market for the treatment of various kinds of cancers, they are often found to be inefficient because of their poor selectivity, side effects, and multidrug resistance. High cost is another limitation of the present anticancer drugs that a common man cannot afford. Hence, it is essential to develop new, cheaper, and more efficient chemotherapeutic agents with fewer side effects.

Anthraquinone systems have earned their place in medicinal chemistry owing to their outstanding pharmacological activities. A large number of anthraquinone derivatives were synthesized by many researchers and screened for their cytotoxic activity against various tumor cell lines. Some anthraquinone-based emodin derivatives occur in plants, and their biocompatibility increases their use as anticancer agents. It is believed that some small structural changes in a molecule increase their medicinal activity. In this regard, this review article may be helpful for medicinal chemists to design and develop cheaper, more efficient, and less toxic anticancer agents possessing anthraquinone scaffolds.

## LIST OF ABBREVIATIONS

| | |
|---|---|
| **Akt** | Ak Strain Transforming |
| **ADM** | Adriamycin |
| **ALL** | Acute Lymphoblastic Leukemia |
| **AQ** | Anthraquinone |
| **ATP** | Adenosine Triphosphate |
| **Bcl-2** | B-Cell Lymphoma 2 |
| **BHAQ** | 3-Bromo-1-hydroxy-9,10-anthraquinone |
| **CCK-8** | Cell Counting Kit-8 |
| **c-Met** | Mesenchymal-Epithelial Transition |
| **c-Myc** | Cellular Myc |
| **CNS** | Central Nervous System |
| **Comfa** | Comparative Molecular Field Analysis |
| **CoMSIA** | Comparative Molecular Similarity Index Analysis |
| **COX-2** | Cyclooxygenase 2 |
| *ct-DNA* | *Circulating tumor DNA* |
| **DNA** | Deoxyribonucleic Acid |
| **DOX** | Doxorubicin |
| **DPPH** | 2,2-diphenyl-1-picryl hydrazyl |

| | |
|---|---|
| **$ED_{50}$** | Median Effective Dose |
| **EGFR** | Epidermal Growth Factor Receptor |
| **HDAC** | Histone Deacetylase |
| **HPV** | Human Papillomavirus |
| **HSA** | Human Serum Albumin |
| **h-TERT** | Human Telomerase Reverse Transcriptase |
| **$IC_{50}$** | Half-maximal inhibitory concentration |
| **$ID_{50}$** | Minimum Infectious Dose |
| **$IG_{50}$** | Growth inhibition expressed in $IG_{50}$ values |
| **JNK** | Jun N-terminal Kinase |
| **LLC** | Lewis lung carcinoma cells |
| **MDM2** | Mouse Double Minute 2 Homolog |
| **MMP2** | Matrix Metalloproteinase-2 |
| **MMP9** | Matrix metallopeptidase 9 |
| **mTOR** | Mammalian target of rapamycin |
| **MTT** | 3-(4,5-Dimethylthiazol-2-yl)-2,5- Diphenyltetrazolium Bromide |
| **MTX** | Methotrexate |
| **NSCLC** | Non-Small Cell Lung Cancer |
| **P13K** | Phosphoinositide 3-Kinases |
| **PARP** | Poly (ADP-ribose) polymerase |
| **Pgp** | P-glycoprotein |
| **PLK1** | Polo-Like Kinase 1 |
| **RHL** | Rhein Lysinate |
| **ROS** | Reactive Oxygen Species |
| **SAR** | Structure–Activity Relationship |

| **STAT1** | Signal Transducer and Activator of Transcription 1 |
| **STAT3** | Signal Transducer and Activator of Transcription 3 |
| **tNOX** | tumor-associated NADH oxidase |
| **VEGF** | Vascular Endothelial Growth Factor |
| **WHO** | World Health Organisation |

## REFERENCES

[1]   Sung H, Ferlay J, Siegel RL, *et al.* Global cancer statistics 2020: GLOBOCAN estimates of incidence and mortality worldwide for 36 cancers in 185 countries. CA Cancer J Clin 2021; 71(3): 209-49.
      http://dx.doi.org/10.3322/caac.21660 PMID: 33538338

[2]   Lustberg MB, Kuderer NM, Desai A, Bergerot C, Lyman GH. Mitigating long-term and delayed adverse events associated with cancer treatment: implications for survivorship. Nat Rev Clin Oncol 2023; 20(8): 527-42.
      http://dx.doi.org/10.1038/s41571-023-00776-9 PMID: 37231127

[3]   Hanauske A-R. The development of new chemotherapeutic agents. Anticancer Drugs 1996; 7(Suppl. 2): 29-32.
      http://dx.doi.org/10.1097/00001813-199608002-00008 PMID: 8862708

[4]   Wang P, Wei J, Hua X, *et al.* Plant anthraquinones: Classification, distribution, biosynthesis, and regulation. J Cell Physiol 2023; jcp.31063.
      http://dx.doi.org/10.1002/jcp.31063 PMID: 37393608

[5]   Hafez Ghoran S, Taktaz F, Ayatollahi SA, Kijjoa A. Anthraquinones and their analogues from marine-derived fungi: Chemistry and biological activities. Mar Drugs 2022; 20(8): 474.
      http://dx.doi.org/10.3390/md20080474 PMID: 35892942

[6]   Cheemalamarri C, Batchu UR, Thallamapuram NP, Katragadda SB, Reddy Shetty P. A review on hydroxy anthraquinones from bacteria: crosstalk's of structures and biological activities. Nat Prod Res 2022; 36(23): 6186-205.
      http://dx.doi.org/10.1080/14786419.2022.2039920 PMID: 35175877

[7]   Singh R, Geetanjali , Chauhan SMS. 9,10-Anthraquinones and other biologically active compounds from the genus Rubia. Chem Biodivers 2004; 1(9): 1241-64.
      http://dx.doi.org/10.1002/cbdv.200490088 PMID: 17191903

[8]   Wang D, Wang XH, Yu X, *et al.* Pharmacokinetics of anthraquinones from medicinal plants. Front Pharmacol 2021; 12: 638993.
      http://dx.doi.org/10.3389/fphar.2021.638993 PMID: 33935728

[9]   Xin D, Li H, Zhou S, Zhong H, Pu W. Effects of anthraquinones on immune responses and inflammatory diseases. Molecules 2022; 27(12): 3831.
      http://dx.doi.org/10.3390/molecules27123831 PMID: 35744949

[10] Deng T, Du J, Yin Y, *et al.* Rhein for treating diabetes mellitus: A pharmacological and mechanistic overview. Front Pharmacol 2023; 13: 1106260.
http://dx.doi.org/10.3389/fphar.2022.1106260 PMID: 36699072

[11] Raghuveer D, Pai VV, Murali TS, Nayak R. Exploring anthraquinones as antibacterial and antifungal agents. ChemistrySelect 2023; 8(6): e202204537.
http://dx.doi.org/10.1002/slct.202204537

[12] Wood S, Huffman J, Weber N, *et al.* Antiviral activity of naturally occurring anthraquinones and anthraquinone derivatives. Planta Med 1990; 56(6): 651-2.
http://dx.doi.org/10.1055/s-2006-961304

[13] Osman CP, Ismail NH, Ahmad R, Ahmat N, Awang K, Jaafar FM. Anthraquinones with antiplasmodial activity from the roots of *Rennellia elliptica Korth.* (Rubiaceae). Molecules 2010; 15(10): 7218-26.
http://dx.doi.org/10.3390/molecules15107218 PMID: 20966871

[14] Gan KH, Teng CH, Lin HC, *et al.* Antiplatelet effect and selective binding to cyclooxygenase by molecular docking analysis of 3-alkylaminopropoxy-9,10-anthraquinone derivatives. Biol Pharm Bull 2008; 31(8): 1547-51.
http://dx.doi.org/10.1248/bpb.31.1547 PMID: 18670087

[15] Nowak-Perlak M, Ziółkowski P, Woźniak M. A promising natural anthraquinones mediated by photodynamic therapy for anti-cancer therapy. Phytomedicine 2023; 119: 155035.
http://dx.doi.org/10.1016/j.phymed.2023.155035 PMID: 37603973

[16] Briggs LH, Jacombs FE, Nicholls GA. 612. The colouring matters of the bark of Rhamnus alaternus L. J Chem Soc 1953; 3069-72.
http://dx.doi.org/10.1039/jr9530003069

[17] Jayasuriya H, Koonchanok NM, Geahlen RL, McLaughlin JL, Chang CJ. Emodin, a protein tyrosine kinase inhibitor from *Polygonum cuspidatum.* J Nat Prod 1992; 55(5): 696-8.
http://dx.doi.org/10.1021/np50083a026 PMID: 1517743

[18] Murti PBR, Seshadri TR. Chemical composition of Indian senna leaves (Cassia angustifolia). Proc Indian Acad Sci Sect A Phys Sci 1939; 10(2): 96-103.
http://dx.doi.org/10.1007/BF03170994

[19] Ng T, Liu F, Lu Y, Cheng CHK, Wang Z. Antioxidant activity of compounds from the medicinal herb Aster tataricus. Comp Biochem Physiol C Toxicol Pharmacol 2003; 136(2): 109-15.
http://dx.doi.org/10.1016/S1532-0456(03)00170-4 PMID: 14559292

[20] Wells JM, Cole RJ, Kirksey JW. Emodin, a toxic metabolite of *Aspergillus wentii* isolated from weevil-damaged chestnuts. Appl Microbiol 1975; 30(1): 26-8.
http://dx.doi.org/10.1128/am.30.1.26-28.1975 PMID: 1147616

[21] Agosti G, Birkinshaw JH, Chaplen P. Studies in the biochemistry of micro-organisms. 112. Anthraquinone pigments of strains of *Cladosporium fulvum* Cooke. Biochem J 1962; 85(3): 528-30.
http://dx.doi.org/10.1042/bj0850528 PMID: 14011254

[22]   Wang FQ, Jiang J, Ma HR, Cheng L, Zhang G. Study on secondary metabolites of endophytic Chaetomium sp. Chin Tradit Herbal Drugs 2017; 48(7): 1298-301.

[23]   Kiyoshi K, Taketoshi K, Hideki M, Jiro K, Yoshinori N. A comparative study on cytotoxicities and biochemical properties of anthraquinone mycotoxins emodin and skyrin from *Penicillium islandicum* sopp. Toxicol Lett 1984; 20(2): 155-60.
       http://dx.doi.org/10.1016/0378-4274(84)90141-3 PMID: 6320499

[24]   Sharifi-Rad J, Herrera-Bravo J, Kamiloglu S, *et al.* Recent advances in the therapeutic potential of emodin for human health. Biomed Pharmacother 2022; 154: 113555.
       http://dx.doi.org/10.1016/j.biopha.2022.113555 PMID: 36027610

[25]   Stompor-Gorący M. The health benefits of emodin, a natural anthraquinone derived from rhubarb-a summary update. Int J Mol Sci 2021; 22(17): 9522.
       http://dx.doi.org/10.3390/ijms22179522 PMID: 34502424

[26]   Hsu SC, Chung JG. Anticancer potential of emodin. Biomedicine (Taipei) 2012; 2(3): 108-16.
       http://dx.doi.org/10.1016/j.biomed.2012.03.003 PMID: 32289000

[27]   Wei WT, Lin SZ, Liu DL, Wang ZH. The distinct mechanisms of the antitumor activity of emodin in different types of cancer (Review). Oncol Rep 2013; 30(6): 2555-62.
       http://dx.doi.org/10.3892/or.2013.2741 PMID: 24065213

[28]   Radha MH, Laxmipriya NP. Evaluation of biological properties and clinical effectiveness of Aloe vera: A systematic review. J Tradit Complement Med 2015; 5(1): 21-6.
       http://dx.doi.org/10.1016/j.jtcme.2014.10.006 PMID: 26151005

[29]   Dong X, Zeng Y, Liu Y, *et al.* Aloe-emodin: A review of its pharmacology, toxicity, and pharmacokinetics. Phytother Res 2020; 34(2): 270-81.
       http://dx.doi.org/10.1002/ptr.6532 PMID: 31680350

[30]   Chen R, Zhang J, Hu Y, Wang S, Chen M, Wang Y. Potential antineoplastic effects of Aloe-emodin: a comprehensive review. Am J Chin Med 2014; 42(2): 275-88.
       http://dx.doi.org/10.1142/S0192415X14500189 PMID: 24707862

[31]   Sanders B, Ray AM, Goldberg S, *et al.* Anti-cancer effects of aloe-emodin: a systematic review. J Clin Transl Res 2017; 3(3): 283-96.
       PMID: 30895270

[32]   Şeker Karatoprak G, Küpeli Akkol E, Yücel Ç, Bahadır Acıkara Ö, Sobarzo-Sánchez E. Advances in understanding the role of aloe emodin and targeted drug delivery systems in cancer. Oxid Med Cell Longev 2022; 2022.
       http://dx.doi.org/10.1155/2022/7928200

[33]   Zhang H, Ma L, Kim E, *et al.* Rhein induces oral cancer cell apoptosis and ROS *via* suppresse AKT/mTOR signaling pathway *in vitro* and *in vivo*. Int J Mol Sci 2023; 24(10): 8507.
       http://dx.doi.org/10.3390/ijms24108507 PMID: 37239855

[34]   Zhang H, Yi JK, Huang H, *et al.* Rhein suppresses colorectal cancer cell growth by inhibiting the mTOR pathway *in vitro* and *in vivo*. Cancers (Basel) 2021; 13(9): 2176.
       http://dx.doi.org/10.3390/cancers13092176 PMID: 33946531

[35] Henamayee S, Banik K, Sailo BL, *et al.* Therapeutic emergence of rhein as a potential anticancer drug: A review of its molecular targets and anticancer properties. Molecules 2020; 25(10): 2278.
http://dx.doi.org/10.3390/molecules25102278 PMID: 32408623

[36] Wu C, Cao H, Zhou H, *et al.* Research progress on the antitumor effects of rhein: literature review. Anticancer Agents Med Chem 2017; 17(12): 1624-32.
PMID: 26419468

[37] Cheng L, Chen Q, Pi R, Chen J. A research update on the therapeutic potential of rhein and its derivatives. Eur J Pharmacol 2021; 899: 173908.
http://dx.doi.org/10.1016/j.ejphar.2021.173908 PMID: 33515540

[38] Ren L, Li Z, Dai C, *et al.* Chrysophanol inhibits proliferation and induces apoptosis through NF-κB/cyclin D1 and NF-κB/Bcl-2 signaling cascade in breast cancer cell lines. Mol Med Rep 2018; 17(3): 4376-82.
http://dx.doi.org/10.3892/mmr.2018.8443 PMID: 29344652

[39] Ni CH, Yu CS, Lu HF, *et al.* Chrysophanol-induced cell death (necrosis) in human lung cancer A549 cells is mediated through increasing reactive oxygen species and decreasing the level of mitochondrial membrane potential. Environ Toxicol 2014; 29(7): 740-9.
http://dx.doi.org/10.1002/tox.21801 PMID: 22848001

[40] Lee MS, Cha EY, Sul JY, Song IS, Kim JY. Chrysophanic acid blocks proliferation of colon cancer cells by inhibiting EGFR/mTOR pathway. Phytother Res 2011; 25(6): 833-7.
http://dx.doi.org/10.1002/ptr.3323 PMID: 21089180

[41] Lim W, Yang C, Bazer FW, Song G. Chrysophanol induces apoptosis of choriocarcinoma through regulation of ROS and the AKT and ERK1/2 pathways. J Cell Physiol 2017; 232(2): 331-9.
http://dx.doi.org/10.1002/jcp.25423 PMID: 27171670

[42] Wang MY, West BJ, Jensen CJ, *et al.* Morinda citrifolia (Noni): a literature review and recent advances in Noni research. Acta Pharmacol Sin 2002; 23(12): 1127-41.
PMID: 12466051

[43] Woradulayapinij W, Pothiluk A, Nualsanit T, *et al.* Acute oral toxicity of damnacanthal and its anticancer activity against colorectal tumorigenesis. Toxicol Rep 2022; 9: 1968-76.
http://dx.doi.org/10.1016/j.toxrep.2022.10.015 PMID: 36518435

[44] Nualsanit T, Rojanapanthu P, Gritsanapan W, Lee SH, Lawson D, Baek SJ. Damnacanthal, a noni component, exhibits antitumorigenic activity in human colorectal cancer cells. J Nutr Biochem 2012; 23(8): 915-23.
http://dx.doi.org/10.1016/j.jnutbio.2011.04.017 PMID: 21852088

[45] Tosa H, Iinuma M, Asai F, *et al.* Anthraquinones from *Neonauclea calycina* and their inhibitory activity against DNA topoisomerase II. Biol Pharm Bull 1998; 21(6): 641-2.
http://dx.doi.org/10.1248/bpb.21.641 PMID: 9657055

[46] Nualsanit T, Rojanapanthu P, Gritsanapan W, Kwankitpraniti T, Min KW, Joon Baek S. Damnacanthal-induced anti-inflammation is associated with inhibition of NF-κB activity. Inflamm Allergy Drug Targets 2011; 10(6): 455-63.
http://dx.doi.org/10.2174/187152811798104908 PMID: 21999179

[47]   Zengin G, Degirmenci NS, Alpsoy L, Aktumsek A. Evaluation of antioxidant, enzyme inhibition, and cytotoxic activity of three anthraquinones (alizarin, purpurin, and quinizarin). Hum Exp Toxicol 2016; 35(5): 544-53.
http://dx.doi.org/10.1177/0960327115595687 PMID: 26178874

[48]   Sachithanandam V, Lalitha P, Parthiban A, *et al.* A comprehensive *in silico* and *in vitro* studies on quinizarin: a promising phytochemical derived from *Rhizophora mucronata Lam.* J Biomol Struct Dyn 2022; 40(16): 7218-29.
http://dx.doi.org/10.1080/07391102.2021.1894983 PMID: 33682626

[49]   Fotia C, Avnet S, Granchi D, Baldini N. The natural compound Alizarin as an osteotropic drug for the treatment of bone tumors. J Orthop Res 2012; 30(9): 1486-92.
http://dx.doi.org/10.1002/jor.22101 PMID: 22411621

[50]   Badria FA, Ibrahim AS. Evaluation of natural anthracene-derived compounds as antimitotic agents. Drug Discov Ther 2013; 7(2): 84-9.
PMID: 23715507

[51]   Malik EM, Müller CE. Anthraquinones as pharmacological tools and drugs. Med Res Rev 2016; 36(4): 705-48.
http://dx.doi.org/10.1002/med.21391 PMID: 27111664

[52]   Martins-Teixeira MB, Carvalho I. Antitumour anthracyclines: progress and perspectives. ChemMedChem 2020; 15(11): 933-48.
http://dx.doi.org/10.1002/cmdc.202000131 PMID: 32314528

[53]   Li Y, Guo F, Chen T, *et al.* Design, synthesis, molecular docking, and biological evaluation of new emodin anthraquinone derivatives as potential antitumor substances. Chem Biodivers 2020; 17(9): e2000328.
http://dx.doi.org/10.1002/cbdv.202000328 PMID: 32627416

[54]   Wang W, Bai Z, Zhang F, Wang C, Yuan Y, Shao J. Synthesis and biological activity evaluation of emodin quaternary ammonium salt derivatives as potential anticancer agents. Eur J Med Chem 2012; 56(56): 320-31.
http://dx.doi.org/10.1016/j.ejmech.2012.07.051 PMID: 22921966

[55]   Yang K, Jin MJ, Quan ZS, Piao HR. Design and synthesis of novel anti-proliferative emodin derivatives and studies on their cell cycle arrest, apoptosis pathway and migration. Molecules 2019; 24(5): 884.
http://dx.doi.org/10.3390/molecules24050884 PMID: 30832378

[56]   Narender T, Sukanya P, Sharma K, Bathula SR. Apoptosis and DNA intercalating activities of novel emodin derivatives. RSC Advances 2013; 3(17): 6123-31.
http://dx.doi.org/10.1039/c3ra23149f

[57]   Narender T, Sukanya P, Sharma K, Bathula SR. Preparation of novel antiproliferative emodin derivatives and studies on their cell cycle arrest, caspase dependent apoptosis and DNA binding interaction. Phytomedicine 2013; 20(10): 890-6.
http://dx.doi.org/10.1016/j.phymed.2013.03.015 PMID: 23669265

[58]   Tan JH, Zhang QX, Huang ZS, *et al.* Synthesis, DNA binding and cytotoxicity of new pyrazole emodin derivatives. Eur J Med Chem 2006; 41(9): 1041-7.
http://dx.doi.org/10.1016/j.ejmech.2006.04.006 PMID: 16716458

[59]   Wang XD, Gu LQ, Wu JY. Apoptosis-inducing activity of new pyrazole emodin derivatives in human hepatocellular carcinoma HepG2 cells. Biol Pharm Bull 2007; 30(6): 1113-6.
        http://dx.doi.org/10.1248/bpb.30.1113 PMID: 17541163

[60]   Teich L, Daub KS, Krügel V, Nissler L, Gebhardt R, Eger K. Synthesis and biological evaluation of new derivatives of emodin. Bioorg Med Chem 2004; 12(22): 5961-71.
        http://dx.doi.org/10.1016/j.bmc.2004.08.024 PMID: 15498672

[61]   Shao J, Zhang F, Bai Z, Wang C, Yuan Y, Wang W. Synthesis and antitumor activity of emodin quaternary ammonium salt derivatives. Eur J Med Chem 2012; 56: 308-19.
        http://dx.doi.org/10.1016/j.ejmech.2012.07.047 PMID: 22901410

[62]   Zheng Y, Zhu L, Fan L, *et al.* Synthesis, SAR and pharmacological characterization of novel anthraquinone cation compounds as potential anticancer agents. Eur J Med Chem 2017; 125: 902-13.
        http://dx.doi.org/10.1016/j.ejmech.2016.10.012 PMID: 27769031

[63]   Koerner SK, Hanai J, Bai S, *et al.* Design and synthesis of emodin derivatives as novel inhibitors of ATP-citrate lyase. Eur J Med Chem 2017; 126: 920-8.
        http://dx.doi.org/10.1016/j.ejmech.2016.12.018 PMID: 27997879

[64]   Cui XR, Takahashi K, Shimamura T, Koyanagi J, Komada F, Saito S. Preparation of 1,8-di-O-alkylaloe-emodins and 15-amino-, 15-thiocyano-, and 15-selenocyanochrysophanol derivatives from aloe-emodin and studying their cytotoxic effects. Chem Pharm Bull (Tokyo) 2008; 56(4): 497-503.
        http://dx.doi.org/10.1248/cpb.56.497 PMID: 18379097

[65]   Thimmegowda NR, Park C, Shwetha B, *et al.* Synthesis and antitumor activity of natural compound aloe emodin derivatives. Chem Biol Drug Des 2015; 85(5): 638-44.
        http://dx.doi.org/10.1111/cbdd.12448 PMID: 25323822

[66]   Chen C, Cao T, Li Y, Hu Y, Yang H, Yin S. Synthesized derivatives of aloe-emodin as proliferation inhibitors for human breast adenocarcinoma, human nonsmall cell lung carcinoma, and human cervix carcinoma. Chem Nat Compd 2020; 56(1): 30-3.
        http://dx.doi.org/10.1007/s10600-020-02937-z

[67]   Kumar GD, Siva B, Vadlamudi S, Bathula SR, Dutta H, Suresh Babu K. Design, synthesis, and biological evaluation of pyrazole-linked aloe emodin derivatives as potential anticancer agents. RSC Med Chem 2021; 12(5): 791-6.
        http://dx.doi.org/10.1039/D0MD00315H PMID: 34124677

[68]   Dileep Kumar G, Siva B, Ashwini K, *et al.* Design, synthesis, cytotoxic, and anti-inflammatory activities of some novel analogues of aloe-emodin isolated from the rhizomes of *Rheum emodi*. Nat Prod Res 2023; 37(9): 1511-7.
        http://dx.doi.org/10.1080/14786419.2021.2024531 PMID: 35021945

[69]   Shang H, Hu Y, Li J, *et al.* The synthesis and biological evaluation of aloe-emodin-coumarin hybrids as potential antitumor agents. Molecules 2022; 27(19): 6153.
        http://dx.doi.org/10.3390/molecules27196153 PMID: 36234685

[70]   Zhang Q, Wang J, Lan F, *et al.* Synthesis and DNA interaction of aloe-emodin α-amino phosphate derivatives. J Mol Struct 2023; 1279: 134950.
        http://dx.doi.org/10.1016/j.molstruc.2023.134950

[71]   Yang X, Sun G, Yang C, Wang B. Novel rhein analogues as potential anticancer agents. ChemMedChem 2011; 6(12): 2294-301.
       http://dx.doi.org/10.1002/cmdc.201100384 PMID: 21954017

[72]   Wang PL, Cheng YT, Xu K, *et al.* Synthesis and antitumor evaluation of one novel tetramethylpyrazine-rhein derivative. Asian J Chem 2013; 25(8): 4885-8.
       http://dx.doi.org/10.14233/ajchem.2013.14135

[73]   Ye MY, Yao GY, Wei JC, Pan YM, Liao ZX, Wang HS. Synthesis, cytotoxicity, DNA binding and apoptosis of rhein-phosphonate derivatives as antitumor agents. Int J Mol Sci 2013; 14(5): 9424-39.
       http://dx.doi.org/10.3390/ijms14059424 PMID: 23629673

[74]   Yao G, Ye M, Huang R, *et al.* Synthesis and antitumor activities of novel rhein α-aminophosphonates conjugates. Bioorg Med Chem Lett 2014; 24(2): 501-7.
       http://dx.doi.org/10.1016/j.bmcl.2013.12.030 PMID: 24378217

[75]   Liu J, Zhang K, Zhen YZ, *et al.* Antitumor activity of rhein lysinate against human glioma U87 cells *in vitro* and *in vivo*. Oncol Rep 2016; 35(3): 1711-7.
       http://dx.doi.org/10.3892/or.2015.4518 PMID: 26707131

[76]   Huang J, Zhang Z, Huang P, He L, Ling Y. Design, synthesis and biological evaluation of rhein derivatives as anticancer agents. MedChemComm 2016; 7(9): 1812-8.
       http://dx.doi.org/10.1039/C6MD00252H

[77]   Li X, Liu Y, Zhao Y, *et al.* Rhein derivative 4F inhibits the malignant phenotype of breast cancer by downregulating Rac1 protein. Front Pharmacol 2020; 11: 754.
       http://dx.doi.org/10.3389/fphar.2020.00754 PMID: 32547389

[78]   Tian W, Li J, Su Z, *et al.* Novel anthraquinone compounds induce cancer cell death through paraptosis. ACS Med Chem Lett 2019; 10(5): 732-6.
       http://dx.doi.org/10.1021/acsmedchemlett.8b00624 PMID: 31097991

[79]   Su Z, Li Z, Wang C, *et al.* A novel Rhein derivative: Activation of Rac1/NADPH pathway enhances sensitivity of nasopharyngeal carcinoma cells to radiotherapy. Cell Signal 2019; 54: 35-45.
       http://dx.doi.org/10.1016/j.cellsig.2018.11.015 PMID: 30463023

[80]   Liang D, Su Z, Tian W, *et al.* Synthesis and screening of novel anthraquinone-quinazoline multitarget hybrids as promising anticancer candidates. Future Med Chem 2020; 12(2): 111-26.
       http://dx.doi.org/10.4155/fmc-2019-0230 PMID: 31718309

[81]   Quaglio D, Infante P, Cammarone S, *et al.* Exploring the potential of anthraquinone-based hybrids for identifying a novel generation of antagonists for the smoothened receptor in HH-dependent tumour. Chemistry 2023; 29(62): e202302237.
       http://dx.doi.org/10.1002/chem.202302237 PMID: 37565343

[82]   Koyama M, Kelly TR, Watanabe KA. Novel type of potential anticancer agents derived from chrysophanol and emodin. Some structure-activity relationship studies. J Med Chem 1988; 31(2): 283-4.
       http://dx.doi.org/10.1021/jm00397a002 PMID: 3339601

[83]   Darzynkiewicz Z, Carter SP, Kapuscinski J, Watanabe KA. Effect of derivatives of chrysophanol, a new type of potential antitumor agents of anthraquinone family, on growth and cell cycle of L1210 leukemic cells. Cancer Lett 1989; 46(3): 181-7.
http://dx.doi.org/10.1016/0304-3835(89)90128-6 PMID: 2766258

[84]   Saha K, Lam KW, Abas F, *et al.* Synthesis of damnacanthal, a naturally occurring 9,10-anthraquinone and its analogues, and its biological evaluation against five cancer cell lines. Med Chem Res 2013; 22(5): 2093-104.
http://dx.doi.org/10.1007/s00044-012-0197-5

[85]   Yao G, Dai W, Ye M, *et al.* Synthesis and antitumor properties of novel alizarin analogs. Med Chem Res 2014; 23(12): 5031-42.
http://dx.doi.org/10.1007/s00044-014-1062-5

[86]   Hu X, Cao Y, Yin X, *et al.* Design and synthesis of various quinizarin derivatives as potential anticancer agents in acute T lymphoblastic leukemia. Bioorg Med Chem 2019; 27(7): 1362-9.
http://dx.doi.org/10.1016/j.bmc.2019.02.041 PMID: 30827866

[87]   Zhao LM, Cao FX, Jin HS, Zhang JH, Szwaya J, Wang G. One-pot synthesis of 1,4-dihydroxy-2-(( E )-1-hydroxy-4-phenylbut-3-enyl)anthracene-9,10-diones as novel shikonin analogs and evaluation of their antiproliferative activities. Bioorg Med Chem Lett 2016; 26(11): 2691-4.
http://dx.doi.org/10.1016/j.bmcl.2016.04.006 PMID: 27080175

[88]   Liu Y, Liang Y, Jiang J, Qin Q, Wang L, Liu X. Design, synthesis and biological evaluation of 1,4-dihydroxyanthraquinone derivatives as anticancer agents. Bioorg Med Chem Lett 2019; 29(9): 1120-6.
http://dx.doi.org/10.1016/j.bmcl.2019.02.026 PMID: 30846253

[89]   Jin GZ, Jin HS, Jin LL. Synthesis and antiproliferative activity of 1,4-bis(dimethylamino)-9,10-anthraquinone derivatives against P388 mouse leukemic tumor cells. Arch Pharm Res 2011; 34(7): 1071-6.
http://dx.doi.org/10.1007/s12272-011-0704-0 PMID: 21811913

[90]   Tu HY, Huang AM, Teng CH, *et al.* Anthraquinone derivatives induce G2/M cell cycle arrest and apoptosis in NTUB1 cells. Bioorg Med Chem 2011; 19(18): 5670-8.
http://dx.doi.org/10.1016/j.bmc.2011.07.021 PMID: 21852140

[91]   Lin L, Hutzen B, Li PK, *et al.* A novel small molecule, LLL12, inhibits STAT3 phosphorylation and activities and exhibits potent growth-suppressive activity in human cancer cells. Neoplasia 2010; 12(1): 39-IN5.
http://dx.doi.org/10.1593/neo.91196 PMID: 20072652

[92]   Bhasin D, Etter JP, Chettiar SN, Mok M, Li PK. Antiproliferative activities and SAR studies of substituted anthraquinones and 1,4-naphthoquinones. Bioorg Med Chem Lett 2013; 23(24): 6864-7.
http://dx.doi.org/10.1016/j.bmcl.2013.09.098 PMID: 24176397

[93]   Abu N, Akhtar M, Ho W, Yeap S, Alitheen N. 3-Bromo-1-hydroxy-9,10-anthraquinone (BHAQ) inhibits growth and migration of the human breast cancer cell lines MCF-7 and MDA-MB231. Molecules 2013; 18(9): 10367-77.

http://dx.doi.org/10.3390/molecules180910367 PMID: 23985955

[94]   Marković V, Janićijević A, Stanojković T, *et al.* Synthesis, cytotoxic activity and DNA-interaction studies of novel anthraquinone–thiosemicarbazones with tautomerizable methylene group. Eur J Med Chem 2013; 64: 228-38.
http://dx.doi.org/10.1016/j.ejmech.2013.03.071 PMID: 23644206

[95]   Castro-Castillo V, Suárez-Rozas C, Castro-Loiza N, Theoduloz C, Cassels BK. Annulation of substituted anthracene-9,10-diones yields promising selectively antiproliferative compounds. Eur J Med Chem 2013; 62: 688-92.
http://dx.doi.org/10.1016/j.ejmech.2013.01.049 PMID: 23454511

[96]   Lee CC, Huang KF, Lin PY, *et al.* Synthesis, antiproliferative activities and telomerase inhibition evaluation of novel asymmetrical 1,2-disubstituted amidoanthraquinone derivatives. Eur J Med Chem 2012; 47(1): 323-36.
http://dx.doi.org/10.1016/j.ejmech.2011.10.059 PMID: 22100139

[97]   Lee YR, Yu DS, Liang YC, *et al.* New approaches of PARP-1 inhibitors in human lung cancer cells and cancer stem-like cells by some selected anthraquinone-derived small molecules. PLoS One 2013; 8(2): e56284.
http://dx.doi.org/10.1371/journal.pone.0056284 PMID: 23451039

[98]   Liang Z, Ai J, Ding X, *et al.* Anthraquinone derivatives as potent inhibitors of c-Met kinase and the extracellular signaling pathway. ACS Med Chem Lett 2013; 4(4): 408-13.
http://dx.doi.org/10.1021/ml4000047 PMID: 24900685

[99]   Cogoi S, Shchekotikhin AE, Membrino A, Sinkevich YB, Xodo LE. Guanidino anthrathiophenediones as G-quadruplex binders: uptake, intracellular localization, and anti-Harvey-Ras gene activity in bladder cancer cells. J Med Chem 2013; 56(7): 2764-78.
http://dx.doi.org/10.1021/jm3019063 PMID: 23458775

[100]  Sangthong S, Sangphech N, Palaga T, *et al.* Anthracene-9, 10-dione derivatives induced apoptosis in human cervical cancer cell line (CaSki) by interfering with HPV E6 expression. Eur J Med Chem 2014; 77: 334-42.
http://dx.doi.org/10.1016/j.ejmech.2014.02.006 PMID: 24657570

[101]  Almutairi M, Hegazy G, Haiba M, Ali H, Khalifa N, Soliman A. Synthesis, docking and biological activities of novel hybrids celecoxib and anthraquinone analogs as potent cytotoxic agents. Int J Mol Sci 2014; 15(12): 22580-603.
http://dx.doi.org/10.3390/ijms151222580 PMID: 25490139

[102]  Kolundžija B, Marković V, Stanojković T, *et al.* Novel anthraquinone based chalcone analogues containing an imine fragment: Synthesis, cytotoxicity and anti-angiogenic activity. Bioorg Med Chem Lett 2014; 24(1): 65-71.
http://dx.doi.org/10.1016/j.bmcl.2013.11.075 PMID: 24332490

[103]  Marković V, Debeljak N, Stanojković T, *et al.* Anthraquinone–chalcone hybrids: Synthesis, preliminary antiproliferative evaluation and DNA-interaction studies. Eur J Med Chem 2015; 89: 401-10.
http://dx.doi.org/10.1016/j.ejmech.2014.10.055 PMID: 25462255

[104] Zhao LM, Ma FY, Jin HS, Zheng S, Zhong Q, Wang G. Design and synthesis of novel hydroxyanthraquinone nitrogen mustard derivatives as potential anticancer agents *via* a bioisostere approach. Eur J Med Chem 2015; 102: 303-9.
http://dx.doi.org/10.1016/j.ejmech.2015.08.006 PMID: 26291039

[105] Yang L, Fu Z, Niu X, Zhang G, Cui F, Zhou C. Probing the interaction of anthraquinone with DNA by spectroscopy, molecular modeling and cancer cell imaging technique. Chem Biol Interact 2015; 233: 65-70.
http://dx.doi.org/10.1016/j.cbi.2015.03.026 PMID: 25834985

[106] Fu Z, Cui Y, Cui F, Zhang G. Modeling techniques and fluorescence imaging investigation of the interactions of an anthraquinone derivative with HSA and ctDNA. Spectrochim Acta A Mol Biomol Spectrosc 2016; 153: 572-9.
http://dx.doi.org/10.1016/j.saa.2015.09.011 PMID: 26436845

[107] Lee YR, Chen TC, Lee CC, *et al.* Ring fusion strategy for synthesis and lead optimization of sulfur-substituted anthra[1,2-c][1,2,5]thiadiazole-6,11-dione derivatives as promising scaffold of antitumor agents. Eur J Med Chem 2015; 102: 661-76.
http://dx.doi.org/10.1016/j.ejmech.2015.07.052 PMID: 26344783

[108] Yeap S, Akhtar MN, Lim KL, *et al.* Synthesis of an anthraquinone derivative (DHAQC) and its effect on induction of G2/M arrest and apoptosis in breast cancer MCF-7 cell line. Drug Des Devel Ther 2015; 9: 983-92.
PMID: 25733816

[109] Tikhomirov AS, Shchekotikhin AE, Lee YH, *et al.* Synthesis and characterization of 4,11-diaminoanthra[2,3- *b* ]furan-5,10-diones: Tumor cell apoptosis through tNOX-modulated NAD $^+$ /NADH Ratio and SIRT1. J Med Chem 2015; 58(24): 9522-34.
http://dx.doi.org/10.1021/acs.jmedchem.5b00859 PMID: 26633734

[110] Cogoi S, Zorzet S, Shchekotikhin AE, Xodo LE. Potent apoptotic response induced by chloroacetamidine anthrathiophenediones in bladder cancer cells. J Med Chem 2015; 58(14): 5476-85.
http://dx.doi.org/10.1021/acs.jmedchem.5b00409 PMID: 26057859

[111] Juhan SF, Sukari MA, Abdul SS. New Synthesised Aminoanthraquinone Derivatives and Its Antimicrobial and Anticancer Activities (Route II). Int J Contemp Med Res 2016; 3: 18-33.

[112] Choi HK, Ryu H, Son A, *et al.* The novel anthraquinone derivative IMP1338 induces death of human cancer cells by p53-independent S and G2/M cell cycle arrest. Biomed Pharmacother 2016; 79: 308-14.
http://dx.doi.org/10.1016/j.biopha.2016.02.034 PMID: 27044842

[113] Stanojković T, Marković V, Matić IZ, *et al.* Highly selective anthraquinone-chalcone hybrids as potential antileukemia agents. Bioorg Med Chem Lett 2018; 28(15): 2593-8.
http://dx.doi.org/10.1016/j.bmcl.2018.06.048 PMID: 29970309

[114] Lozynskyi A, Sabadakh O, Luchkevich E, *et al.* The application of anthraquinone-based triazenes as equivalents of diazonium salts in reaction with methylene active compounds. Phosphorus Sulfur Silicon Relat Elem 2018; 193(7): 409-14.
http://dx.doi.org/10.1080/10426507.2018.1452236

[115]   Sweidan K, Zalloum H, Sabbah DA, Idris G, Abudosh K, Mubarak MS. Synthesis, characterization, and anticancer evaluation of some new *N* 1-(anthraquinon-2-yl) amidrazone derivatives. Can J Chem 2018; 96(12): 1123-8.
http://dx.doi.org/10.1139/cjc-2018-0145

[116]   Volodina YL, Dezhenkova LG, Tikhomirov AS, *et al.* New anthra[2,3-b]furancarboxamides: A role of positioning of the carboxamide moiety in antitumor properties. Eur J Med Chem 2019; 165: 31-45.
http://dx.doi.org/10.1016/j.ejmech.2018.12.068 PMID: 30659997

[117]   Huang K, Jiang L, Liang R, *et al.* Synthesis and biological evaluation of anthraquinone derivatives as allosteric phosphoglycerate mutase 1 inhibitors for cancer treatment. Eur J Med Chem 2019; 168: 45-57.
http://dx.doi.org/10.1016/j.ejmech.2019.01.085 PMID: 30798052

[118]   Siyamak S, Masoome S, Radwan A, Liudmila F, Evgenij D. Antitumor and antioxidant activities of the new anthraquinone derivatives: Synthesis, DPPH, ABTS, biological and DFT Investigations. Jiegou Huaxue 2019; 38(10): 1673-90.

[119]   Niedziałkowski P, Czaczyk E, Jarosz J, *et al.* Synthesis and electrochemical, spectral, and biological evaluation of novel 9,10-anthraquinone derivatives containing piperidine unit as potent antiproliferative agents. J Mol Struct 2019; 1175: 488-95.
http://dx.doi.org/10.1016/j.molstruc.2018.07.070

[120]   Awasthi P, Vatsal M, Sharma A. Structural and biological study of synthesized anthraquinone series of compounds with sulfonamide feature. J Biomol Struct Dyn 2019; 37(17): 4465-80.
http://dx.doi.org/10.1080/07391102.2018.1552198 PMID: 30489230

[121]   Okumura N, Mizutani H, Ishihama T, *et al.* Study on redox properties and cytotoxicity of anthraquinone derivatives to understand antitumor active anthracycline substances. Chem Pharm Bull (Tokyo) 2019; 67(7): 717-20.
http://dx.doi.org/10.1248/cpb.c19-00103 PMID: 31257327

[122]   Sirazhetdinova NS, Savelyev VA, Frolova TS, *et al.* 1-Hydroxyanthraquinones containing aryl substituents as potent and selective anticancer agents. Molecules 2020; 25(11): 2547.
http://dx.doi.org/10.3390/molecules25112547 PMID: 32486108

[123]   Tikhomirov AS, Litvinova VA, Andreeva DV, *et al.* Amides of pyrrole- and thiophene-fused anthraquinone derivatives: A role of the heterocyclic core in antitumor properties. Eur J Med Chem 2020; 199: 112294.
http://dx.doi.org/10.1016/j.ejmech.2020.112294 PMID: 32428792

[124]   Oliveira LA, Nicolella HD, Furtado RA, *et al.* Design, synthesis, and antitumor evaluation of novel anthraquinone derivatives. Med Chem Res 2020; 29(9): 1611-20.
http://dx.doi.org/10.1007/s00044-020-02587-4

[125]   Li Y, Guo F, Guan Y, *et al.* Novel anthraquinone compounds inhibit colon cancer cell proliferation *via* the reactive oxygen species/JNK pathway. Molecules 2020; 25(7): 1672.
http://dx.doi.org/10.3390/molecules25071672 PMID: 32260423

[126] Anifowose A, Agbowuro AA, Tripathi R, *et al.* Inducing apoptosis through upregulation of p53: structure–activity exploration of anthraquinone analogs. Med Chem Res 2020; 29(7): 1199-210.
http://dx.doi.org/10.1007/s00044-020-02563-y PMID: 32719577

[127] Lozynskyi A, Holota S, Yushyn I, *et al.* Synthesis and biological Aactivity evaluation of polyfunctionalized Anthraquinonehydrazones. Lett Drug Des Discov 2021; 18(2): 199-209.
http://dx.doi.org/10.2174/1570180817999200802032844

[128] Awasthi P, Sharma A, Vatsal M. Spectroscopic, viscometric and computational binding study of 1 and 2 substituted anthraquinone analogs to be potential anti-cancer agents. J Mol Struct 2021; 1223: 129293.
http://dx.doi.org/10.1016/j.molstruc.2020.129293

[129] Lin S, Zhang Y, Wang Z, *et al.* Preparation of novel anthraquinone-based aspirin derivatives with anti-cancer activity. Eur J Pharmacol 2021; 900: 174020.
http://dx.doi.org/10.1016/j.ejphar.2021.174020 PMID: 33741381

[130] Lin KW, Lin WH, Su CL, Hsu HY, Lin CN. Design, synthesis and antitumour evaluation of novel anthraquinone derivatives. Bioorg Chem 2021; 107: 104395.
http://dx.doi.org/10.1016/j.bioorg.2020.104395 PMID: 33384144

[131] Volodina YL, Tikhomirov AS, Dezhenkova LG, *et al.* Thiophene-2-carboxamide derivatives of anthraquinone: A new potent antitumor chemotype. Eur J Med Chem 2021; 221: 113521.
http://dx.doi.org/10.1016/j.ejmech.2021.113521 PMID: 34082225

[132] Mohamadzadeh M, Zarei M. Anticancer activity and evaluation of apoptotic genes expression of 2-azetidinones containing anthraquinone moiety. Mol Divers 2021; 25(4): 2429-39.
http://dx.doi.org/10.1007/s11030-020-10142-x PMID: 32944866

[133] Sirazhetdinova NS, Savelyev VA, Baev DS, *et al.* Synthesis, characterization and anticancer evaluation of nitrogen-substituted 1-(3-aminoprop-1-ynyl)-4-hydroxyanthraquinone derivatives. Med Chem Res 2021; 30(8): 1541-56.
http://dx.doi.org/10.1007/s00044-021-02754-1

[134] Singh M, Malhotra L, Haque MA, *et al.* Heteroarene-fused anthraquinone derivatives as potential modulators for human aurora kinase B. Biochimie 2021; 182: 152-65.
http://dx.doi.org/10.1016/j.biochi.2020.12.024 PMID: 33417980

[135] Zou Y, Cao Z, Wang J, *et al.* A series of novel HDAC inhibitors with anthraquinone as a cap group. Chem Pharm Bull (Tokyo) 2020; 68(7): 613-7.
http://dx.doi.org/10.1248/cpb.c20-00206 PMID: 32611998

[136] Tikhomirov AS, Tsvetkov VB, Volodina YL, *et al.* Heterocyclic ring expansion yields anthraquinone derivatives potent against multidrug resistant tumor cells. Bioorg Chem 2022; 127: 105925.
http://dx.doi.org/10.1016/j.bioorg.2022.105925 PMID: 35728293

[137] Morgan I, Wessjohann LA, Kaluđerović GN. *In vitro* anticancer screening and preliminary mechanistic study of A-ring substituted Anthraquinone derivatives. Cells 2022; 11(1): 168.
http://dx.doi.org/10.3390/cells11010168 PMID: 35011730

[138]   Tikhomirov AS, Sinkevich YB, Dezhenkova LG, *et al.* Synthesis and antitumor activity of cyclopentane-fused anthraquinone derivatives. Eur J Med Chem 2024; 265: 116103.
http://dx.doi.org/10.1016/j.ejmech.2023.116103 PMID: 38176358

[139]   Olszewski M, Stasevych M, Zvarych V, Maciejewska N. 9,10-Dioxoanthracenyldithiocarbamates effectively inhibit the proliferation of non-small cell lung cancer by targeting multiple protein tyrosine kinases. J Enzyme Inhib Med Chem 2024; 39(1): 2284113.
http://dx.doi.org/10.1080/14756366.2023.2284113 PMID: 38078360

[140]   Wu S, Zhou X, Li F, Sun W, Zheng Q, Liang D. Novel anthraquinone-based benzenesulfonamide derivatives and their analogues as potent human carbonic Anhydrase Inhibitors with Antitumor Activity: Synthesis, Biological Evaluation, and *In Silico* Analysis. Int J Mol Sci 2024; 25(6): 3348.
http://dx.doi.org/10.3390/ijms25063348 PMID: 38542320

[141]   Andreeva DV, Tikhomirov AS, Shchekotikhin AE. Synthesis and antiproliferative activity of thiazole-fused anthraquinones. Org Biomol Chem 2024; 22(42): 8493-504.
http://dx.doi.org/10.1039/D4OB01284D PMID: 39344399

[142]   Andreeva DV, Vedekhina TS, Gostev AS, *et al.* Thiadiazole-, selenadiazole- and triazole-fused anthraquinones as G-quadruplex targeting anticancer compounds. Eur J Med Chem 2024; 268: 116222.
http://dx.doi.org/10.1016/j.ejmech.2024.116222 PMID: 38387333

[143]   Malik MS, Alsantali RI, Jassas RS, *et al.* Journey of anthraquinones as anticancer agents – a systematic review of recent literature. RSC Advances 2021; 11(57): 35806-27.
http://dx.doi.org/10.1039/D1RA05686G PMID: 35492773

[144]   Siddamurthi S, Gutti G, Jana S, Kumar A, Singh SK. Anthraquinone: a promising scaffold for the discovery and development of therapeutic agents in cancer therapy. Future Med Chem 2020; 12(11): 1037-69.
http://dx.doi.org/10.4155/fmc-2019-0198 PMID: 32349522

[145]   Pang H, Li X, Zhao Y, *et al.* Confirming whether novel rhein derivative 4a induces paraptosis-like cell death by endoplasmic reticulum stress in ovarian cancer cells. Eur J Pharmacol 2020; 886: 173526.
http://dx.doi.org/10.1016/j.ejphar.2020.173526 PMID: 32890460

[146]   Li Y, Yan W, Qin Y, Zhang L, Xiao S. The anthraquinone derivative C2 enhances Oxaliplatin-induced cell death and triggers autophagy *via* the PI3K/AKT/mTOR pathway. Int J Mol Sci 2024; 25(12): 6468.
http://dx.doi.org/10.3390/ijms25126468 PMID: 38928176

[147]   Qiao S, Zhang W, Jiang Y, Su Y. Sennoside A induces autophagic death of prostate cancer *via* inactivation of PI3K/AKT/mTOR axis. J Mol Histol 2023; 54(6): 645-54.
http://dx.doi.org/10.1007/s10735-023-10156-3 PMID: 37740843

[148]   Zhai H, Wang C, Li J, *et al.* Novel anthraquinone derivatives trigger endoplasmic reticulum stress response and induce apoptosis. Future Med Chem 2023; 15(2): 129-45.
http://dx.doi.org/10.4155/fmc-2022-0217 PMID: 36799271

[149] Kabir A, Tilekar K, Upadhyay N, Ramaa CS. Novel anthraquinone derivatives as dual inhibitors of topoisomerase 2 and casein kinase 2: in silico studies, synthesis and biological evaluation on leukemic cell lines. Anticancer Agents Med Chem 2019; 18(11): 1551-62.
http://dx.doi.org/10.2174/1871520618666180423111309 PMID: 29683096

[150] Wang S, Yan WW, He M, Wei D, Long ZJ, Tao YM. Aloe emodin inhibits telomerase activity in breast cancer cells: transcriptional and enzymological mechanism. Pharmacol Rep 2020; 72(5): 1383-96.
http://dx.doi.org/10.1007/s43440-020-00062-w PMID: 32207090

[151] Jamshidi S, Rostami A, Shojaei S, Taherkhani A, Taherkhani H. Exploring natural anthraquinones as potential MMP2 inhibitors: A computational study. Biosystems 2024; 235: 105103.
http://dx.doi.org/10.1016/j.biosystems.2023.105103 PMID: 38123060

[152] Pommier Y. DNA topoisomerase I inhibitors: chemistry, biology, and interfacial inhibition. Chem Rev 2009; 109(7): 2894-902.
http://dx.doi.org/10.1021/cr900097c PMID: 19476377

[153] Smith PJ, Blunt NJ, Desnoyers R, Giles Y, Patterson LH. DNA topoisomerase II-dependent cytotoxicity of alkylaminoanthraquinones and their N-oxides. Cancer Chemother Pharmacol 1997; 39(5): 455-61.
http://dx.doi.org/10.1007/s002800050598 PMID: 9054961

[154] McKeown SR, Hejmadi MV, McIntyre IA, McAleer JJA, Patterson LH. AQ4N: an alkylaminoanthraquinone N-oxide showing bioreductive potential and positive interaction with radiation *in vivo*. Br J Cancer 1995; 72(1): 76-81.
http://dx.doi.org/10.1038/bjc.1995.280 PMID: 7599069

[155] Patterson LH. Rationale for the use of aliphatic N-oxides of cytotoxic anthraquinones as prodrug DNA binding agents: a new class of bioreductive agent. Cancer Metastasis Rev 1993; 12(2): 119-34.
http://dx.doi.org/10.1007/BF00689805 PMID: 8375016

[156] Pors K, Paniwnyk Z, Teesdale-Spittle P, *et al.* Alchemix: a novel alkylating anthraquinone with potent activity against anthracycline- and cisplatin-resistant ovarian cancer. Mol Cancer Ther 2003; 2(7): 607-10.
PMID: 12883032

[157] Son JK, Jung SJ, Jung JH, *et al.* Anticancer constituents from the roots of *Rubia cordifolia* L. Chem Pharm Bull (Tokyo) 2008; 56(2): 213-6.
http://dx.doi.org/10.1248/cpb.56.213 PMID: 18239313

[158] Bielawska A, Kosk K, Bielawski K. L-proline analogues of anthraquinone-2-carboxylic acid: cytotoxic activity in breast cancer MCF-7 cells and inhibitory activity against topoisomerase I and II. Pol J Pharmacol 2001; 53(3): 283-7.
PMID: 11785929

[159] Huang HS, Chen IB, Huang KF, *et al.* Synthesis and human telomerase inhibition of a series of regioisomeric disubstituted amidoanthraquinones. Chem Pharm Bull (Tokyo) 2007; 55(2): 284-92.
http://dx.doi.org/10.1248/cpb.55.284 PMID: 17268103

[160]   Lin YH, Chuang SM, Wu PC, *et al.* Selective recognition and stabilization of new ligands targeting the potassium form of the human telomeric G-quadruplex DNA. Sci Rep 2016; 6(1): 31019.
http://dx.doi.org/10.1038/srep31019 PMID: 27511133

[161]   Huang HS, Huang KF, Li CL, *et al.* Synthesis, human telomerase inhibition and anti-proliferative studies of a series of 2,7-bis-substituted amido-anthraquinone derivatives. Bioorg Med Chem 2008; 16(14): 6976-86.
http://dx.doi.org/10.1016/j.bmc.2008.05.072 PMID: 18571928

[162]   Chen CL, Chang DM, Chen TC, *et al.* Structure-based design, synthesis and evaluation of novel anthra[1,2-d]imidazole-6,11-dione derivatives as telomerase inhibitors and potential for cancer polypharmacology. Eur J Med Chem 2013; 60: 29-41.
http://dx.doi.org/10.1016/j.ejmech.2012.11.032 PMID: 23279865

[163]   Cairns D, Michalitsi E, Jenkins TC, Mackay SP. Molecular modelling and cytotoxicity of substituted anthraquinones as inhibitors of human telomerase. Bioorg Med Chem 2002; 10(3): 803-7.
http://dx.doi.org/10.1016/S0968-0896(01)00337-6 PMID: 11814869

[164]   Huang HS, Chiou JF, Fong Y, *et al.* Activation of human telomerase reverse transcriptase expression by some new symmetrical bis-substituted derivatives of the anthraquinone. J Med Chem 2003; 46(15): 3300-7.
http://dx.doi.org/10.1021/jm0204921 PMID: 12852760

[165]   Perry PJ, Gowan SM, Reszka AP, *et al.* 1,4- and 2,6-disubstituted amidoanthracene-9,10-dione derivatives as inhibitors of human telomerase. J Med Chem 1998; 41(17): 3253-60.
http://dx.doi.org/10.1021/jm9801105 PMID: 9703471

[166]   Cheng HC, Qi RZ, Paudel H, Zhu HJ. Regulation and function of protein kinases and phosphatases. Enzyme Res 2011; 2011: 1-3.
http://dx.doi.org/10.4061/2011/794089 PMID: 22195276

[167]   Bhullar KS, Lagarón NO, McGowan EM, *et al.* Kinase-targeted cancer therapies: progress, challenges and future directions. Mol Cancer 2018; 17(1): 48.
http://dx.doi.org/10.1186/s12943-018-0804-2 PMID: 29455673

[168]   Paul MK, Mukhopadhyay AK. Tyrosine kinase – Role and significance in Cancer. Int J Med Sci 2004; 1(2): 101-15.
http://dx.doi.org/10.7150/ijms.1.101 PMID: 15912202

[169]   Yan YY, Zheng LS, Zhang X, *et al.* Blockade of Her2/neu binding to Hsp90 by emodin azide methyl anthraquinone derivative induces proteasomal degradation of Her2/neu. Mol Pharm 2011; 8(5): 1687-97.
http://dx.doi.org/10.1021/mp2000499 PMID: 21812426

[170]   Zhang L, Lau YK, Xi L, *et al.* Tyrosine kinase inhibitors, emodin and its derivative repress HER-2/neu-induced cellular transformation and metastasis-associated properties. Oncogene 1998; 16(22): 2855-63.
http://dx.doi.org/10.1038/sj.onc.1201813 PMID: 9671406

[171]   Faltynek CR, Schroeder J, Mauvais P, *et al.* Damnacanthal is a highly potent, selective inhibitor of p56lck tyrosine kinase activity. Biochemistry 1995; 34(38): 12404-10.

http://dx.doi.org/10.1021/bi00038a038 PMID: 7547985

[172]   Shi Y, Wang CH, Gong XG. Apoptosis-inducing effects of two anthraquinones from *Hedyotis diffusa* WILLD. Biol Pharm Bull 2008; 31(6): 1075-8.
http://dx.doi.org/10.1248/bpb.31.1075 PMID: 18520033

[173]   Sträter N. Ecto-5′-nucleotidase: Structure function relationships. Purinergic Signal 2006; 2(2): 343-50.
http://dx.doi.org/10.1007/s11302-006-9000-8 PMID: 18404474

[174]   Venugopala KN, Buccioni M. Current understanding of the role of adenosine receptors in cancer. Molecules 2024; 29(15): 3501.
http://dx.doi.org/10.3390/molecules29153501 PMID: 39124905

[175]   Baqi Y. Anthraquinones as a privileged scaffold in drug discovery targeting nucleotide-binding proteins. Drug Discov Today 2016; 21(10): 1571-7.
http://dx.doi.org/10.1016/j.drudis.2016.06.027 PMID: 27373759

[176]   Minotti G, Menna P, Salvatorelli E, Cairo G, Gianni L. Anthracyclines: molecular advances and pharmacologic developments in antitumor activity and cardiotoxicity. Pharmacol Rev 2004; 56(2): 185-229.
http://dx.doi.org/10.1124/pr.56.2.6 PMID: 15169927

[177]   Gewirtz D. A critical evaluation of the mechanisms of action proposed for the antitumor effects of the anthracycline antibiotics adriamycin and daunorubicin. Biochem Pharmacol 1999; 57(7): 727-41.
http://dx.doi.org/10.1016/S0006-2952(98)00307-4 PMID: 10075079

[178]   Tacar O, Sriamornsak P, Dass CR. Doxorubicin: an update on anticancer molecular action, toxicity and novel drug delivery systems. J Pharm Pharmacol 2013; 65(2): 157-70.
http://dx.doi.org/10.1111/j.2042-7158.2012.01567.x PMID: 23278683

[179]   Brown JP, Brown RJ. Mutagenesis by 9,10-anthraquinone derivatives and related compounds in *Salmonella typhimurium.* Mutat Res Genet Toxicol Test 1976; 40(3): 203-24.
http://dx.doi.org/10.1016/0165-1218(76)90046-X PMID: 785247

[180]   Brown JP, Dietrich PS. Mutagenicity of anthraquinone and benzanthrone derivatives in the salmonella/microsome test: Activation of anthraquinone glycosides by enzymic extracts of rat cecal bacteria. Mutat Res Genet Toxicol Test 1979; 66(1): 9-24.
http://dx.doi.org/10.1016/0165-1218(79)90003-X PMID: 370585

[181]   Brown JP. A review of the genetic effects of naturally occurring flavonoids, anthraquinones and related compounds. Mutat Res Rev Genet Toxicol 1980; 75(3): 243-77.
http://dx.doi.org/10.1016/0165-1110(80)90029-9 PMID: 6770263

[182]   Müller SO, Eckert I, Lutz WK, Stopper H. Genotoxicity of the laxative drug components emodin, aloe-emodin and danthron in mammalian cells: Topoisomerase II mediated? Mutat Res Genet Toxicol Test 1996; 371(3-4): 165-73.
http://dx.doi.org/10.1016/S0165-1218(96)90105-6 PMID: 9008718

[183]   Mueller SO, Stopper H. Characterization of the genotoxicity of anthraquinones in mammalian cells. Biochim Biophys Acta, Gen Subj 1999; 1428(2-3): 406-14.
http://dx.doi.org/10.1016/S0304-4165(99)00064-1 PMID: 10434060

[184]   Müeller SO, Lutz WK, Stopper H. Factors affecting the genotoxic potency ranking of natural anthraquinones in mammalian cell culture systems. Mutat Res Genet Toxicol Environ Mutagen 1998; 414(1-3): 125-9.
http://dx.doi.org/10.1016/S1383-5718(98)00047-3 PMID: 9630566

[185]   Liu Y, Mapa MST, Sprando RL. Liver toxicity of anthraquinones: A combined *in vitro* cytotoxicity and in silico reverse dosimetry evaluation. Food Chem Toxicol 2020; 140: 111313.
http://dx.doi.org/10.1016/j.fct.2020.111313 PMID: 32240702

[186]   Semwal RB, Semwal DK, Combrinck S, Viljoen A. Emodin - A natural anthraquinone derivative with diverse pharmacological activities. Phytochemistry 2021; 190: 112854.
http://dx.doi.org/10.1016/j.phytochem.2021.112854 PMID: 34311280

[187]   Wang Y, Zhao M, Li B, Geng X. Advances in the mechanism of emodin-induced hepatotoxicity. Heliyon 2024; 10(13): e33631.
http://dx.doi.org/10.1016/j.heliyon.2024.e33631 PMID: 39027614

[188]   He Q, Liu K, Wang S, Hou H, Yuan Y, Wang X. Toxicity induced by emodin on zebrafish embryos. Drug Chem Toxicol 2012; 35(2): 149-54.
http://dx.doi.org/10.3109/01480545.2011.589447 PMID: 21834668

CHAPTER 5

# Role of Nanotechnology in Anthraquinones-Mediated Disease Management

**Pratima P. Pandey**[1] and **Maushmi S. Kumar**[1,*]

[1] *Somaiya Institute for Research and Consultancy, Somaiya Vidyavihar University, Vidyavihar (East), Mumbai – 400077, India*

**Abstract:** To maintain therapeutic efficacy while reducing toxicity and biocompatibility, innovative safety delivery techniques with improved nanotechnology are developed. Nanotechnology develops delivery systems that enhance the solubilizing and release quality of drugs, as well as their circulation time, which is important to enhance the therapeutic efficacy of drugs. Anthraquinones are organic substances that may be found in various plants, animals, and even some marine life. They are simple anthrones or bianthrones chemically speaking. They are used as pigments, dyes, and pharmaceuticals. Anthraquinone glycosides possess biological qualities including laxative, anti-inflammatory, anti-cancer, and antioxidants. Anthraquinones have certain drawbacks, such as their poor solubility in aqueous media, which restricts the routes of administration and lowers their bioavailability while also exhibiting a lower degree of selectivity for target tissues. It is speculated that anthraquinones and nanostructures work together. This chapter describes the application of nanotechnology in the treatment of anthraquinone-mediated diseases. The utilization of anthraquinone-loaded nanoparticles, nanocapsules, and nanocarriers in the treatment of various illnesses is highlighted.

**Keywords:** Anti-bacterial, Anthraquinones, Age macular diseases, Angiogenesis, Cancer, Chitosan, Diabetic nephropathy, Doxorubicin, Emodin, Liponanoparticles, Natural products, Nanocarriers, Nanoparticles, Nanotechnology, Peripheral vascular disease, Photosensitizers, Photodynamic, Rhein, Reactive oxygen species.

## INTRODUCTION

For millennia, natural products have been a primary source of therapeutic agents. Nevertheless, the use of biologically active natural metabolites in pharmaceutical products and drug discovery is still a viable option. Living organisms in a variety of environments can develop various secondary metabolites, which can be useful

* **Corresponding author Maushmi S. Kumar:** Somaiya Institute for Research and Consultancy, Somaiya Vidyavihar University, Vidyavihar (East), Mumbai – 400077, India; E-mails: maushmiskumar@gmail.com; maushmi@somaiya.edu

**Pardeep Kaur, Ajay Kumar, Robin, Tarunpreet Singh Thind & Kamaljit Kaur (Eds.)**

to the organism and may have many applications for human beings Anthraquinone (AQ) is one of the main metabolites produced by various plants, marine organisms, and microorganisms, used in a broad range of applications, *e.g.*, colouring agents for foodstuffs as well as textile products which are therapeutic to different diseases [1,2]. Anthraquinones (Fig. **1**) originate from the compound known as 9,10-anthracenedione. The introduction of hydroxyl, methyl, carboxyl, and methoxy functional groups onto 9,10-anthracenedione leads to the generation of several anthraquinone derivatives, exhibiting various therapeutic properties (Fig. **2**) [1]. These are anthracene derivatives comprising three benzene rings and one or more hydroxyl groups that can combine with sugar molecules. Therefore, they occur in nature as anthraquinone glycosides. Anthracene compounds exist in oxidized (anthraquinone) or reduced (anthrones, anthranols) and dimer (dianthrones) forms in nature [3].

More than 75 naturally occurring AQs are identified from various natural sources which include algae, marine organisms, fungi, and medicinal plants belonging to distinct family groups. Researchers are interested in the AQ scaffold because of its broad spectrum of biological actions, which include antitumor, anti-inflammatory, anticancer, antimutagenic, anti-fungal, anti-viral, anti-malarial, anti-microbial, anti-platelet, antidiabetic, neuroprotective, antioxidant, anti-bacterial, laxative, *etc* (Table **1**) [4]. Many more biological properties have been recorded with different effects [5]. In addition to its biological activities, many natural and synthetic anthraquinones are finding applications in textiles, electronic goods, biochips, food, cosmetics, medicine, and imaging photocleavage protection groups [3].

In addition, the design of new techniques has shown that an effective nanoformulation can be produced by the combination of anthraquinones and nanostructures. A potential area of research in nanotechnology involves the creation of nanocarriers that encapsulate hydrophobic and lipophilic bioactive medicines. This will improve bioavailability and broaden the spectrum of delivery strategies. Because nanocarriers are bio-compatible and bio-degradable, they offer greater opportunities for innovation and early detection of many diseases [6].

Nanotechnology is the study of nanomaterials or structures less than 100 nm with high surface density and volume ratios that could change physical chemistry, and biological parameters in chemical compositions. Nanoscience has gained worldwide interest because of its potential for applications in pharmaceuticals, diagnostics, and disease treatment [7]. Nanomaterials have a variety of qualities and characteristics, such as the required size, greater solubility, easier passage over biological barriers or an improvement in reactivity. New ambitions to tackle today's human challenges have arisen from the use of nanotechnology. The use of nanotechnology also benefits the pharmaceutical and medical industry, resulting in new products being launched on the market [8].

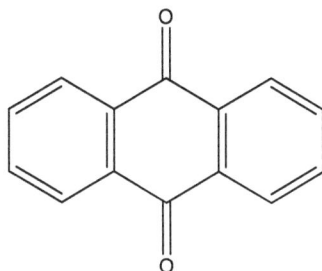

**Fig. (1).** Anthraquinone structure.

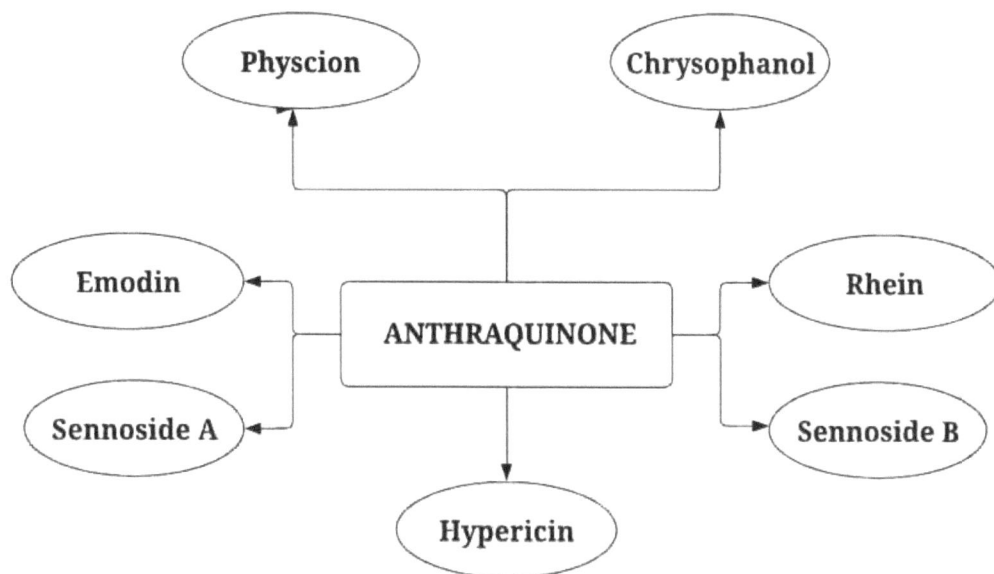

**Fig. (2).** Derivatives of Anthraquinone.

## Toxicity of Anthraquinone

Anthraquinones are recognized for their diverse toxicological effects, which pose risks to both human health and environmental integrity. The degree of toxicity associated with these compounds can vary markedly based on the specific anthraquinone derivative and the levels of exposure involved. Certain derivatives have been classified as potential carcinogens, indicating their possible role in cancer development in living organisms, thereby underscoring the necessity for thorough safety evaluations. The toxicological mechanisms attributed to anthraquinones include the induction of oxidative stress and the production of reactive oxygen species, both of which can result in cellular damage and enhance their carcinogenic potential. Given their toxic and carcinogenic characteristics, anthraquinones are subjected to regulatory oversight, which involves assessments by health and safety agencies to establish safe exposure thresholds and appropriate safety measures. The adverse effects of anthraquinones extend beyond human health, impacting ecological systems as well. Their environmental presence can lead to detrimental consequences for both wildlife and plant species [15]. Current research highlights several toxicity profiles associated with anthraquinones, including hepatotoxicity, nephrotoxicity, genotoxicity, reproductive toxicity, and phototoxicity [16]. Among these, hepatotoxicity and nephrotoxicity are particularly concerning. Investigations have indicated that both direct DNA damage (through incorporation into DNA base pairs) and indirect DNA damage (mediated by reactive oxygen species) can lead to significant apoptosis or inflammatory responses, which are critical factors in the liver and kidney toxicity observed with high doses of anthraquinones. Furthermore, prolonged exposure to these compounds has been associated with colonic toxicity, attributed to the accumulation of harmful metabolites that may trigger apoptosis and autophagy in colonic epithelial cells.

**Table 1. Pharmacological activities of anthraquinone derivatives.**

| Anthraquinone Derivatives | Pharmacological Activities | References |
|---|---|---|
| Emodin | Anti-inflammatory, hepatoprotective, Anti-ulcer antimicrobial, muscle relaxant, immunosuppressive, neuroprotective and anti-fibrotic activities. | [9] |
| Chrysophanol | Anticancer, hepatoprotective, anti-inflammatory, antiobesity, hypolipidemic, pulmonary injury, antidiabetic, neuroprotective, and anti-CVDs. | [1,10] |
| Physcion | Hepatoprotective, anti-microbial, anti-inflammatory, and anti-proliferative effects. | [11] |
| Rhein | Hepatoprotective, anti-inflammatory, anticancer, nephroprotective, antidiabetic, lipid-lowering activity, antimicrobial activity, purgative, and antioxidant activity. | [12] |

*(Table 1) cont.....*

| Hypericin | Antibacterial, antiviral, and neuroprotective effects, photosensitizing activity, antimicrobial, antitumor, antidepressant, anti-inflammatory, and antipsoriatic properties. | [13,14] |
|---|---|---|

In recent years, aloe-emodin has been identified as having phototoxic properties, primarily evidenced by its detrimental effects on skin fibroblasts when exposed to ultraviolet radiation. Numerous studies indicate that a photochemical mechanism, which includes the generation of singlet oxygen and direct photo-oxidative damage to nucleic acids such as DNA or RNA, may underlie this phototoxicity. Consequently, the development of AQ as a therapeutic agent should prioritize achieving high efficacy while minimizing toxicity. As a result, the strategy of utilizing anthraquinone encapsulated in nanoparticles has garnered significant interest in the management of various diseases (Table **2**) [17].

## ANTHRAQUINONE NANOPARTICLES (AQNPs) IN DISEASE MANAGEMENT

### Diabetic Nephropathy

Nephropathy, a serious microvascular disorder that is prevalent in both types of diabetes, is the primary contributor to renal failure. Additionally, it can accelerate the progression of cardiovascular disease and exacerbate macrovascular issues. An important characteristic of Diabetic Nephropathy (DN) is the accumulation of (Extra cellular matrix) ECM proteins, including collagens, which can lead to fibrosis and hypertrophy in the glomerulus and tubules of the kidneys, both of which contribute to renal failure in type 2 diabetes. Currently available treatments for DN do not offer a complete therapeutic effect, indicating that a more detailed understanding of molecular pathways underlying the development of these diseases is needed to improve disease management [18].

Rhein (RH) is an anthraquinone derivative extracted from *Rheum palmatum, Rheum tanguticum*, having varied pharmacological actions on DN. RH has received a lot of interest in Asian nations for its anti-diabetic properties. Pharmacological studies have shown that RH lowers glomerular mesangial cell proliferation, glomerular hypertrophy, and the production as well as deposition of intercellular matrix in DN patients. It can also maintain kidney intrinsic function by suppressing mRNA transcription, TSP-1 and TGF-1 expression inside renal tubular epithelial cells, and lowering nephritis. However, its clinical use is limited due to low solubility and bioavailability, reduced renal distribution, and side effects [19,20].

To improve RH absorption into the kidneys and DN's therapeutic effectiveness, nanotechnology has been employed by Chen and his co-workers to synthesize and utilize polyethyleneglycol-co-polycaprolactone-co-polyethylenimine (PPP) triblock amphiphilic polymers to create polyethyleneglycol-co-polycaprolactoneco-polyethylenimine -rhein -nanoparticles (PPP-RH-NPs) for the purpose of drug delivery to the kidneys. Injecting intravenous PPP-RH-NPs into the body of mice without hair resulted in a high level of fluorescence, hair should be eliminated because it significantly affects fluorescence. As time went on, the drug was found to be digested in the mice, resulting in a decrease in fluorescence intensity. After 6 hours, most of the drug was found in the chest and epigastrium, with the kidney being the most fluorescing organ. The lungs and liver were the next most fluorescing organs, with the kidney having the highest fluorescence intensity. This suggested that the drug may be able to target and reach the maximum concentration of an infected kidney, providing strong evidence for direct medication delivery to a diseased kidney [21].

Wang and his colleagues conducted a similar investigation by developing kidney-targeted RH-loaded Liponanoparticles (KLPPR) with a yolk-shell structure consisting of polycaprolactone-polyethyleneimine (PCL-PEI) -based cores and lipid layers with kidney-targeting peptide (KTP) modifications (Fig. **3**) [22].

Due to metabolism and clearance, Liponanoparticles (LPPR) and KLPPR were broadly dispersed throughout the mice's bodies and accumulated in the primary organs. After 8 hours, the majority of LPPR and KLPPR were discovered in the mice's abdomen and chest. KLPPR fluoresced more strongly in the kidney, whereas LPPR fluoresced significantly in the bladder and urine. Rapid urine excretion was demonstrated by a 2.4-fold increase in urine fluorescence intensity in LPPR-treated animals compared to control mice. The most common sites of *peripheral vascular disease* (PVD) were the intestines and kidneys, followed by the liver and lungs. In particular, the kidneys of mice treated with KLPPR showed fluorescence intensity 2.5 times higher than that of the liver, lungs, and colon, and 7.5 times higher than that of the heart, spleen, and brain. In contrast, the fluorescence intensity of KLPPR in kidneys was 2.6 times that of LPPR. These results indicate that KTP-mediated may increase renal cellular uptake of KLPPR, resulting in increased renal distribution and accumulation while reducing urine excretion [20].

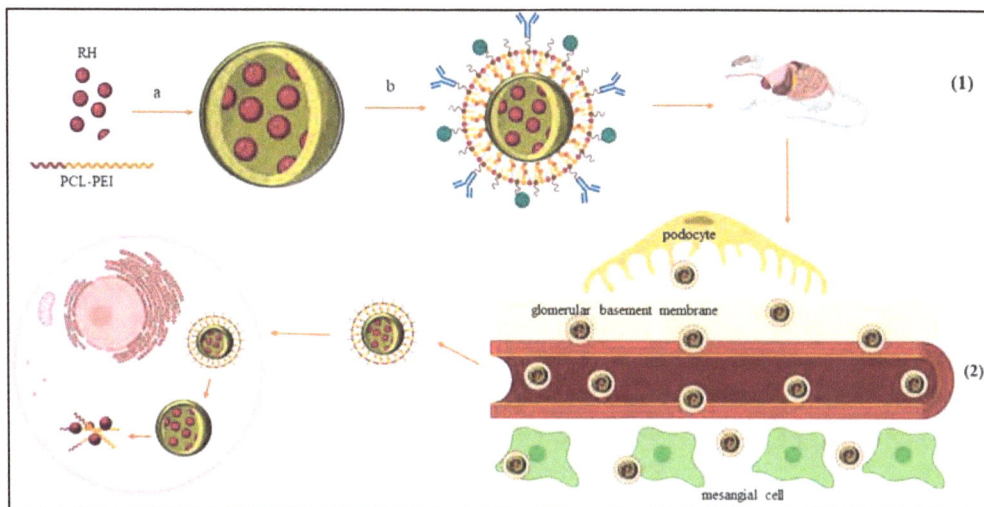

**Fig. (3).** Diagrammatic illustration of kidney-targeted medication delivery using KLPPR Liponanoparticles. **(1)** KLPPR Liponanoparticles consisted of a RH-loaded PCL-PEI nanoparticular core and a KTP-modified lipid layer **(a)** Independently assembled RH-loaded PCL-PEI nanoparticle **(b)** Lipid coating of KTP **(2)** The glomerular filtration membrane allows KLPPR to enter the kidney, by facilitating membrane fusion, KTP adorning enhanced KLPPR kidney retention by promoting renal cellular uptake and internalization.

## Cancer Management

Cancer is a global, complex, and diverse disease, becoming a primary source of illness and mortality worldwide as the population rises [5]. In the upcoming decades, every region of the world will experience a significant increase in the incidence and mortality of cancer [23]. Therefore, there is a desperate need for research into the development of cancer cures [24]. There are a lot of techniques to treat cancer [25]. Chemotherapy, surgery, and radiation are the most prevalent forms of cancer therapies accessible today [26]. The disadvantages and the side effects of these treatments include high treatment costs, acute toxicity, tumour recurrence, adverse medication responses, and treatment failures [27]. Numerous initiatives have been taken to reduce the negative pharmacological side effects associated with cancer treatment [28].

The group of anticancer drugs belonging to the AQs class that is one of the most used classes, includes daunorubicin, epirubicin, doxorubicin, and mitoxantrone. These drugs are inhibitory to the growth of cancer cells, in conditions such as leukaemia. The mechanism of these anti-cancer agents is to intricate deoxyribose nucleic acid (DNA) by placing themselves between two DNA strands, for its separation. DNA damage is caused by the formation of free radicals, particularly reactive oxygen species (ROS), which react with topoisomerases to inhibit them and induce apoptosis through the inhibition of p53 and the inhibition of ROS-induced topoisomerases. Additionally, anthraquinones stimulate the apoptosis of cells *via* mitochondrial pathways, as well as through the N-terminal kinase (Akt/PKB) and the C-Jun N-terminal kinase [29].

Nanoparticles (NPs) due to their small size, have been found to be beneficial for sensitive cancer detection capacities and successful treatment outcomes. Nanotechnology-based treatments have been shown to provide benefits in several ways, *e.g.*, improved absorption, dose-response, and targeted effectiveness, as well as fewer side effects than standard chemotherapies [30]. Both *in vitro* tests and *in vivo* investigations using biologically active compounds have shown high selectivity and cytotoxicity in suppressing various cancer cells [31]. The hormone-dependent human breast adenocarcinoma cell line MCF-7 was used to test the anticancer potential of AQNPs and anthraquinone bulk compounds. Briefly, the NPs of anthraquinone showed moderate anticancer activity comparable to corresponding bulk substances with $GI_{50}$ values that range from 0.31 to 0.52 M [27]. Another study by Redah *et al.* synthesized chitosan nanoparticles (CS-NPs) and loaded them with AQ to increase the medication's effectiveness and decrease its side effects [29]. To assess the cytotoxicity of manufactured nanoparticles, MTT assays were used. The proliferation of tumour cells was shown to be suppressed when exposed to CS-NPs, AQ nanoparticles, and chitosan anthraquinone CS-AQ. The capacity of human acute myeloid leukemia (HL-60) cells for growth has been investigated in both concentration and time-dependent methods. The increased concentration of CS-AQ nanoparticles in the medium had an effect on HL-60 cell viability for 24 hours, resulting in an accelerated rate of cell death. The propensity to die became more evident as the incubation period continued, indicating an extended medication release from the CS-NPs. Cell viability decreased from approximately 82.1% to 12.6% for CS-AQ NPs. CSNPs suppressed cellular proliferation at least as strongly as AQ in terms of dose dependency. It has been demonstrated that CS-AQ nanoparticles possess a synergistic anticancer effect by inhibiting the proliferation of HL-60 cells, thus demonstrating the potential of biodegradable nanomaterials as an ideal therapeutic drug for the purpose of treating cancer. Research was undertaken to determine the influence of CS-AQ NPs on the growth of HL-60 cells. The results indicated that, after 24 hours of treatment, the percentage of lactate dehydrogenase (LDH) released to the culture medium increased significantly, rising from 20.13% to 96.58%. The pinnacle of LDH accumulation in the media was noted when the cells were treated with 80 mg/ml of CS-AQ NPs. The results of this investigation indicated that CS-AQ NPs had a favourable impact on the cells' development. These studies demonstrated a promising effect of anthraquinone-entrapped chitosan nanoparticles on *in vitro* apoptotic and cytotoxic tests [32].

Emodin (EMO) is a naturally occurring AQ that serves several biological purposes in addition to being a common laxative. To investigate EMO's anticancer efficacy, Wang *et al.* prepared Emodin-solid lipid nanoparticles (EMO-SLNs). This could significantly reduce the cell proliferation in human breast cancer cell lines (MCF-7, MCF-10A, and MDA-MB-231) in a dose- and time-dependent manner. Compared to free EMO, EMO-SLNs significantly outperform it in terms of inhibiting MCF-7 and MDA-MB-231 cells at the same concentration. Further investigation demonstrated that EMO-SLNs had a greater ability than bulk EMO solution to considerably arrest MCF-7 cells at the gap phase 2/meiotic phase (G2/M phase). Studies on apoptosis also reveal that EMO-SLNs increase the rate of apoptosis in MCF-7 cells. These findings led researchers to draw the conclusion that the EMO-SLNs could be a promising drug delivery method, particularly for moderately insoluble anticancer drugs [33]. Liu *et al.* conducted a comprehensive study to investigate the cytotoxic efficacy of polymer-lipid hybrid nanoparticles that encapsulate emodin against MCF-7 cells. These nanoparticles, referred to as E-PLNs, were synthesized using a sophisticated nanoprecipitation technique, which allows for the controlled formation of nanoparticles with specific characteristics. When the cytotoxic effects of the E-PLNs were compared to those of free emodin, the E-PLNs exhibited significantly enhanced cytotoxicity against MCF-7 cells, a widely used breast cancer cell line. This increased cytotoxic effect can be attributed to the improved bioavailability and absorption of emodin when delivered in the nanoparticle form. The E-PLNs facilitate a more efficient uptake of emodin by the cancer cells, which is essential for maximizing its therapeutic effects. Moreover, the study revealed that the enhanced cytotoxicity of the E-PLNs was associated with the activation of apoptotic pathways within the MCF-7 cells [34,35]. Another investigation assessed the antiproliferative properties of rhein and emodin, which were conjugated with ZnO-CuO nanoparticles, against two cancer cell lines: Panc-1 (pancreatic cancer) and OVCAR-3 (ovarian cancer). Both compounds demonstrated significant cytotoxic effects on the cancer cells examined. Notably, their nanoparticle-conjugated forms exhibited enhanced cytotoxicity and selectivity towards Panc-1 and OVCAR-3 cells. Among these, emodin-conjugated ZnO-CuO nanoparticles recorded the lowest IC50 value for Panc-1 cells, compared to doxorubicin. Further, the annexin V/PI staining analysis revealed that emodin-conjugated ZnO-CuO nanoparticles induce cytotoxicity in Panc-1 cells primarily through the activation of apoptosis, as evidenced by an 8.3-fold increase in the population of apoptotic cells compared to the untreated control [36].

Because of the lipophilic nature of AQ, it is at risk for ROS absorption and therefore does not reach a therapeutic level. To increase the bioavailability of antioxidant and water solubility *in vivo*. A variety of drug carriers including polymer nanoparticles, lipid-based nanospheres as well as biodegradable microspheres have been used [37]. The use of chitosan (CS) nanoparticles and their derivatives in drug delivery vehicles has been given a lot of attention recently [38]. An innovative cationic polylactic acid polymers (PLA) nanoparticle formulation coated with chitosan-encapsulated anthraquinone formulation that can increase anthraquinone solubility and inhibit ROS uptake is described by D. Jeevitha *et al.* in an uncomplicated way. Without the use of organic solvents or surfactants, chitosan polylactic acid (CS-PLA) nanoparticles are produced under natural circumstances. These nanoparticles are stable in neutral and acidic environments between pH 4 and 8, and they agglomerate at pH levels above 9. According to preliminary findings from research on the size of chitosan-polylactic acid anthraquinone -CS/PLA-AQ NPs, AQ loading, and release appeared to be good vehicles for the chemo preventing agents' administration in hepatoblastoma cell line HepG2 cells. Because CS-PLA nanoparticles are biodegradable and rarely harmful, their therapeutic/clinical significance depends on this characteristic. These encouraging results therefore suggest that the idea of nano-chemoprevention holds a great advantage and incentive to perform additional in-depth human disease prevention studies on relevant animal models [39].

**Photodynamic Therapy**

Photodynamic therapy (PDT) has become widely used in the treatment of skin problems; however, the response rate to PDT treatment varies greatly. The low penetration depth of photosensitizers in the tissue affects the penetration depth of PDT, which clearly has an impact on the therapeutic efficacy [40].

Nanotechnology integration with photosensitizers (PSs), has shown to be a very efficient technique to boost therapeutic efficacy while researching new techniques to improve PDT. To enhance the results in terms of precise targeting, high drug loading, multifunctional integration, enhanced hydrophobic PS solubility, sustaining a steady PS delivery rate, and decreased harmful effects on healthy cells, nanomaterials have lately become a significant part of photodynamic therapy (PDT). Poly (D, L-lactide-co-glycoside (PLGA) due to its biodegradability and simplicity of formulation, has gained interest for PSs encapsulation. PLGA nanoparticles loaded with PSs excel conventional PDT medicines in terms of photoactivity [41].

In one of the studies, anthraquinone-PLGA nanocapsules were used in photodynamic treatment investigations in NIH-3T3 cells in 24-well plates, and NIH/3T3 cells were grown. These were incubated for 3 h at 37°C in a humid air atmosphere containing 5% $CO_2$ and 50 g of free red dye (RD) and red dye nanocapsules (RD/NC). Before the cells were exposed to radiation, any free RD or RD/NC that had not been absorbed by them was removed from each well through a PBS wash. The irradiation was conducted using a 24-well light-emitting diode (LED) table lighting system with 28 mW of optical power and an emission wavelength of 460 nm. Each plate was subjected to energy densities ranging from 1.5 to 2.5, and 5.0 to 10.0 J $cm^2$. Cells were cultured under controlled conditions for 24 hours after PDT. Consequently, the viability of NIH-3T3 cells treated with RD/NC decreased significantly across all irradiation levels employed in this investigation. Even with all irradiation dosages applied, NIH-3T3 cells treated with free RD showed just a little reduction in cell viability in comparison to RD/NC. The results of the photodynamic test showed that, at the same concentration and energy dose, NCs induced a greater proportion of cell death than RD-Free, which may suggest that NC promotes cell uptake. This work established the novel anthraquinone derivative's photodynamic potential and opened the door for further investigation and use of RD as a biotechnological photosensitizer nanostructured in PDT [6].

## Antibacterial Agent

The use of antibiotics is widely and effectively used in the fight against infections. However, accidental, uncontrolled, and improper antibiotic use contributes to the emergence of antimicrobial resistance which poses one of the main threats to Public Health [42]. The search for new antibacterial agents has been triggered by this situation [43].

Naturally occurring compounds have been demonstrated to alter bacteria's pathogenicity islands, proving their antibacterial effectiveness against disease. The antibacterial and antibiofilm properties of anthraquinones, an extensive variety of naturally existing constituents that include dantron, 1,2 hydroxyanthraquinone, alizarin, and purpurin, demonstrated against important fungal, opportunistic bacterial pathogens [44].

By altering the physicochemical properties of substances, nanotechnology opens up new possibilities for biological applications. To overcome resistance to antibiotics and enhance their effectiveness, NPs may be mixed with antibacterial agents. Furthermore, these agents may decrease the dose and toxicity of antibiotics to be used [43]. It is hard for microorganisms to develop resistance because NPs behave in various ways, acting on bacteria by several targets and mechanisms [45]. As a result, one potential method to satisfy this need is to employ a combination of medications and NPs that target the same unique pathogenicity islands of infections by antimicrobials.

Various investigations have demonstrated that copper nanoparticles (CuO NPs) display broad-spectrum antibacterial action against both Gram-positive and Gram-negative bacteria, including, *Escherichia* spp., *Bacillus* spp., *Staphylococcus* spp., *Pseudomona* spp., *Proteus* spp. [44,46]. CuO NPs and AQ's combination effect on enhanced antibacterial activity in opposition to *Staphylococci* spp., was described by Srivastava and co-authors. According to their results, copper (CuO) nanoparticles and AQ showed effective synergistic activity by the suppression of biofilm development and alteration of cell morphology as key mechanisms. *Staphylococcus* spp., in particular *Staphylococcus aureus* (*S. aureus*), is the majorly frequent species discovered in infections linked to biofilms, encompassing superficial skin infections and long-term injuries. Since *S. aureus* exopolysaccharide-protected architecture encourages bacterial aggregation, biofilm populations exhibit higher metabolic activity and less antibiotic diffusion in the matrix. Combination treatment is frequently employed to improve the empiric protection against Gram-positive bacterial infections provided by two or more antimicrobials with distinct activity profiles. By inhibiting the growth of biofilms and using changes in cell shape as a primary mechanism, CuO NPs and AQ displayed significant synergistic effectiveness *in vitro*. For *S. aureus* infections, it will be used to develop an effective combination therapy because this is the first report of a synergistic combination of CuO NPs with AQ. Further studies on the synergy's mechanisms revealed that the combination's primary effects on the phenotype's manifestation were the inhibition of biofilm formation and the reduction of cytoplasmic volume as cells shrank. This work demonstrated unequivocally that combining AQ with CuO NPs would be an inventive method of curing *S. aureus* infections (Fig. **4**) [44].

PLAs are the most widely studied and environmentally friendly polymers available, offering superior biodegradability, processability, and reduced energy dependency. Biodegradable polymers are used in the preparation of these beads, cylinders, discs, microspheres, and absorbable nanospheres for sustained drug release. However, PLAs have some drawbacks that limit their application in certain contexts, such as their crystalline composition and hydrophobic nature [47]. To lessen PLA's adverse effects and boost its efficacy, Jeevitha D *et al.*, and colleagues designed anthraquinone-loaded chitosan nanoparticles, AQ-CS-PLA. Hydrophobic PLA and hydrophilic CS were utilized in combination to establish a solid link between amino acids and hydroxyl groups. Anthraquinone was employed to create AQ-CS-PLA nanoparticles here by mixing it with chitosan-CS-lactic acid nanoparticles. When the first investigation was conducted, chitosan and PLA were simply combined. *Pseudomonas auginosa, Kiebellia pneumoniae, Porcine vulgaris,* and *E. coli* were among the bacteria that the anthraquinone-loaded CS- PLA nanoparticles were found to be effective against [48].

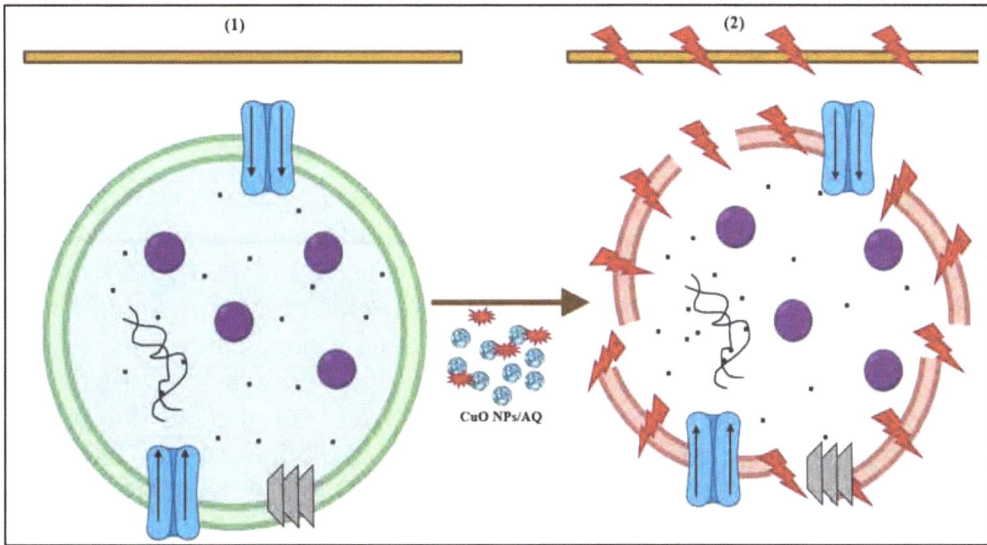

**Fig. (4).** CuO NPs/AQ combined action on the morphology of *S. aureus* **(1)** Stable biofilm- A bacterial cell embedded in a biofilm **(2)** degradation of biofilm- suppression of biofilms and shrinking of cells with loss of cytoplasmic contents.

## ANTHRAQUINONE-LOADED NANOPARTICLES EFFECTS ON VASCULAR ENDOTHELIAL GROWTH FACTOR

In recent years, it has been widely reported that anti-vascular endothelial growth factor therapy is used for the prevention of angiogenesis in ophthalmic fields. However, in practice, its delivery is hard to achieve owing to the hydrophobic nature of several agents. Therefore, studies are being carried out to introduce a new form of hydrophobic drug carriers, using nanoparticles [49].

The synthesis, characterization, and cytotoxicity of magnesium silicate hollow spheres were investigated in retina capillary endothelial cells. The researchers investigated the loading and releasing capacity of the spheres, as well as the effect of emodin on vascular endothelial growth factor (VEGF) expression at both the gene and protein levels. Furthermore, the angiogenesis of fertilized chicken eggs was studied, and the results revealed that the nanoparticles of magnesium silicate ($MgSiO_3$) were not hazardous. Emodin-loaded $MgSiO_3$ was discovered to be efficient in suppressing the expression of VEGF-related genes and proteins. In addition, chicken angiogenesis was drastically decreased [50].

Several medications, including doxorubicin, are characterized as Hypoxia-inducible factor –(HIF-1) inhibitors. HIF-1 regulates the expression of various angiogenic factors, including VEGF. Doxorubicin (DOX) stops HIF-1 from binding to DNA, which prevents it from upregulating angiogenic factors, as opposed to lowering the quantity of HIF-1 that is present directly. Anthracycline antibiotic DOX has been used in the treatment of cancer [51].

Kelly *et al.* developed DOX-containing polymer nanoparticles to limit the growth of HIF-1, a chemical that interferes with the levels of some angiogenic hormones. However, because DOX is hazardous, encapsulating it in NPs will facilitate the delivery of the medication while lowering its toxicity without sacrificing its efficacy. The PLGA ones were manufactured using nanoprecipiters and single or double emulsion dosing (SE, DE), and the chitosan was created using nanoprecipiters. Of all the formulations evaluated, all the NP formulations exhibited the lowest zeta potential (ZP) and the quickest drug release. The Fourier transform infrared (FTIR) verified the existence of DOX on the SE NP's surface, hence it indicates that nanoparticle capsule (NPC) is the most efficient delivery method for age macular diseases (AMD) [50].

**Table 2. Role of AQNPs in disease management.**

| Condition | Characteristics | Therapeutic Agent | Combination Therapy | Mechanism of Action |
|---|---|---|---|---|
| Cancer | Increasing prevalence and mortality rates represent a significant global health concern. | CS-NPs with AQ | Inhibits the proliferation of HL-60 cells | Intercalates DNA, generates ROS, inhibits topoisomerases, induces apoptosis. |
| DN | A microvascular condition that significantly contributes to renal failure and | PPP-RH-NPs, KLPPR | Nanoparticles for targeted delivery | Reduces mesangial cell proliferation, glomerular hypertrophy, |

*(Table 2) cont.....*

| | | | | ECM deposition, and suppresses TSP-1 and TGF-1 expressions. |
|---|---|---|---|---|
| PDT | Two-stage treatment that combines light energy with a medicine called a photosensitizer. | Anthraquinone-PLGA nanocapsules | Integration of nanotechnology with PS. | Enhances cell uptake; induces cell death upon irradiation. |
| Bacterial Infections | The rise of antimicrobial resistance necessitates the development of novel therapeutic agents. | Anthraquinon.es, CuO nanoparticles | CuO NPs combined with anthraquinones for synergy | Alters pathogenicity islands, inhibits biofilm development, alters cell morphology. |
| Anti- VEGF | Prevention of angiogenesis in ophthalmic fields. | Emodin-loaded MgSiO3, Encapsulated DOX in polymer nanoparticles. | Facilitates the delivery of the medication while lowering its toxicity. | Decreases the expression of VEGF-related genes and proteins. |

## CONCLUSION

Nanotechnology has the possibility of being a powerful tool for the advancement of drug-resistant nanocarriers. These nanocarriers can expand the route of administration and enhance bioavailability, while also being biodegradable and biocompatible. Furthermore, they have the potential to develop new forms of medication and diagnosis for a variety of illnesses. Polymeric nanoparticles (nano-spheres and nano-capsules), solid lipoplastics, nanostructured lipid systems, and liposomes are among the many nanocarriers that have emerged in recent years. The spherical form and smaller particle size of nano-capsules (NCs) boost their attractiveness because they increase the surface area of the target tissues and improve their interactions with one another. Anthraquinone, a secondary metabolite, has a wide range of pharmacological activities, however, due to some disadvantages, anthraquinone may be coupled with nanocarriers and can be used to treat most diseases with fewer side effects.

## ACKNOWLEDGEMENTS

We acknowledge the facility provided and the help extended by Somaiya Institute for Research and Consultancy, Somaiya Vidyavihar University, Mumbai in writing this chapter.

## LIST OF ABBREVIATIONS

| | |
|---|---|
| **AMD** | Age Macular Diseases |
| **AQ** | Anthraquinone |
| **AQ-CS-NP** | Anthraquinone-Loaded Chitosan Nanoparticles |
| **AQ-CS-PLA** | Anthraquinone -Chitosan-Polylactic Acid |
| **AQNP's** | Anthraquinone Nanoparticles |
| **CO₂** | Carbon Dioxide |
| **CSAQ** | Chitosan Anthraquinone |
| **CS-AQ NPs** | Chitosan Loaded Anthraquinone Nanoparticles |
| **CSNPs** | Chitosan Nanoparticles |
| **CS-PLA** | Chitosan Polylactic Acid |
| **CuO NPs** | Copper Oxide Nanoparticles |
| **DN** | Diabetic Nephropathy |
| **DNA** | Deoxyribose Nucleic Acid |
| **DOX** | Doxorubicin |
| **ECM** | Extracellular Matrix |
| **EMO** | Emodin |
| **E-SLNs** | Emodin-Solid Lipid Nanoparticles |
| **FTIR** | Fourier Transform Infrared |

| **HepG2** | Hepatoblastoma Cell Line |
|---|---|
| **HIF-1** | Hypoxia Inducible Factor -1 |
| **HL-60** | Human Acute Myeloid Leukemia |
| **KLPPR** | Kidney-Targeted Rhein (RH)-Loaded Lipo Nanoparticles |
| **KTP** | Kidney Targeting Protein |
| **LDH** | Lactate Dehydrogenase |
| **LED** | Light-Emitting Diode |
| **MCF-10A** | Michigan Cancer Foundation -10A |
| **MCF-7** | Michigan Cancer Foundation -7 |
| **MDA-MB-231** | Epithelial, Human Breast Cancer Cell Line |
| **MgSiO₃** | Magnesium Silicate |
| **mRNA** | Messenger RNA |
| **MTT** | 3-[4,5-dimethylthiazol-2-yl]-2,5 Diphenyl Tetrazolium Bromide |
| **NCs** | Nanocapsules |
| **NP's** | Nanoparticles |
| **OVCAR-3** | Ovarian Cancer |
| **Panc-1** | Pancreatic Cancer |
| **PCL-PEI** | Polycaprolactone-polyethyleneimine |
| **PDT** | Photodynamic Therapy |
| **PLA** | Polylactic Acid Polymers |
| **PLGA** | Poly (D, L-lactide-co-glycoside) |
| **PPP** | Polyethyleneglycol-co-polycaprolactone-co-polyethyleneimine |
| **PPP-RH-NP's** | polyethyleneglycol-co-polycaprolactoneco-polyethyleneimine - Rhein -Nanoparticles |

| **PS** | Photosensitizer |
|---|---|
| **RD** | Red Dye |
| **RD/NC** | Red dye / Nanocapsules |
| **RH** | Rhein |
| **ROS** | Reactive Oxygen Species |
| **SE NP's** | Single emulsion diffusion |
| **SLN** | Solid Lipid Nanoparticles |
| **TGF-1** | Transforming Growth Factor-beta1 |
| **TSP-1** | Thrombospondin-1 |
| **VEGF** | Vascular Endothelial Growth Factor |
| **ZP** | Zeta Potential |

## REFERENCES

[1]    Prateeksha, Yusuf MA, Singh BN, *et al.* Chrysophanol: A natural anthraquinone with multifaceted biotherapeutic potential. Biomolecules. 2019; 9(2).

[2]    Preet G, Gomez-Banderas J, Ebel R, Jaspars M. A structure-activity relationship analysis of anthraquinones with antifouling activity against marine biofilm-forming bacteria. Frontiers in Natural Products 2022; 1(Oct): 990822.
http://dx.doi.org/10.3389/fntpr.2022.990822

[3]    Stompor-Gorący M. The health benefits of emodin, a natural anthraquinone derived from rhubarb—a summary update. Int J Mol Sci 2021; 22(17): 9522.
http://dx.doi.org/10.3390/ijms22179522 PMID: 34502424

[4]    Siddamurthi S, Gutti G, Jana S, Kumar A, Singh SK. Anthraquinone: a promising scaffold for the discovery and development of therapeutic agents in cancer therapy. Future Med Chem 2020; 12(11): 1037-69.
http://dx.doi.org/10.4155/fmc-2019-0198 PMID: 32349522

[5]    Cheng W, Li Y, Yang W, *et al.* Simultaneous determination of 13 constituents of radix polygoni multiflori in rat plasma and its application in a pharmacokinetic study. International Journal of Analytical Chemistry. 2020; 2020.

[6]    Amantino CF, de Baptista-Neto Á, Badino AC, Siqueira-Moura MP, Tedesco AC, Primo FL. Anthraquinone encapsulation into polymeric nanocapsules as a new drug from biotechnological origin designed for photodynamic therapy. Photodiagn Photodyn Ther 2020; 31(Sep): 101815.
http://dx.doi.org/10.1016/j.pdpdt.2020.101815 PMID: 32407889

[7]     Koul B, Poonia AK, Yadav D, Jin JO. Microbe-mediated biosynthesis of nanoparticles: Applications and future prospects. Biomolecules 2021; 11(6): 886.
        http://dx.doi.org/10.3390/biom11060886 PMID: 34203733

[8]     Rezaei R, Safaei M, Mozaffari HR, *et al.* The role of nanomaterials in the treatment of diseases and their effects on the immune system. Open Access Maced J Med Sci 2019; 7(11): 1884-90.
        http://dx.doi.org/10.3889/oamjms.2019.486 PMID: 31316678

[9]     Semwal RB, Semwal DK, Combrinck S, Viljoen A. Emodin - A natural anthraquinone derivative with diverse pharmacological activities. Phytochemistry 2021; 190(Oct): 112854.
        http://dx.doi.org/10.1016/j.phytochem.2021.112854 PMID: 34311280

[10]    Su S, Wu J, Gao Y, Luo Y, Yang D, Wang P. The pharmacological properties of chrysophanol, the recent advances. Biomed Pharmacother 2020; 125(May): 110002.
        http://dx.doi.org/10.1016/j.biopha.2020.110002 PMID: 32066044

[11]    Pang M, Yang Z, Zhang X, Liu Z, Fan J, Zhang H. Physcion, a naturally occurring anthraquinone derivative, induces apoptosis and autophagy in human nasopharyngeal carcinoma. Acta Pharmacol Sin 2016; 37(12): 1623-40.
        http://dx.doi.org/10.1038/aps.2016.98 PMID: 27694907

[12]    Zhou YX, Xia W, Yue W, Peng C, Rahman K, Zhang H. Rhein: A review of pharmacological activities. evidence-based complementary and alternative medicine. 2015; 2015.
        http://dx.doi.org/10.1155/2015/578107

[13]    Amrit D, Singh P. Hypericin-A Napthodianthrone from Hypericum perforatum 2003.

[14]    Oglah MK, Kahtan Bashir M, Mustafa YF. Hypericin And Its Analogues: A review of their biological activities structure-based 3D-QSAR and molecular docking approach of n-benzylpiperidine derivatives as acetylcholinesterase inhibitors to design anti-alzheimer agents view project medicinal chemistry view project. article in turkish journal of field crops [Internet]. 2021; Available from: https://www.researchgate.net/publication/354717849

[15]    Sendelbach LE. A review of the toxicity and carcinogenicity of anthraquinone derivatives. Toxicology 1989; 57(3): 227-40.
        http://dx.doi.org/10.1016/0300-483X(89)90113-3 PMID: 2667196

[16]    Dong X, Fu J, Yin X, *et al.* Emodin: A review of its pharmacology, toxicity and pharmacokinetics. Phytother Res 2016; 30(8): 1207-18.
        http://dx.doi.org/10.1002/ptr.5631 PMID: 27188216

[17]    Xin D, Li H, Zhou S, Zhong H, Pu W. effects of anthraquinones on immune responses and inflammatory diseases. Molecules 2022; 27(12): 3831.
        http://dx.doi.org/10.3390/molecules27123831 PMID: 35744949

[18]    Kato M, Natarajan R. Diabetic nephropathy—emerging epigenetic mechanisms. Nat Rev Nephrol 2014; 10(9): 517-30.
        http://dx.doi.org/10.1038/nrneph.2014.116 PMID: 25003613

[19]    Zeng CC, Liu X, Chen GR, *et al.* The molecular mechanism of rhein in diabetic nephropathy. Evidence-based Complementary and Alternative Medicine 2014; 2014.
        http://dx.doi.org/10.1155/2014/487097

[20]   Wang G, Li Q, Chen D, *et al.* Kidney-targeted rhein-loaded liponanoparticles for diabetic nephropathy therapy *via* size control and enhancement of renal cellular uptake. Theranostics 2019; 9(21): 6191-208.
http://dx.doi.org/10.7150/thno.37538 PMID: 31534545

[21]   Chen D, Han S, Zhu Y, Hu F, Wei Y, Wang G. Kidney-targeted drug delivery *via* rhein-loaded polyethyleneglycol-co-polycaprolactone-co-polyethyleneimine nanoparticles for diabetic nephropathy therapy. Int J Nanomedicine 2018; 13: 3507-27.
http://dx.doi.org/10.2147/IJN.S166445 PMID: 29950832

[22]   Cheng L, Chen Q, Pi R, Chen J. A research update on the therapeutic potential of rhein and its derivatives. Eur J Pharmacol 2021; 899(May): 173908.
http://dx.doi.org/10.1016/j.ejphar.2021.173908 PMID: 33515540

[23]   Bray F, Jemal A, Grey N, Ferlay J, Forman D. Global cancer transitions according to the Human Development Index (2008–2030): a population-based study. Lancet Oncol 2012; 13(8): 790-801.
http://dx.doi.org/10.1016/S1470-2045(12)70211-5 PMID: 22658655

[24]   Pucci C, Martinelli C, Ciofani G. Innovative approaches for cancer treatment: Current perspectives and new challenges. ecancermedicalscience. 2019; 13.

[25]   Iqbal J, Abbasi BA, Mahmood T, *et al.* Plant-derived anticancer agents: A green anticancer approach. Asian Pac J Trop Biomed 2017; 7(12): 1129-50.
http://dx.doi.org/10.1016/j.apjtb.2017.10.016

[26]   Zhu R, Zhang F, Peng Y, Xie T, Wang Y, Lan Y. Current progress in cancer treatment using nanomaterials. Front Oncol 2022; 12(Jul): 930125.
http://dx.doi.org/10.3389/fonc.2022.930125 PMID: 35912195

[27]   Bansode PA, Patil PV, Birajdar AR, Somasundaram I, Bachute MT, Rashinkar GS. Anticancer, antioxidant and antiangiogenic activities of nanoparticles of bioactive dietary nutraceuticals. ChemistrySelect 2019; 4(47): 13792-6.
http://dx.doi.org/10.1002/slct.201903946

[28]   Vinogradov S, Wei X. Cancer stem cells and drug resistance: the potential of nanomedicine. Nanomedicine (Lond) 2012; 7(4): 597-615.
http://dx.doi.org/10.2217/nnm.12.22 PMID: 22471722

[29]   Zivarpour P, Hallajzadeh J, Asemi Z, Sadoughi F, Sharifi M. Chitosan as possible inhibitory agents and delivery systems in leukemia. Cancer Cell Int 2021; 21(1): 544.
http://dx.doi.org/10.1186/s12935-021-02243-w PMID: 34663339

[30]   Quader S, Kataoka K. Nanomaterial-enabled cancer therapy. Mol Ther 2017; 25(7): 1501-13.
http://dx.doi.org/10.1016/j.ymthe.2017.04.026 PMID: 28532763

[31]   Hashemzaei M, Far AD, Yari A, *et al.* Anticancer and apoptosis-inducing effects of quercetin *in vitro* and *in vivo*. Oncol Rep 2017; 38(2): 819-28.
http://dx.doi.org/10.3892/or.2017.5766 PMID: 28677813

[32]   Redah Alassaif F, Redah Alassaif E, Rani Chavali S, Dhanapal J. Suppressing the growth of HL-60 acute myeloid leukemia cells by chitosan coated anthraquinone nanoparticles *in vitro*. Int J Polym Mater 2019; 68(14): 819-26.

http://dx.doi.org/10.1080/00914037.2018.1509340

[33]   Wang S, Chen T, Chen R, Hu Y, Chen M, Wang Y. Emodin loaded solid lipid nanoparticles: Preparation, characterization and antitumor activity studies. Int J Pharm 2012; 430(1-2): 238-46.
       http://dx.doi.org/10.1016/j.ijpharm.2012.03.027 PMID: 22465546

[34]   Liu H, Zhuang Y, Wang P, *et al.* Polymeric lipid hybrid nanoparticles as a delivery system enhance the antitumor effect of emodin *in vitro* and *in vivo*. J Pharm Sci 2021; 110(8): 2986-96.
       http://dx.doi.org/10.1016/j.xphs.2021.04.006 PMID: 33864779

[35]   Okon E, Gaweł-Bęben K, Jarzab A, Koch W, Kukula-Koch W, Wawruszak A. Therapeutic potential of 1,8-dihydroanthraquinone derivatives for breast cancer. Int J Mol Sci 2023; 24(21): 15789. http://dx.doi.org/10.3390/ijms242115789 PMID: 37958772

[36]   Abdelhameed RFA, Eltahawy NA, Nafie MS, *et al.* Rhein and Emodin anthraquinones of Cassia fistula leaves: HPTLC concurrent estimation, green synthesis of bimetallic ZnO-CuO NPs and anticancer activity against Panc-1 and OVCAR-3 cancer cells. Biomass Conversion and       Biorefinery       [Internet].       2024;       Available       from: https://www.researchgate.net/publication/380031909_Rhein_and_Emodin_anthraquinones _of_Cassia_fistula_leaves_HPTLC_concurrent_estimation_green_synthesis_of_bimetallic_ ZnO-CuO_NPs_and_anticancer_activity_against_Panc-1_and_OVCAR-3_cancer_cells

[37]   Liang J, Li F, Fang Y, *et al.* Synthesis, characterization and cytotoxicity studies of chitosan-coated tea polyphenols nanoparticles. Colloids Surf B Biointerfaces 2011; 82(2): 297-301.
       http://dx.doi.org/10.1016/j.colsurfb.2010.08.045 PMID: 20888740

[38]   Li G, Zhuang Y, Mu Q, Wang M, Fang Y. Preparation, characterization and aggregation behavior of amphiphilic chitosan derivative having poly (l-lactic acid) side chains. Carbohydr Polym 2008; 72(1): 60-6.
       http://dx.doi.org/10.1016/j.carbpol.2007.07.042

[39]   Jeevitha D, Amarnath K. Chitosan/PLA nanoparticles as a novel carrier for the delivery of anthraquinone: Synthesis, characterization and *in vitro* cytotoxicity evaluation. Colloids Surf B Biointerfaces 2013; 101: 126-34.
       http://dx.doi.org/10.1016/j.colsurfb.2012.06.019 PMID: 22796782

[40]   Liu D, Zhao S, Li J, Chen M, Wu L. The application of physical pretreatment in photodynamic therapy for skin diseases. Lasers Med Sci 2021; 36(7): 1369-77.
       http://dx.doi.org/10.1007/s10103-020-03233-6 PMID: 33404884

[41]   Niculescu AG, Grumezescu AM. therapy—an up-to-date review. Applied Sciences (Switzerland). 2021; 11(8).

[42]   Makowski M, Silva ÍC, Pais do Amaral C, Gonçalves S, Santos NC. Advances in lipid and metal nanoparticles for antimicrobial peptide delivery. Pharmaceutics 2019; 11(11): 588.
       http://dx.doi.org/10.3390/pharmaceutics11110588 PMID: 31717337

[43]   Hutchings MI, Truman AW, Wilkinson B. Antibiotics: past, present and future. Curr Opin Microbiol 2019; 51: 72-80.
       http://dx.doi.org/10.1016/j.mib.2019.10.008 PMID: 31733401

[44] Srivastava P, Kim Y, Cho H, Kim K. Synergistic action between copper oxide (CuO) nanoparticles and anthraquinone-2-carboxylic acid (AQ) against *Staphylococcus aureus.* Journal of Composites Science 2023; 7(4): 135.
http://dx.doi.org/10.3390/jcs7040135

[45] Lee NY, Ko WC, Hsueh PR. Nanoparticles in the treatment of infections caused by multidrug-resistant organisms. Front Pharmacol 2019; 10: 1153.
http://dx.doi.org/10.3389/fphar.2019.01153 PMID: 31636564

[46] Bai B, Saranya S, Dheepaasri V, *et al.* Biosynthesized copper oxide nanoparticles (CuO NPs) enhances the anti-biofilm efficacy against K. pneumoniae and S. aureus. J King Saud Univ Sci 2022; 34(6): 102120.
http://dx.doi.org/10.1016/j.jksus.2022.102120

[47] Ikada Y, Tsuji H. Biodegradable polyesters for medical and ecological applications. Macromol Rapid Commun 2000; 21(3): 117-32.
http://dx.doi.org/10.1002/(SICI)1521-3927(20000201)21:3<117::AID-MARC117>3.0.CO;2-X

[48] Jeevitha D, Br M, Ps P, Srividya S. Antibacterial activity of anthraquinone encapsulated chitosan/poly(lactic acid) NANOPARTICLES. Int J Pharm Bio Sci 2014; 5(4): 20-8.

[49] Ren H, Zhu C, Li Z, Yang W, Song E. Emodin-loaded magnesium silicate hollow nanocarriers for anti-angiogenesis treatment through inhibiting VEGF. Int J Mol Sci 2014; 15(9): 16936-48.
http://dx.doi.org/10.3390/ijms150916936 PMID: 25250911

[50] Ganapathy D, Shanmugam R, Sekar D. Current status of nanoparticles loaded medication in the management of diabetic retinopathy. J Evol Med Dent Sci 2020; 9(22): 1713-8.
http://dx.doi.org/10.14260/jemds/2020/376

[51] Kelly SJ, Halasz K, Smalling R, Sutariya V. Nanodelivery of doxorubicin for age-related macular degeneration. Drug Dev Ind Pharm 2019; 45(5): 715-23.
http://dx.doi.org/10.1080/03639045.2019.1569024 PMID: 30704311

# CHAPTER 6

# Anthraquinone-Based Nanomaterials: Emerging Strategies in Cancer Therapy

**Shilpa[1,#], Manoj Kumar[2,#], Ajay Kumar[3], Simrandeep Kaur[4], Sonika[1], Gulshan Kumar[5,\*] and Sukhvinder Dhiman[6,\*]**

[1]*Department of Chemistry, Maharishi Markandeshwar Engineering College, Maharishi Markandeshwar (Deemed to be University), Mullana, Haryana-133207, India*

[2]*Department of Microbiology, Guru Nanak Dev University, Amritsar-143005, Punjab, India*

[3]*University Centre for Research and Development (UCRD), Chandigarh University, Mohali-140413, Punjab, India*

[4]*Department of Botanical and Environmental Sciences, Guru Nanak Dev University, Amritsar, Punjab-143005, India*

[5]*Department of Chemistry, Banasthali University, Banasthali Newai 304022, Rajasthan, India*

[6]*Institute of Nano Science and Technology, Mohali, Punjab-140306, India*

**Abstract:** Anthraquinone-based nanomaterials (ANMs) have recently garnered considerable attention due to their potential applications in cancer therapy. Anthraquinones are characterized by their tricyclic aromatic structure, which can be modified and incorporated into nanomaterials for various therapeutic purposes in cancer treatment. There are several ways in which ANMs are currently being investigated for cancer therapy such as improved drug delivery systems, photothermal therapy, photodynamic therapy, imaging agents (anti-cancer agents), combination therapy, and biomarker detection. It is important to highlight that ongoing research in the field of nanomedicine is continuously advancing, and the exploration of ANMs for cancer therapy is a rapidly evolving area. Recent studies reported in the literature show that ANMs effectively inhibit cancer by reactive oxygen species formation, paraptosis, autophagy, apoptosis, and various cell signaling pathways. Furthermore, before these ANMs can be extensively utilized in cancer therapy, regulatory approval and clinical trials are mandatory steps in the process. The chapter outlines a comprehensive overview of ANMs, highlighting their potential use for therapeutic, cancer therapy, and various health products.

**\* Corresponding authors Gulshan Kumar and Sukhvinder Dhiman:** Department of Chemistry, Banasthali University, Banasthali Newai 304022, Rajasthan, India; Institute of Nano Science and Technology, Mohali, Punjab-140306, India; E-mails: dr.gulshan.kmr@gmail.com; sukhvinderdhimank@gmail.com
#These authors contributed equally

**Keywords:** Anthraquinone-based nanomaterials, Cancer therapy, Drug delivery systems, Nanomedicine, Photothermal therapy.

## INTRODUCTION

Anthraquinone, derived from Rubia and other higher plant sources, is a tricyclic compound reported for its antioxidant properties [1-3]. It finds a wide range of applications in both industrial and medical fields due to functional activity. These compounds are derivatives of 9,10-anthracenedione, which offer several therapeutic effects in humans exhibiting antibacterial, antitrypanosomal, and antineoplastic activities [4]. Additionally, they also inhibit lipid peroxidation, intestinal motility, and human telomerase activity. Anthraquinone-based nanomaterials (ANMs) have demonstrated hepatoprotective, renal calculi elimination, immunomodulatory, anti-inflammatory, calcium channel antagonistic, antithrombotic, and DNA binding properties in both animals and humans [5]. Recent studies have shown that ANMs isolated from flowers and roots exhibit *in vitro* antitumor effects on human cancer cell lines [6-7]. These findings underscore their potential to activate antiproliferative activity and induce cytotoxic effects on these cancer cells. ANMs are recognized for their ability to induce apoptosis in various human cancer cell lines, including lung adenocarcinoma A549, myelogenous leukemia HL-60, lung squamous carcinoma CH27, cervical carcinoma HeLa cells, neuroblastoma IMR-32, bladder cancer T24, and hepatoma HepG2 cells [8]. Anthraquinone also inhibits the uptake of glucose in tumor cells, leading to alterations in membrane-associated functions that ultimately induce cell death [9].

Cancer is the second leading cause of death, with projections indicating over 21.7 million new cases and 13 million deaths attributed to this disease by 2030 [10,11]. Contemporary anticancer drugs face a lot of challenges, encompassing concerns related to their selectivity, resistance, toxicity, and limited therapeutic window. The focus has shifted to developing highly selective and potent drugs that minimize side effects. These drugs target specific factors, including DNA, topoisomerases, telomerase, MMPs, kinases, ectonucleotidase, and quinone reductase [12]. The pharmaceutical industry is placing a growing focus on targeted therapies, which is propelling the progress of novel drug formulations. The introduction of nanotechnology has brought about a revolution in drug delivery, resulting in enhanced solubility, increased bioavailability, and a reduction in toxicity [13-17]. Both natural and synthetic nanoparticles are gaining recognition and popularity because of their capability to precisely target drug delivery and improve the controlled release of medications. Within the field of cancer treatment, nanoparticle-based delivery systems have made substantial advancements, effectively reducing toxicity and precision-based targeting only cancer cells [18-21]. Doxorubicin (DOX) is a modern Anthraquinone derivative that is being studied intensively in the field of nanotechnology to cure various tumors (Fig. **1**) [22]. A heterobifunctional linker was used to create an artificial recombinant chimeric polypeptide (CP)-based near-monodisperse nanoparticle containing DOX at Cys residues. In solid tumors, CP-DOX nanoparticles are reported to accumulate preferentially.

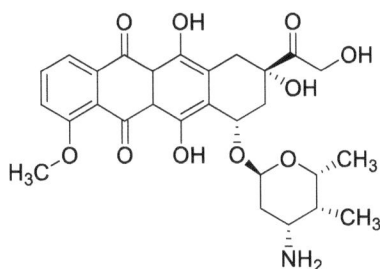

**Fig. (1).** Structure of Doxorubicin.

The scope of ANMs has gained significant attention in the field of cancer therapy due to their unique properties and potential applications due to their planar and aromatic ring structure making it easy to form various functionalized nanoscale structures. These nanomaterials have shown promise in several aspects of cancer therapy, and their scope includes, drug delivery systems, imaging agents, photothermal therapy (PTT), photodynamic therapy (PDT), reactive oxygen species (ROS) modulation, multimodal therapy, targeted therapies, overcoming drug resistance, nanotoxicology and biocompatibility. The potential of ANMs in cancer therapy is continuously evolving, as ongoing research strives to refine their properties and expand their applications. With increasing research efforts and the development of novel derivatives, the potential for these nanomaterials to bring about revolutionary changes in cancer therapy is on a steady path of expansion.

This chapter highlights the medicinal importance of natural anthraquinone-based nanoparticles specifically focusing on their role in anticancer activities. The creation of thoughtfully designed ANMs can pave the way for innovative approaches in chemotherapy. Various diverse nanoparticle formulations have been researched for cancer therapy and its applications in the healthcare sector. In addition to this, ongoing collaboration between scientists, engineers, healthcare professionals, and regulatory agencies will be vital in realizing these future applications.

# ANTHRAQUINONE-BASED NANOMATERIALS: CHEMISTRY AND SYNTHESIS

Anthraquinone-based nanomaterials (ANMs) are a class of nanomaterials that utilize anthraquinone as a core component. These materials have gained significant attention due to their unique properties and diverse applications in fields such as materials science, chemistry, and nanotechnology. Anthraquinone is a tricyclic aromatic compound with a quinone moiety, which imparts redox activity to these materials. This redox activity is one of the key factors that make anthraquinone-based nanomaterials desirable for various applications [22].

## Chemistry and Synthesis of Anthraquinones

Anthraquinone is typically synthesized through the oxidation of anthracene, a hydrocarbon compound. The oxidation process usually involves the use of strong oxidizing agents, such as chromates or permanganates, to convert anthracene into anthraquinone. The synthesis of ANMs involves the preparation of nanoscale structures or nanoparticles that incorporate anthraquinone as a core component. These nanomaterials can be customized with various properties and functionalities depending on the specific application [22-24]. To tailor anthraquinone for specific applications, it can be functionalized by introducing different chemical groups, such as alkyl, aryl, or functional moieties. This step is critical for enhancing solubility and compatibility with other materials. For the synthesis of organic material-based nanomaterials, different methods were adopted in the literature *e.g.* chemical reduction, sol-gel process, and surfactant-assisted synthesis, which are briefly discussed hereafter. In the chemical reduction process, we generally use a reducing agent with the anthraquinone precursor in a solvent. Apply heat or other energy sources to initiate the reduction reaction. Control reaction conditions to obtain nanoparticles of the desired size and morphology. In case of the sol-gel process, anthraquinone precursor is combined with a sol-gel solution, often containing metal alkoxides, allowing the solution to undergo hydrolysis and condensation reactions, and forming a solid gel. The gel can then be processed to obtain anthraquinone-based nanomaterials. In surfactant-assisted synthesis, surfactants are used to stabilize the anthraquinone nanoparticles and control their size and shape. The nanoparticles are precipitated from the solution by altering pH, and temperature, or adding a counter-ion. After the synthesis using the above-mentioned methods, it can be purified using centrifugation, filtration, or dialysis to remove impurities and unreacted components. Also, the handling and storage of anthraquinone-based nano-particles require appropriate conditions to prevent aggregation or degradation. Proper storage and handling are essential to maintain their properties over time.

# Chemical Properties and Structure of Anthraquinone-Based Nanomaterials

The chemical properties and structures of anthraquinone-based nanomaterials are critical to understanding their applications because, as compared to other nanomaterials, anthraquinone-based materials have distinct chemical and structural characteristics from their bulk counterparts. [23].

## Chemical Properties

Anthraquinone-based nanomaterials (ANMs) have been reported for their varied properties. ANMs exhibit redox activity, undergoing reversible two-electron/two-proton redox reactions, transforming between their reduced form (hydroquinone) and oxidized form (quinone). This property is utilized in diverse applications, such as energy storage and electrocatalysis. Based on their particular structure and synthesis methods, certain ANMs can display electrical conductivity. This feature is valuable for applications in electronic devices and sensors. ANMs are typically stable under normal conditions but can be sensitive to extreme chemical environments. The stability of these materials may be influenced by the presence of functional groups. The solubility of ANMs varies depending on their structure and functionalization. Some derivatives are soluble in common solvents, while others may require specialized solvents for dispersion. [24-26]

## Structure of Anthraquinone-Based Nanomaterials

### Core Structure

Anthraquinone consists of a tricyclic aromatic ring system with a quinone moiety. The core structure of ANMs is retained, and modifications are made to this core structure through functionalization or incorporation into nanocomposites. Anthraquinone can be functionalized by introducing various chemical groups, such as alkyl, aryl, or heteroatom-containing groups, to the core structure. These functional groups can influence the chemical reactivity and properties of the nanomaterials. The specific nanostructure of ANMs can vary. Depending on the synthesis method, these materials can exist as nanoparticles, nanowires, nanosheets, or other nanostructures. The size and morphology of these structures can be tailored for specific applications. Anthraquinone can be incorporated into polymer matrices to form nanocomposites. In these materials, anthraquinone nanoparticles or functionalized derivatives are dispersed within a polymer matrix, leading to unique structural properties and enhanced functionalities. In some cases, ANMs can exhibit crystalline structures with well-defined lattice arrangements. The crystal structure can impact their electrical and optical properties. The surface of ANMs may be decorated with functional groups that can interact with other molecules or surfaces. These functional groups can be used to modify the materials for specific applications. [24-26]

### *Anthraquinone-Based Nanomaterials*

Anthraquinones are organic compounds with a unique tricyclic aromatic structure, consisting of three interconnected benzene rings. This cyclic six-membered carbon structure with alternating single and double bonds gives them their aromatic properties. One of the benzene rings in anthraquinones contains a quinone functional group, responsible for their redox activity. The tricyclic aromatic system forms a conjugated system, resulting in the delocalization of π-electrons across the entire system, influencing its optical and electronic properties. The planar structure of anthraquinones is due to the conjugation of π-electrons, leading to a flat, extended structure. Anthraquinones and their derivatives have unique properties, including redox activity, optical properties, and a high degree of conjugation, making them valuable in various applications, including dyes, pigments, antioxidants, and organic synthesis [25, 26].

Here are a few examples of anthraquinones' tricyclic aromatic structure, such as 9,10-antracenedione, chrysophenol, *etc*. These substances are present in a variety of foods that humans eat, including peas, cabbage, lettuce, and beans. These foods contain approximately 0.04, 3.6, 5.9, and 36 mg of anthraquinones per kilogram of fresh vegetables, along with chrysophenol, aloe, and emodin, which serve as natural dyes and possess biological properties such as antibacterial, antifungal, antitumor, and diuretic effects. [27] Some of the other examples based on anthraquinones' tricyclic aromatic structure are given in Fig. (**2**).

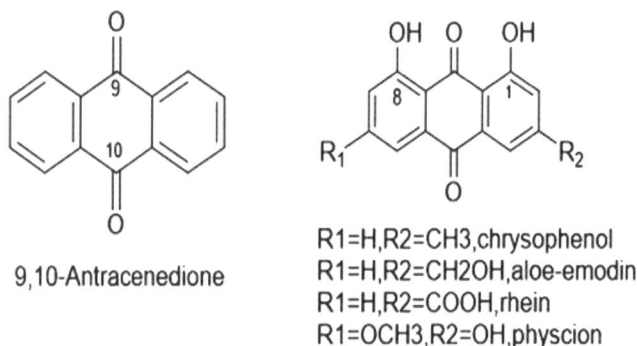

9,10-Antracenedione

R1=H,R2=CH3,chrysophenol
R1=H,R2=CH2OH,aloe-emodin
R1=H,R2=COOH,rhein
R1=OCH3,R2=OH,physcion

**Fig. (2).** Examples of anthraquinones based tricyclic aromatic structure.

## Anthraquinone-Based Nanomaterials: Modification and Incorporation into Nanomaterials

ANMs have gained significant attention in recent years due to their unique properties and potential applications in various fields, including electronics, energy storage, catalysis, and environmental remediation. Modification and incorporation of ANMs into other nanomaterials can further enhance their properties and expand their utility. Here's an overview of the modification and incorporation of ANMs into other nanomaterials [22, 23].

### Synthesis of Anthraquinone-Based Nanomaterials

The synthesis of ANMs can be achieved through various methods, including chemical reduction, hydrothermal synthesis, solvothermal methods, and electrochemical techniques. The ANMs can be modified through the attachment of functional groups, such as -OH, -COOH, -NH$_2$, *etc.*, which can enhance their solubility, stability, and reactivity. Also, the introduction of dopant atoms (*e.g.*, nitrogen, sulfur) into the anthraquinone structure can modify their electronic properties and improve their performance in various applications [22-26].

### *Incorporation into Nanomaterials*

ANMs can be incorporated into other nanomaterials to form composite materials. For example, they can be combined with graphene, carbon nanotubes, or metal oxides to create hybrid materials with unique properties. ANMs can be coated onto the surface of other materials to provide specific functionalities. This is often used in energy storage applications, such as lithium-ion batteries and supercapacitors. By blending ANMs with polymers or other matrices, nanocomposites can be created with improved mechanical, electrical, or thermal properties.

### Anthraquinone-Based Nanomaterials: Structural Characteristics for Therapeutic Adaptation

ANMs can be tailored for therapeutic applications by modifying their structural characteristics. These modifications are essential to ensure that these nanomaterials are well-suited for drug delivery, therapy, and related medical applications [28-33]. Some key structural properties, characteristics, and applications of anthraquinone-based nanomaterials are summarised in Table **1**.

**Table 1. Structural properties, characteristics, and applications of anthraquinone-based nanomaterials.**

| S. No. | Structural Property | Characteristics | Applications | References |
|---|---|---|---|---|
| 1. | Particle size and shape | Nanoparticles with diameters in the range of 1-100 nm for drug delivery due to their ability to efficiently transport drugs to target sites. The shapes of nanomaterials such as spheres, rods, or discs can influence how they interact with biological systems. | Therapeutic properties and antitumor activity | [34] |
| 2. | Surface chemistry and functionalization | Anthraquinone nanoparticles' surface can be tailored through functionalization with specific ligands, targeting molecules, or biomimetic coatings improving biocompatibility, enhanced cellular uptake, and targeted drug delivery to specific tissues or cells. | Pharmaceutical and therapeutic applications in drug delivery | [35] |
| 3. | Biodegradability | Anthraquinone nanoparticles having biodegradable properties can be designed to break down into non-toxic byproducts, reducing concerns about long-term accumulation in the body. | Pharmacological and therapeutic applications | [34, 38] |
| 4. | Drug loading and release mechanism | Anthraquinone nanoparticles can be designed to efficiently load on drugs or biomolecules to control the release rate by incorporating porous structures or responsive materials that release drugs in response to specific stimuli, such as pH, temperature, or enzymes. | Application in drug delivery related properties in treatment of various diseases | [36, 37] |

*(Table 1) cont.....*

| | | | | |
|---|---|---|---|---|
| 5. | Stability and aggregation | Anthraquinones nanoparticles should be stable in biological fluids, preventing aggregation and ensuring drug delivery efficiency. Surface modifications, like the addition of steric stabilizers or biocompatible coatings, can help maintain stability. | Application in drug stability and delivery | [37,38,39] |
| 6. | Surface charge | The surface charge of nanomaterials can affect their interaction with cells and biological membranes to improve the cellular uptake and biodistribution of therapeutic nanomaterials. | Application in drug delivery | [34,37] |
| 7. | Biocompatibility and toxicity | Anthraquinone nanoparticles should have low toxicity, high biocompatibility, and functional groups to be well-tolerated by biological systems. | Better therapeutic property and compatibility in therapeutic applications | [34,38] |
| 8. | Targeting and homing | Anthraquinone nanoparticles should have specific targeting capabilities, such as ligands or antibodies that recognize and bind to receptors or biomarkers on the target cells. This helps ensure that therapeutic agents are delivered precisely to the intended location. | Application in drug delivery and its potent activity | [34, 35, 38] |
| 9. | Imaging and monitoring | Adding contrast agents or fluorescent molecules can facilitate real-time monitoring and imaging of the therapeutic process, enabling physicians to track treatment progress. | Drug monitoring and tracking | [39] |

# CURRENT INVESTIGATION OF CANCER THERAPY

Current studies in the field of cancer therapy are actively investigating nanoparticle and nanocarriers-based novel ways in drug delivery systems to improve the targeted delivery of anti-cancer medications and reduce side effects. To increase accuracy and efficacy, researchers are honing light and heat in photodynamic treatment and photothermal therapy. Furthermore, personalized cancer therapy can be guided by unique genetic and molecular signs, and the identification of biomarkers is key to customizing treatments for individual patients. As will be discussed below, these multidisciplinary studies provide the potential for more effective and patient-centered cancer treatments. The various terminologies used in this section of the chapter has been thoroughly covered below.

## Drug Delivery Systems

Cancer treatment is a global challenge, with traditional chemotherapy causing harmful side effects by killing both cancer and healthy cells. Targeted therapies have improved precision, but challenges like side effects and drug resistance persist. Cancer is still a leading cause of death, spurring ongoing research for more precise and effective treatments [33]. In recent decades, nanotechnology has made significant strides in the field of medicine, particularly in the diagnosis, treatment, and targeted therapy of cancer. Nanoparticle (NP)-based drug delivery systems offer numerous benefits, including improved pharmacokinetics, precise tumor cell targeting, reduced side effects, and overcoming drug resistance [40, 41]. NPs are carefully chosen or designed based on their size and surface properties enhancing permeability and retention to match tumor characteristics. These nano-carriers deliver drugs to tumor cells, including traditional chemotherapy agents and nucleic acids, enabling both cytotoxic and gene therapy [42]. NPs also help encapsulate poorly soluble drugs and enhance drug half-life while protecting normal cells [43]. Furthermore, NP drug delivery systems have applications in immunotherapy and cancer ablation, potentially improving immunotherapy outcomes and reversing the tumor's immunosuppressive environment [44]. Drug delivery systems (DDS), particularly in the context of nanomedicine, have demonstrated a remarkable ability to overcome many of the physiological challenges associated with the delivery of hydrophobic chemotherapeutic agents to the tumor site while reducing associated toxicity and sparing healthy tissues [45].

NPs used in medical treatment must be carefully tailored in terms of size, shape, and surface properties to optimize drug delivery and therapeutic effectiveness [46]. NPs in the 10-100 nm diameter range are ideal for cancer therapy, effectively delivering drugs and leveraging the enhanced permeability and retention (EPR) effect [47]. Smaller particles risk leakage and clearance, while larger ones can be removed by immune cells. Surface characteristics, like hydrophilic coatings such as polyethylene glycol (PEG), enhance NP bioavailability, prolong circulation time, and improve tumor penetration. These features collectively influence NPs' therapeutic impact in cancer management [48]. Different kinds of NPs (*viz.* organic and inorganic nanoparticles) have made significant contributions to drug delivery in cancer therapy [49]. Liposomes, composed of an outer lipid layer and a drug-containing core, mimic cell characteristics and have undergone generations of development [50]. They effectively deliver anti-tumor drugs and nucleic acids, reducing cardiotoxicity and enhancing therapeutic outcomes. Polymer-based NPs, like polylactic-co-glycolic acid (PLGA) and dendrimers, offer biocompatible carriers with EPR effects. Polymeric micelles self-assemble and enable efficient delivery of anticancer drugs while increasing drug stability and circulation time. These organic NPs have expanded the range of options for cancer treatment [51, 52].

In the fight against cancer, photothermal therapy (PTT) is used, which transforms light energy into heat energy [53]. To improve the prognosis of cancer patients and reduce side effects on other body parts, research on PTTAs with high selectivity and precise therapy is still crucial. Owing to the body's systemic distribution of photothermal agents (PTAs) and improper laser irradiation during treatment, normal tissue may be harmed [54]. Strong photothermal conversion capacities were demonstrated when $Fe_3O_4$@PDA particles were introduced into the tumors cells of rats and further exposed them to light. The tumor surface heated up to 59.7°C instantly [55]. The simplest and most effective way to eradicate cancer cells without endangering healthy cells is to increase the quantity of PTAs at the tumor site [56]. Both raising the temperature of the tumor site and creating a concentration difference between normal and malignant tissue is feasible [57]. In photodynamic therapy, a material called a photosensitizer, sometimes referred to as a photosensitizing agent, is utilized to kill cancer cells in the presence of light. The source of the light could be a laser source, such as an LED. Photosensitizers produce oxygen radicals a type of oxygen that, when exposed to, destroys cells after absorbing specific light wavelengths [58, 59]. Photodynamic therapy can cause damage to the tumor's blood vessels, preventing the tumor from getting the blood it needs to grow. It may also trigger the immune system to combat tumor cells in other bodily areas. The photosensitizers in photodynamic therapy shield healthy cells from harm. Photodynamic therapy is beneficial for people with skin cancer. All three of the main cell death pathways *viz.* apoptosis, necrosis, and autophagy-associated cell death induced by PDT. In response to photodynamic therapy, apoptosis is commonly observed as a major mechanism of cell death [60-62]. Biological activities in living systems at the cellular and subcellular levels are seen and measured by molecular imaging. With its ability to measure target expression, targeted molecular imaging is a vital diagnostic and therapeutic tool. Targeted and non-targeted agents are the two types of imaging agents. Drugs that target a particular protein on or near tumor cells usually consist of three fundamental parts: a carrier to enhance pharmacokinetics, a targeting moiety, and a signaling segment for detection. Another approach to divide this category is into categories called "always-on" and "activatable". The more traditional "always-on" agents, which work by attaching a fluorescent fragment to peptides and antibodies, are easier to assemble and analyze. Activatable fluorescent agents can be roughly divided into two groups: (1) agents that are unique to a target cell and sensitive to external proteases, and (2) agents that are quenched until they are internalized and activated by endo lysosomal processing (lowered pH) within the target cell. There are no targeted agents that can only contain a signaling dye or that can be pure optical dye [63, 64]. The cornerstone of cancer treatment is combination therapy, a therapeutic approach that combines two or more therapeutic drugs [65]. This strategy may reduce medication resistance in addition to offering therapeutic anti-cancer effects such as stopping mitotically active cells, reducing the amount of cancer stem cells, causing apoptosis, and lowering tumor growth and spreading potential. Drugs that target different pathways can be combined to provide a potentiation or synergistic effect that has a notable anti-cancer impact. Combination therapy has the potential to kill cancer cells while preventing the harmful effects on healthy cells [66]. Combination medicine has its risks because multiple pathways will be targeted if chemotherapeutic drugs are used, and the toxicity will be greatly reduced. Combination therapy results in a more successful treatment response in fewer cycles [67]. Genes, proteins, or other substances that can be tested to provide crucial information about an individual's cancer diagnosis are known as cancer biomarkers [68, 69]. For some people, biomarker testing may not be beneficial. However, biomarker testing may offer helpful information that could influence therapy choices for individuals with a variety of cancer types [70]. Every cancer cell is unique. Different gene alterations that promote cancer cell growth can exist in cancer cells even in individuals with the same type of cancer (*e.g.*, lung or breast cancer) [71]. These alterations may also affect how the cancer reacts to particular types of treatment, such as immunotherapy and targeted drug therapy, which are most effective when cancer cells possess unique properties that distinguish them from healthy cells. The most effective cancer treatment options can be identified with the use of various biomarker testing, including some tests that evaluate specific proteins or other types of indicators, many cancer biomarker tests search for gene changes in the cancer cells. Many biomarker tests are designed specifically for people with certain forms of cancer [72-74].

Hereafter are a few examples of anthraquinone-based derivates for the utilization of cancer treatment. The drug delivery method utilizing amino-anthraquinone compounds was synthesized by Arrousse and colleagues utilizing **1-5** (Fig. **3**). Their research has revealed that various anthraquinones possess the ability to induce cell cycle arrest and apoptosis in prostate cancer cells, effectively impeding their proliferation and growth. Notably, these anthraquinones exhibit minimal harm to normal cells, rendering them an attractive candidate for the development of innovative anti-cancer pharmaceuticals. An analysis of the drugs' pharmacokinetics indicates that the modeled compounds fall within an acceptable parameter range and exhibit significant bioactivity, with one exception - product **2** due to its toxicity. Furthermore, the drug selection process involved the criteria set forth by Lipinski, Ghose, Egan, and Veber [75]. The PC-3 Cell line exhibits the same sequence of reactivity in molecular docking investigations.

**Fig. (3).** Structure of anthraquinone-based derivates 1-5.

The efficacy of photothermal therapy (PTT), a treatment method that employs light to generate heat and destroy cancer cells, relies on the disparity in the ability of tumor tissue and normal tissue to convert light into heat. This difference in photothermal conversion capacity dictates the selective effect of PTT. It can be categorized into two primary strategies: first, by increasing the concentration difference between normal tissues and tumor sites, and second, by conferring upon photothermal agents (PTAs) the capability to autonomously regulate photothermal conversion. These strategies play a crucial role in enhancing the effectiveness of PTT in the context of cancer therapy [76].

Chen and their research team developed a therapeutic and diagnostic agent like doxorubicin (6) (Fig. **1**) and others have been employed for treating various cancers such as breast and ovarian cancer and prostate cancer, respectively [77]. These agents serve a dual purpose by both diagnosing and treating cancer.

Perlak and colleagues have developed an anthraquinone-based photodynamic therapy for its potential as an anticancer agent [78]. Their research delves into various photosensitizing drugs (Fig. **4**), including emodin (7) [79] (utilized in gall bladder treatment), aloe-emodin (8) [80] (applied in skin cancer therapy), rubiadin (9) [81] (targeting breast cancer cells), and hypericin (10) [82] (used for breast, cervical, and other cancers). Notably, among the anthraquinone derivatives, hypericin is undergoing clinical investigations. It is worth highlighting that anthraquinones are currently under active investigation for potential clinical applications, and there is growing anticipation that they, including hypericin, may find future utility as natural photosensitizers in cancer treatment. This underlines the ongoing research and potential promise of anthraquinones and hypericin in the field of cancer therapy.

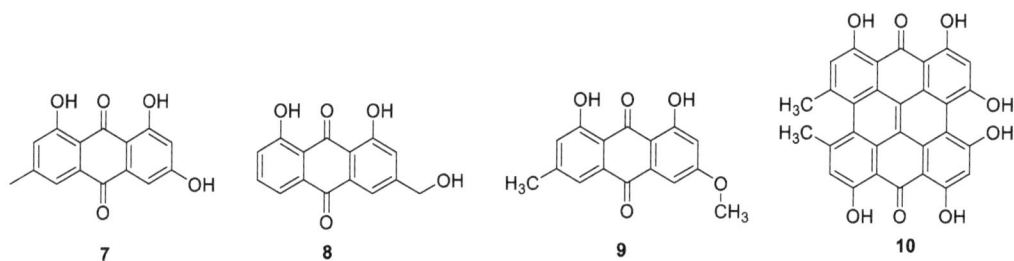

**Fig. (4).** Structure of anthraquinone-based derivates 7-10.

Malik and colleagues have developed anticancer and imaging drugs using anthraquinone as their foundational component [83]. The anthraquinone structure forms the basis of several anticancer drugs that have proven effective in halting the spread of cancer. These compounds primarily exert their effects by targeting essential cellular proteins. Chemical agents based on anthraquinones play a major role in the composition of chemotherapy medications, which are commonly employed for cancer treatment.

Notable anthraquinone-derived drugs include daunorubicin (11) (used for HeLa and A549 cell lines), idarubicin (12) (used for treating acute myelogenous leukemia, a form of white blood cell cancer), epirubicin (13) (effective against breast cancer), valrubicin (14) (applied in the treatment of bladder cancer), and pixantrone (15) (used for YC-murine lymphoma and L1210-murine leukemia) [84]. Understanding the profile and potential of these novel anthraquinone-based compounds involved conducting comprehensive studies, such as toxicity assessments and investigations into pharmacokinetics and pharmacodynamics (Fig. **5**).

Opportunities for advancing the field of anthraquinone-based anticancer medications are vast and encompass targeted drug delivery, synergistic combinations with other anticancer drugs, and the utilization of computational tools for enhanced drug development. These anthraquinone-based anticancer medications hold significant promise for integration into combination therapies alongside recently approved anticancer drugs that target distinct biological receptors, paving the way for more effective cancer treatments.

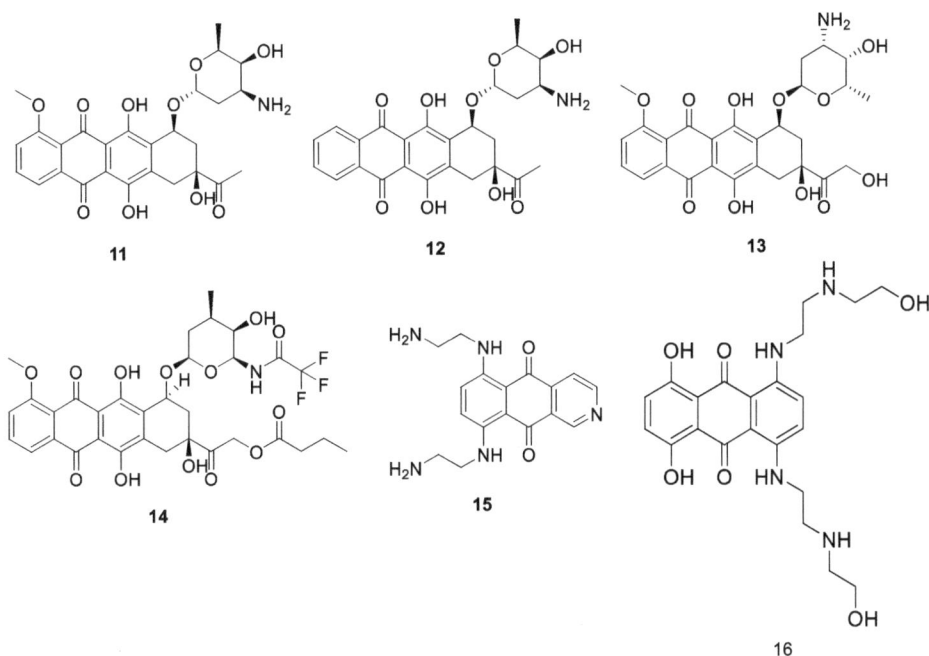

**Fig. (5)**. Structure of anthraquinone-based derivates 11-16.

Nisiewicz and colleagues have developed a gravimetric and voltammetric responsive glycine-leucine-anthraquinone sensor for the detection of active-matrix metalloproteinase (MMP)-2 in plasma through the cleavage of glycine-leucine peptide bond and antigen-antibody recognition. The sensors have a 3D structure involving a cationic polymer that ensures effective immobilization of molecules for biosensing. They use quartz crystal microbalance and voltammetry for detection. The results show great stability, a wide detection range (from 2.0 pg mL$^{-1}$ to 5.0 mg mL$^{-1}$), and a low detection limit (around 10 fg mL$^{-1}$). They perform well in real sample analysis, making them promising for early cancer diagnosis [85]. In cancer diagnosis and treatment, endopeptidases play a vital role in degrading the extracellular matrix (ECM) and serve as biomarkers. For the highly sensitive detection of active-matrix metalloproteinase (MMP)-2 in plasma, both gravimetric and voltammetric sensors are employed. Voltammetric techniques were utilized to detect the active form of MMP-2 due to its ability to enzymatically cleave the glycine-leucine peptide bond. Conversely, gravimetric detection of MMP-2 relies on the interaction between the antigen and the antibody. Active MMP-2, a significant cancer biomarker employed in the diagnosis and treatment of prostate and breast cancer, is analogous to MMP-9, another biomarker used for the treatment of breast and lung cancer [86, 87].

Babichev and colleagues have identified a promising and selective therapeutic approach for leiomyosarcoma through a systematic cell line screening process. This approach involves the utilization of PI3K and mTOR protein inhibitors, either as stand-alone treatments or in combination with standard therapies. Notably, within these pathways, there are two specific inhibitors, BEZ235, and BKM120, which exhibit the potential to impede the growth of leiomyosarcoma in *in vitro* settings. Furthermore, the use of doxorubicin (DOX) and phosphatidylinositol-3-kinase protein inhibitors in ovarian and breast cancer, as well as the combination of emodin with paclitaxel in the context of breast cancer cells, presents additional combination therapies of interest in the field of cancer treatment [88].

One of the benefits of employing nanoparticles is that they can be made to have specific chemical functions and can enter tumor cells instead of healthy cells. Numerous nanoparticle compositions have been evaluated for simultaneous *in vivo* imaging and enhanced cancer therapy efficacy. In the field of nanotechnology, the modern anthraquinone derivative DOX (6) is being studied to treat a variety of cancers [89]. Tf-loaded DOX-modified lipid-coated poly D, L-lactic-co-glycolic acid nanoparticles (PLGA-NP) were used to prevent lung cancer. The inhibitory effect of DOX-loaded TF-LPs on A549 cells and the inhibition of tumor spheroid formation were greater than those of DOX-loaded lipid-coated PLGA-NPs and PLGA-NPs [90]. Additionally, the inhibitory effect was improved on tumor growth in A549 *in vivo* tumor-bearing mice. Several amphiphilic PEG derivatives were produced to load polyelectrolyte nanoparticles with mitoxantrone [91]. The largest-sized CHP nanoparticles loaded with mitoxantrone (16) were the most harmful to bladder cancer cells [91].

The most significant naturally occurring dyes found in the class of anthraquinones are physicon, Damnacanthal, Nordamnacanthal, Chrysophanol, Danthron, Rubiadin, Quinalizarin, N-Ethylhydroxy-doxorubicin, CP-DOX, and DOX-loaded TF-LPs, as mentioned in Table **2**. Physicon, present in several plants such as *Rheum palmatum*, has exhibited antiproliferative ability against human breast cancer cells and colorectal cancer cells HCT116 [92], whereas DAM (Damnacanthal) exhibited cytotoxic action against small-cell lung cancer and breast cancer cell lines. DAM (Damnacanthal) and NDAM (Nordamnacanthal) are unique from other anthraquinones that have been described in terms of their ability to function as anti-HIV agents and their ability to have an anti-cancer impact on human B-lymphoblastoid cells. (Jasril *et al.*, 2003). Similarly, chrysophanol and rubiadin have been shown to have a number of advantageous biological characteristics over the years, such as their anticancer in MCF-7 cell lines, HeLa cells, hypolipidemic, and hepatoprotective characteristics, *etc.* [93] and all the treatments like thyroid carcinoma, metastatic gastric carcinoma, acute myeloblastic leukemia, *etc.* can be curable with doxorubicin. [94].

**Table 2. Summary of a few anthraquinone-based nanocarriers.**

| S. No. | Compound | Structure | Source | Activity Against Tumor | Mode of Action | References |
|---|---|---|---|---|---|---|
| 1 | Mitoxantrone | | -- | Bladder cancer cell (MB49) | hydrophobic substitution degree in the polymer. | [95] |
| 2 | Physcion | | *Rubia cordifolia* [64]. | HeLa cell line, HCT116 cell line | down-regulating of Bcl-2 expression, up-regulating of Bax expression. | [96] |

*(Table 2) cont.....*

| | | | | | | |
|---|---|---|---|---|---|---|
| 3 | Damnacanthal | | *Morinda citrifolia* [67]. | colorectal cancer cells | Blocking of C/EBPβ by shRNA. | [97] |
| 4 | Nordamnacanthal | | *M. citrifolia* [68]. | MCF-7 | IFN-γ and IL-2contribute. | [98] |
| 5 | Chrysophanol | | Rhubarb [71], *Cassia* sp. [57] | MCF-7 | NF-κB/cyclin D1 and NF-κB/Bcl-2 signaling | [99] |
| 6 | Danthron | | *R. palmatum* [69] | SNU-1 cells | MMP-9 mRNA expression | [100] |
| 7 | Rubiadin | | *R. cordifolia* [78] | HeLa cell lines | enzyme EGFR tyrosine kinase | [101] |
| 8 | DOX-loaded TF-LPs | -- | -- | A549 cells | inhibition of tumor spheroid growth | [102] |
| 9 | Daunorubicin (DNR, 5)-based poly (lactic-co-glycolic acid) (PLGA)-poly-l-lysine (PLL)-polyethylene glycol (PEG)-transferrin (Tf ) | -- | -- | (leukemia K562 cells) | down-regulating of Bcl-2 expression | [103] |

*(Table 2) cont.....*

| 10 | Quinalizarin | | -- | CK2 cell line | protein kinase CK2 | [104] |
|---|---|---|---|---|---|---|
| 11 | N-Ethylhydroxy-doxorubicin | | -- | DNA (PDB ID 385D) | DNA-interacting capacity | [105] |
| 12 | 1,4-Diamino-5,8-dihydroxy anthracene-9,10-dione | | -- | CK2 kinase | protein kinase CK2 | [106] |
| 13 | Tf-DOX | -- | -- | lung cancer | Tf receptor-targeting function | [107] |
| 14 | 9,10-Dioxo-9,10-dihydro anthracene-2,6,-disulfonic acid | | -- | Cytosolic5' - nucleotidas e cN-II | potential cN-II inhibitors, | [108] |
| 15 | CP-DOX | -- | -- | mild hypertherm ia of solid tumors | -- | [109] |
| 16 | DOX-FA-MNPs | -- | -- | human ovarian cancer cell lines | sharp decreases in the levels of bcl-2 | [110] |

# RECENT ADVANCEMENTS AND PROMISING RESULTS

Recent advancements in the application of anthraquinone derivatives for cancer treatment have gained significant attention, owing to their distinct structural features and potent biological activities.

## Current Trends in Nanomedicine Employed in Cancer Therapy

Over the past few years, nanomedicines have gained a lot of interest, due to their higher efficacy in combating pharmacokinetic challenges and formulating more profound treatments with low toxicity profiles for the treatment of several types of cancer (Fig. **6**). The use of nanomedicine modalities as a drug delivery approach enables ameliorated drug targeting and biodistribution. It also inhibits the drug solubility, drug degradation, and multi-drug resistance, thus, improving the drug performance [111]. Numerous nanomedicines have been designed and approved by EMA (European Medicines Agency) and FDA (US Food and Drug Administration) for drug delivery in the treatment of cancer such as Doxil, NK105, NanoTherm Abraxane 1, *etc.* Doxil – is a doxorubicin PEGylated liposomes FDA-approved drug also named, Lipodox R$^©$, Evacet R$^©$, and Caelyx R$^©$ widely used in cancer therapeutics mainly for ovarian cancer and AIDS-associated Kaposi's sarcoma. It functions by the generation of free radicals and ROS, causing cellular oxidative damage. It also binds to the DNA, leading to dissociation of topoisomerase, thus, inducing the activation of DNA repair mechanisms. This liposomic drug has reduced toxicity and renders fewer side effects, therefore, remains a safer alternative to conventional anticancer drugs [112]. DaunoXome R$^©$ is an anthracycline daunorubicin liposomal FDA-approved chemotherapeutic drug employed in HIV-associated Kaposi's sarcoma and cancer. It is assumed to operate by the enhanced permeability and retention (EPR) effect across tumor tissues but its mode of action is still unclear.

Another drug majorly utilized in treating pancreatic tumors is Onivyde R$^©$, which is also an FDA-approved anticancer drug, nanoformulated with irinotecan liposome, having better biodistribution, targeted accumulation, and lesser adverse effects [113]. A daunorubicin and cytarabine-coated liposomal drug also referred to as Vyxeos R$^©$, is an FDA-approved in the year 2017 used as an antileukemic drug, with certain side effects [114]. Apart from liposomic nanomedicines, certain polymeric nanomedicines are also being utilized in cancer therapeutics such as Neulasta R$^©$, a Polyethylene glycol conjugated filgrastim, also called PEGylated filgrastim used in chemotherapy-induced febrile neutropenia and infections resulting from neutrophil's deprivation [115]. A palliative prostate cancer, slow-releasing drug, Eligard R$^©$ was FDA-approved in 2002 having Atrigel-based polymeric nanosuspension available in injectable form. It is composed of Lupron, also known as Leuprolide acetate, a synthetic variant of Gonadotrophin releasing a hormone that modulates the gonadotropin levels as well as other associated hormone levels, and Atrigel which is a polymeric mixture of polylactic and polyglycolide used as a drug delivery system [116]. Abraxane R$^©$ is a protein-derived, FDA-approved drug (2005, 2012, and 2013) formulated from paclitaxel encapsulated with albumin nanoparticles. It is being broadly utilized in the treatment strategies of pancreatic, breast, and lung carcinoma [117-119]. It functions by hindering the microtubule's movement and influences the process of mitosis in the cancer cells [120].

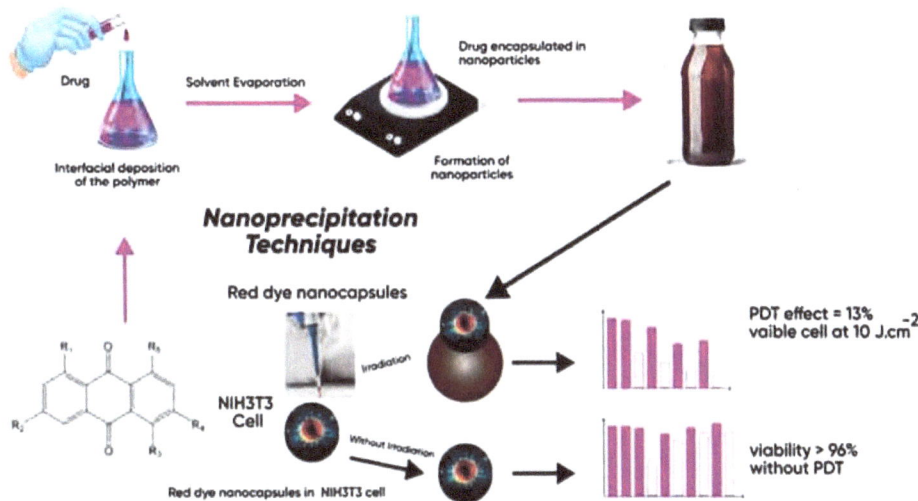

**Fig. (6).** Graphical representation of the mode of action of anthraquinone-nanoparticle structure.

## Evolving Strategies in the Use of Nanomaterials/Nanomedicines

Nanomedicines have been highly investigated for treatment regimens and prognostic clinical uses including heat therapy, functional imaging, and drug and gene deliveries. In combination with synergistic therapeutic techniques, nanomedicines have facilitated the integrated theragnostic methods for tracking the treatment responses, in various preclinical analyses [121]. For enhancing the efficacy of neoadjuvant therapy and complete response to tumor, an active chemoradiotherapy targeted nanosystem was formulated by amalgamating LRP-1 -B5 peptide, 5-FU, and Cy7 fluorophore in the human serum albumin-based nanoparticles. Studies have fabricated a neoadjuvant therapeutic method, for counteracting drug resistance and reversion of tumors, by integrating anti-PD1 (immune checkpoint inhibitor) with an immune-stimulated nanoparticle [122]. Lipid-formulated immune-stimulated nanoparticles were co-embodied with STING agonist and TLR agonist in a definite amount [123]. In addition, targeted gene delivery with the help of nanoparticles was investigated by Yen *et al*. They delineated that poly(c-4-(((2-(piperidine-1-yl) ethyl) amino) methyl) benzyl-L-glutamate) based compact plasmid DNA nanocomplexes encapsulated with hyaluronic acid contributes to 28-36% surge in the transfection of CD44 gene, in the human embryonic stem cells [124]. Furthermore, several studies revealed that tumor-affiliated fibroblasts represent the target site for monitoring the effect of chemotherapeutic drugs. This was supported by research on the combined drug (cisplatin and rapamycin) formulated poly (lactic-co-glycolic acid) nano complexes, which render apoptotic effects in *in vitro* studies on human melanoma cells whereas *in vivo* studies in xenograft cancerous model predicted the growth inhibition potential of nano complexed drugs attributed to the modulated vasculature of tumor and increased permeation of nano complexed drug [125]. Recently, extensive research has focused on CRISPR gene-edited nanomedicines for gene inhibition in cancer therapeutics. Several nanomedicines such as PEG (Polyethylene glycol) encapsulated nanoparticles, lipid nanoparticles, and poly beta-amino esters (PBAE) nanoparticles are principally employed in delivering CRISPR-associated tools in preclinical research [126-129].

## ANTICANCER MECHANISMS AND SIGNALING PATHWAYS OF ANMs

Anthraquinones have been previously reported for their therapeutic potential against different cancers. This property of anthraquinone can be utilized more judiciously when these compounds are incorporated into nanomaterials for targeted drug delivery in cancer therapy [130]. The exact mechanism by which ANMs inhibit or kill cancer cells is still unknown and much research has to be carried out in this area [131]. The use of anthraquinone in combination with nano-materials can enhance targeted drug delivery in tumor tissues because of its enhanced permeability and retention (EPR) effect. This effect reduces the adverse effect of anthraquinone on healthy tissues or cells [132]. In literature, many of the studies were carried out, some anthraquinone-based nano-particles can generate reactive oxygen species (ROS) causing cellular damage leading to cell death and oxidative stress. The targeted delivery of anthraquinone-based nano-particles to cancer cells can cause localized ROS production leading to cytotoxic effects on tumor growth and stress-related pathways leading to apoptosis [133]. Some of the ANMs can be used for specifically delivering to cancer cells causing activation of various cell cycle checkpoints and DNA repair pathways. In case, DNA damage is too severe to repair, then it can lead to programmed cell death or apoptosis in cancer cells, thus inhibiting their growth [130, 134]. Some of these ANMs such as etoposide, can enhance efficacy and delivery causing the inhibition of topoisomerases by blocking

DNA replication as well as repair mechanisms causing DNA strand breaks followed by cell death of tumor cells [135-136].

Some of the reports have also shown immunomodulatory effects enhancing the body's immune response activating T-cells and natural killer cells, and destroying cancer cells or tumor formation. There are many mechanisms by which ANMs act on cancer cells or tumor formation depending on type and individual patient characteristics [130, 137]. Further research and clinical trials are ongoing to better understand and optimize the use of these materials in cancer therapy. The inhibition of cancer by ANMs involves the modulation of several signaling pathways within cancer cells. These pathways play crucial roles in cell survival, proliferation, and metastasis [138].

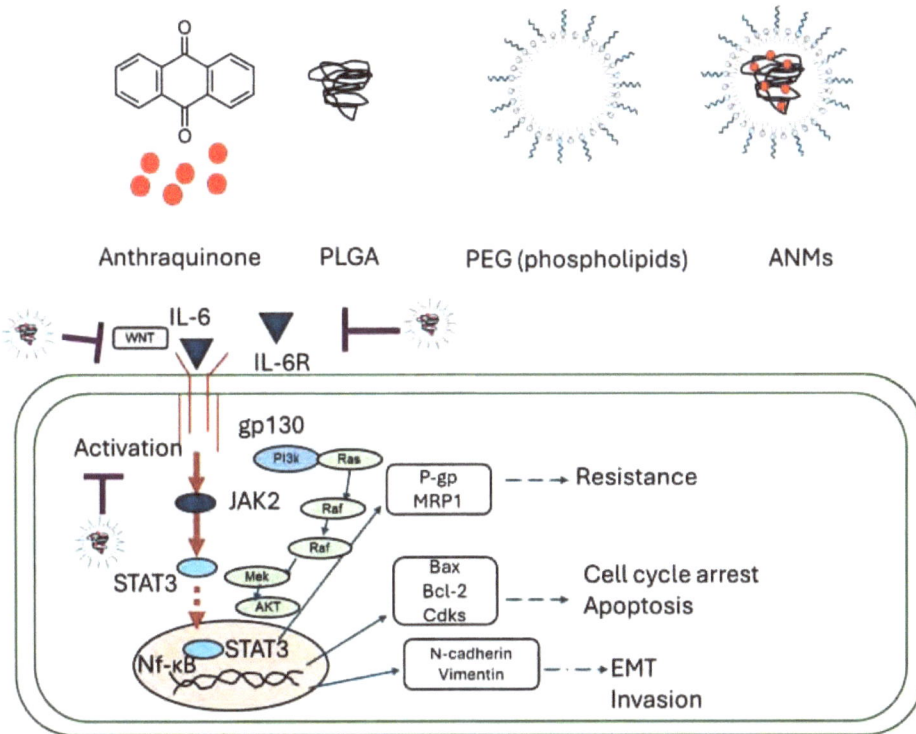

**Fig. (7).** Modulation of different signaling pathways with the treatment of Anthraquinone-Based Nanomaterials (ANMs) in cancer cells.

In literature studies, nuclear factor-kappa B (NF-κB) is reported to play an important role as a transcription factor in inflammation and cell survival, and its suppression using ANMs can decrease cell survival and reduce inflammatory responses in the tumor microenvironment [134, 139]. Similarly, mitogen-activated protein kinase/extracellular signal-regulated kinases (MAPK/ERK) are reported to promote cell proliferation and differentiation of cancer cells, whereas anthraquinone-based nano-particles can interfere in these pathways inhibiting decreased cell proliferation and tumor growth [130, 140]. In addition to this, the protein kinase B (AKT) pathway and phosphoinositide 3-kinase (PI3K) are reported to promote cell survival and growth in cancerous cell formation, whereas ANMs are reported to inhibit protein kinase B (AKT) pathway and phosphoinositide 3-kinase (PI3K). Inhibition of PI3K/AKT signaling can lead to cell cycle arrest and apoptosis in cancer cells (Fig. **7**) [130, 141, 142].

Similarly, the uncontrolled growth in many of the cancer cells is due to dysregulation of the Wnt/β-catenin signaling pathway. ANMs can be used to block the Wnt/β-catenin signaling pathway inhibiting cancer cells by controlling cell growth, differentiation, and apoptosis (Fig. **7**). Similarly, the Janus kinase/signal transducer and activator of transcription (JAK/STAT) pathway are involved in cell growth and immune responses [130, 133, 143]. ANMs can interfere with this pathway, affecting tumor cell proliferation and promoting immune cell-mediated tumor cell death. In addition to these pathways, ANMs can inhibit the process of angiogenesis, mainly responsible for blood vessel formation required for promoting tumor growth and metastasis. This is often achieved by targeting vascular endothelial growth factor (VEGF) and its receptors [144]. Some ANMs can induce DNA damage, activating DNA damage response pathways such as the ATM/ATR pathways and p53. This leads to cell cycle arrest and apoptosis. Also, ANMs can inhibit autophagy, making cancer cells more susceptible to treatment-induced stress and apoptosis [145].

The inhibition of various signaling pathways generally depends on the type of ANMs, their formulation, and the cancer type [130, 146]. Moreover, the precise mechanisms of action are still an area of active research, and ongoing clinical trials aim to establish the effectiveness and safety of these nanoparticles in cancer therapy.

## REGULATORY APPROVAL AND CLINICAL TRIALS

ANMs can be utilized as valuable compounds in pharmaceutical-based industries. They can serve as lead structures for future drug development exhibiting a diverse range of pharmacological properties such as laxative properties, anti-cancer potential, anti-inflammatory and anti-arthritic activities, antifungal and antibacterial capabilities, antiviral effects, and neuroprotective benefits [147-149].

There are various mandatory steps and challenges to be followed for testing anthraquinone-based nanoparticles for pharmaceutical applications.

## Mandatory Steps and Challenges

The use of ANMs for the treatment of various cancers and therapeutic purposes requires proper regulatory approval and clinical trials involving a series of steps and rigorous testing to ensure their safety and efficacy [150]. For clinical trials, extensive preclinical research has to be conducted to study the properties and potential uses of ANMs. The first step is generally the material selection for making ANMs with the desired properties, such as size, surface chemistry, and stability accessing therapeutic or diagnostic purposes. This is followed by preclinical research to evaluate the safety and efficacy of the nanoparticles including *in vitro* and *in vivo* studies, such as cytotoxicity assessments, biodistribution studies, and testing in relevant disease models [151]. Further studies include optimizing the formulation of the nanoparticles enhancing their stability, bioavailability, and targeting capabilities [152]. If the drug passes these preclinical trials then the research institution can file an investigational new drug (IND) application to regulatory authorities such as the US Food and Drug Administration (FDA) [153]. There are different phases of a clinical trial, the first is phase 1 trials involving healthy volunteers to assess the dosage and safety of the new drug to be tested understanding its drug profile, potential side effects, and metabolism. After the successful completion of phase 1, phase 2 trials expand to a larger group of patients aiming at the drug's effectiveness, dosing, and potential side effects in patients with the target condition. This is followed by phase 3 trials experimenting larger group of patients across different sites monitoring the drug's effectiveness and side effects in a broader patient population. This trial mainly accesses the drug's benefit-to-risk ratio. If Phase 3 trials demonstrate the drug's efficacy and safety, the research authority or pharma industry can submit a new drug application (NDA) to the regulatory authority having all data and gathered information during development and clinical trial phases [154]. Furthermore, regulatory bodies review the NDA to determine approval of the drug for marketing after the evaluation of the drug's safety, efficacy, and manufacturing processes [153]. Also, there is a need to submit regulatory applications and interact with regulatory agencies to gain approval for clinical trials and, if successful, for marketing.

The process of making ANMs can face many challenges such as ensuring their biocompatibility and adverse effects on the human body is a significant challenge. Also, the toxicity profile and concerns associated with ANMs are crucial for assessing their long-term effects and accumulation in tissues. The studies related to nano-particles require newer guidelines and regulations [155]. It requires designing effective clinical trials for novel applications and complex diseases requires careful consideration. Furthermore, developing cost-effective manufacturing processes and ensuring the commercial viability of the product can be challenging, especially for niche or highly specialized applications [156]. Collaboration with experts in nanotechnology, medicine, and regulatory affairs can help overcome these challenges and facilitate the safe and effective use of these nanoparticles in various applications.

## FUTURE PROSPECTS IN CLINICAL APPLICATION

ANMs hold promise in various fields as mentioned below. Anthraquinone-based nanoparticles can be used for controlled drug release and targeting plating an important role in the treatment of cancer [157, 158]. These nanoparticles can used for the treatment of other diseases as they can encapsulate and deliver a wide range of therapeutic agents, and improve the bioavailability and pharmacokinetics [158]. Anthraquinones have been reported for their antioxidant and anti-inflammatory properties utilized in developing novel therapies for managing oxidative stress and inflammation in diseases such as autoimmune disorders and neurodegenerative disorders [159, 160]. ANMs could be engineered for enhanced imaging capabilities, enabling more accurate and detailed diagnostics [161, 162].

Some of the nanoparticles can be used to remove environmental pollutants present in water and soil using processes such as degradation, photocatalysis, and adsorption [163]. It can also be used for making sensors to detect specific analytes, such as toxins, pathogens, or chemical substances [164, 165]. In addition to this, it can be used for studying biological processes and drug interactions at the cellular and molecular levels. They may enable advancements in drug discovery and the understanding of complex biological mechanisms [166]. Further, ANMs can be used in healthcare which must be supported by safety assessments and regulatory compliance. In addition to this, suitable coordination is required between scientists, healthcare professionals, and regulatory agencies so that ANMs can be utilized as future therapeutic for cancer therapy.

## CONCLUSION

Anthraquinones are a class of organic compounds that are naturally occurring in plants and have been used for various medicinal purposes for centuries. Recently, the incorporation of anthraquinone-based compounds into nanomaterials can pave the way for enhanced therapeutic applications, making them promising candidates for cancer treatment. ANMs can be designed to encapsulate anticancer drugs, enabling targeted drug delivery to tumor sites reducing systemic exposure of healthy tissues to chemotherapy agents, and minimizing its side effects while enhancing the drug's efficacy against cancer cells. Many anticancer drugs, including some anthraquinone derivatives, have limited solubility in water, which can hinder their bioavailability. Nanomaterials can improve the solubility of these compounds, making them more suitable for intravenous administration and ensuring more efficient drug delivery. ANMs can be engineered to release drugs in a controlled and sustained manner. This allows for a prolonged drug exposure to cancer cells and reduces the frequency of drug administration, improving patient comfort and treatment adherence. ANMs can also be used for cancer imaging and diagnostics. They can be functionalized with imaging agents, such as fluorescent dyes or magnetic nanoparticles, to enable real-time visualization of tumors and monitor treatment response. Anthraquinones have shown anticancer properties through multiple mechanisms, including apoptosis induction, cell cycle arrest, and inhibition of angiogenesis. Combining ANMs with other therapeutic agents or treatments, such as radiation therapy, can potentially lead to synergistic effects and improved cancer outcomes. Resistance to chemotherapy is a significant challenge in cancer treatment. ANMs may offer a solution by enabling the delivery of multiple drugs or drug combinations to overcome drug resistance mechanisms in cancer cells. The potential of ANMs in cancer therapy is promising, their development and clinical translation are ongoing areas of research. Researchers are actively investigating their safety, efficacy, and scalability for clinical use. Additionally, the specific anthraquinone derivatives and nanomaterials used, as well as the treatment strategies employed, can vary depending on the type of cancer and individual patient characteristics. As the field of nanomedicine continues to advance, ANMs may play a significant role in improving cancer treatment in the future.

## LIST OF ABBREVIATIONS

| | |
|---|---|
| **ANMs** | Anthraquinone-Based Nanomaterials |
| **ROS** | Reactive Oxygen Species |
| **PTT** | Photothermal Therapy |
| **PDT** | Photodynamic Therapy |
| **DOX** | Doxorubicin |
| **EPR** | Enhanced Permeability and Retention |

| **DDS** | Drug Delivery Systems |
|---|---|
| **NP** | Nanoparticle |
| **PEG** | Polyethylene Glycol |
| **PLGA** | Polylactic-co-Glycolic Acid |
| **MMPs** | Matrix Metalloproteinases |
| **FDA** | Food and Drug Administration |
| **EMA** | European Medicines Agency |
| **IND** | Investigational New Drug |
| **NDA** | New Drug Application |
| **CRISPR** | Clustered Regularly Interspaced Short Palindromic Repeats |
| **PI3K** | Phosphoinositide 3-Kinase |
| **MMP-2** | Matrix Metalloproteinase-2 |
| **ECM** | Extracellular Matrix |
| **NF-κB** | Nuclear Factor Kappa B |
| **MAPK/ERK** | Mitogen-Activated Protein Kinase/Extracellular Signal-Regulated Kinase |
| **AKT** | Protein Kinase B |
| **JAK/STAT** | Janus Kinase/Signal Transducer and Activator of Transcription |
| **VEGF** | Vascular Endothelial Growth Factor |
| **ATM/ATR** | Ataxia Telangiectasia Mutated/ATM and Rad3 Related |

## REFERENCES

[1]    Hussain H, Al-Harrasi A, Al-Rawahi A, *et al.* A fruitful decade from 2005 to 2014 for anthraquinone patents. Expert Opin Ther Pat 2015; 25(9): 1053-64.
http://dx.doi.org/10.1517/13543776.2015.1050793 PMID: 26036306

[2]    Sendelbach LE. A review of the toxicity and carcinogenicity of anthraquinone derivatives. Toxicology 1989; 57(3): 227-40.

http://dx.doi.org/10.1016/0300-483X(89)90113-3 PMID: 2667196

[3]     Yen G, Din-Der DC, Da-Yon C, *et al.* Antioxidant activity of anthraquinones and anthrone. Food Chem 2000; 70(4): 437-41.

http://dx.doi.org/10.1016/S0308-8146(00)00108-4

[4]     Ko JKS, Leung WC, Ho WK, Chiu P. Herbal diterpenoids induce growth arrest and apoptosis in colon cancer cells with increased expression of the nonsteroidal anti-inflammatory drug-activated gene. Eur J Pharmacol 2007; 559(1): 1-13.

http://dx.doi.org/10.1016/j.ejphar.2006.12.004 PMID: 17258704

[5]     Hou Z, Lambert JD, Chin KV, Yang CS. Effects of tea polyphenols on signal transduction pathways related to cancer chemoprevention. Mutat Res 2004; 555(1-2): 3-19.

http://dx.doi.org/10.1016/j.mrfmmm.2004.06.040 PMID: 15476848

[6]     Orban N, Boldizsar I, Szucs Z, Danos B. Influence of different elicitors on the synthesis of anthraquinone derivatives in Rubia tinctorum L. cell suspension cultures. Dyes Pigments 2008; 77(1): 249-57.

http://dx.doi.org/10.1016/j.dyepig.2007.03.015

[7]     Cichewicz RH, Zhang Y, Seeram NP, Nair MG. Inhibition of human tumor cell proliferation by novel anthraquinones from daylilies. Life Sci 2004; 74(14): 1791-9.

http://dx.doi.org/10.1016/j.lfs.2003.08.034 PMID: 14741736

[8]     Huang Q, Lu G, Shen HM, Chung MCM, Ong CN. Anti-cancer properties of anthraquinones from rhubarb. Med Res Rev 2007; 27(5): 609-30.

http://dx.doi.org/10.1002/med.20094 PMID: 17022020

[9]     Castiglione S, Fanciulli M, Bruno T, *et al.* Rhein inhibits glucose uptake in Ehrlich ascites tumor cells by alteration of membrane-associated functions. Anticancer Drugs 1993; 4(3): 407-14.

http://dx.doi.org/10.1097/00001813-199306000-00019 PMID: 8358069

[10]    Rock CL, Thomson C, Gansler T, *et al.* American Cancer Society guideline for diet and physical activity for cancer prevention. CA Cancer J Clin 2020; 70(4): 245-71.

http://dx.doi.org/10.3322/caac.21591 PMID: 32515498

[11]    Dhadda S, Raigar AK, Saini K, Manju , Guleria A. Benzothiazoles: From recent advances in green synthesis to anti-cancer potential. Sustain Chem Pharm 2021; 24: 100521.

http://dx.doi.org/10.1016/j.scp.2021.100521

[12]    Liu W, Qaed E, Zhu Y, *et al.* Research progress and new perspectives of anticancer effects of emodin. Am J Chin Med 2023; 51(7): 1751-93.

http://dx.doi.org/10.1142/S0192415X23500787 PMID: 37732372

[13]    Emerich DF, Thanos CG. Targeted nanoparticle-based drug delivery and diagnosis. J Drug Target 2007; 15(3): 163-83.

http://dx.doi.org/10.1080/10611860701231810 PMID: 17454354

[14]    Bamrungsap S, Zhao Z, Chen T, *et al.* Nanotechnology in therapeutics: a focus on nanoparticles as a drug delivery system. Nanomedicine (Lond) 2012; 7(8): 1253-71.

http://dx.doi.org/10.2217/nnm.12.87 PMID: 22931450

[15] Tang L, Mei Y, Shen Y, *et al.* Nanoparticle-mediated targeted drug delivery to remodel tumor microenvironment for cancer therapy. Int J Nanomedicine 2021; 16: 5811-29.
http://dx.doi.org/10.2147/IJN.S321416 PMID: 34471353

[16] Lôbo GCNB, Paiva KLR, Silva ALG, Simões MM, Radicchi MA, Báo SN. Nanocarriers used in drug delivery to enhance immune system in cancer therapy. Pharmaceutics 2021; 13(8): 1167.
http://dx.doi.org/10.3390/pharmaceutics13081167 PMID: 34452128

[17] Yao Y, Zhou Y, Liu L, *et al.* Nanoparticle-based drug delivery in cancer therapy and its role in overcoming drug resistance. Front Mol Biosci 2020; 7: 193.
http://dx.doi.org/10.3389/fmolb.2020.00193 PMID: 32974385

[18] Solanki A, Kim JD, Lee KB. Nanotechnology for regenerative medicine: nanomaterials for stem cell imaging. Nanomedicine (Lond) 2008; 3(4): 567-78.
http://dx.doi.org/10.2217/17435889.3.4.567 PMID: 18694318

[19] Bae YH, Park K. Targeted drug delivery to tumors: Myths, reality and possibility. J Control Release 2011; 153(3): 198-205.
http://dx.doi.org/10.1016/j.jconrel.2011.06.001 PMID: 21663778

[20] Wilhelm S, Tavares AJ, Dai Q, *et al.* Analysis of nanoparticle delivery to tumours. Nat Rev Mater 2016; 1(5): 16014.
http://dx.doi.org/10.1038/natrevmats.2016.14

[21] Xin Y, Yin M, Zhao L, Meng F, Luo L. Recent progress on nanoparticle-based drug delivery systems for cancer therapy. Cancer Biol Med 2017; 14(3): 228-41.
http://dx.doi.org/10.20892/j.issn.2095-3941.2017.0052 PMID: 28884040

[22] Gartman JA, Tambar UK. Recent total syntheses of anthraquinone-based natural products. Tetrahedron 2022; 105: 132501.
http://dx.doi.org/10.1016/j.tet.2021.132501 PMID: 35095120

[23] Qi N, Yao B, Sun H, Gao Y, Liu X, Li F. Anthraquinone-based porous organic polymers: From synthesis to applications in electrochemical energy conversion and storage. Arab J Chem 2023; 16(11): 105263.
http://dx.doi.org/10.1016/j.arabjc.2023.105263

[24] Bachman JE, Curtiss LA, Assary RS. Investigation of the redox chemistry of anthraquinone derivatives using density functional theory. J Phys Chem A 2014; 118(38): 8852-60.
http://dx.doi.org/10.1021/jp5060777 PMID: 25159500

[25] Qin M, Qin M, Shi Y, Xu J, Cao J. Redox-active anthraquinone-based π-conjugated polymer anode for high-capacity aqueous organic hybrid flow battery. J Energy Storage 2023; 72: 108642.
http://dx.doi.org/10.1016/j.est.2023.108642

[26] Zhu Y, Li Y, Qian Y, *et al.* Anthraquinone-based anode material for aqueous redox flow batteries operating in nondemanding atmosphere. J Power Sources 2021; 501: 229984.
http://dx.doi.org/10.1016/j.jpowsour.2021.229984

[27] Fouillaud M, Caro Y, Venkatachalam M, Grondin I, Dufossé L. Anthraquinones. In phenolic compounds in food. CRC In: 2018; pp. 131-72.

[28] Vincent MP, Navidzadeh JO, Bobbala S, Scott EA. Leveraging self-assembled nanobiomaterials for improved cancer immunotherapy. Cancer Cell. 2022; 40(3): 255-76.
http://dx.doi.org/10.1016/j.ccell.2022.01.006 PMID: 35148814.

[29] Mund NK, Čellárová E. Recent advances in the identification of biosynthetic genes and gene clusters of the polyketide-derived pathways for anthraquinone biosynthesis and biotechnological applications. Biotechnol Adv 2023; 63: 108104.
http://dx.doi.org/10.1016/j.biotechadv.2023.108104 PMID: 36716800

[30] Mohapatra M, Basak UC. Ethno-medicinal and therapeutic applications of natural anthraquinones: recent trends and advancements. Indian J Pharm Sci 2023; 85(3): 544-54.

[31] Sebak M, Molham F, Greco C, Tammam MA, Sobeh M, El-Demerdash A. Chemical diversity, medicinal potentialities, biosynthesis, and pharmacokinetics of anthraquinones and their congeners derived from marine fungi: a comprehensive update. RSC Advances 2022; 12(38): 24887-921.
http://dx.doi.org/10.1039/D2RA03610J PMID: 36199881

[32] Hafez Ghoran S, Taktaz F, Ayatollahi SA, Kijjoa A. anthraquinones and their analogues from marine-derived fungi: chemistry and biological activities. Mar Drugs 2022; 20(8): 474.
http://dx.doi.org/10.3390/md20080474 PMID: 35892942

[33] Siegel RL, Miller KD, Jemal A. Cancer statistics, 2020. CA Cancer J Clin 2020; 70(1): 7-30.
http://dx.doi.org/10.3322/caac.21590 PMID: 31912902

[34] Jeevitha D, Amarnath K. Chitosan/PLA nanoparticles as a novel carrier for the delivery of anthraquinone: Synthesis, characterization and *in vitro* cytotoxicity evaluation. Colloids Surf B Biointerfaces 2013; 101: 126-34.
http://dx.doi.org/10.1016/j.colsurfb.2012.06.019 PMID: 22796782

[35] Siddamurthi S, Gutti G, Jana S, Kumar A, Singh SK. Anthraquinone: a promising scaffold for the discovery and development of therapeutic agents in cancer therapy. Future Med Chem 2020; 12(11): 1037-69.
http://dx.doi.org/10.4155/fmc-2019-0198 PMID: 32349522

[36] Li J, Ying S, Ren H, *et al.* Molecular dynamics study on the encapsulation and release of anti-cancer drug doxorubicin by chitosan. Int J Pharm 2020; 580: 119241.
http://dx.doi.org/10.1016/j.ijpharm.2020.119241 PMID: 32197982

[37] Liang X, Ding L, Ma J, *et al.* Enhanced mechanical strength and sustained drug release in carrier-free silver-coordinated anthraquinone natural antibacterial anti-inflammatory hydrogel for infectious wound healing. Adv Healthc Mater 2024; 13(23): 2400841.
http://dx.doi.org/10.1002/adhm.202400841 PMID: 38725393

[38] Amantino CF, de Baptista-Neto Á, Badino AC, Siqueira-Moura MP, Tedesco AC, Primo FL. Anthraquinone encapsulation into polymeric nanocapsules as a new drug from biotechnological origin designed for photodynamic therapy. Photodiagn Photodyn Ther 2020; 31: 101815.
http://dx.doi.org/10.1016/j.pdpdt.2020.101815 PMID: 32407889

[39] Zhang W, Li X, Kang M, *et al.* Anthraquinone-centered type I photosensitizer with aggregation-induced emission characteristics for tumor-targeted two-photon photodynamic therapy. ACS Mater Lett 2024; 6(6): 2174-85.

http://dx.doi.org/10.1021/acsmaterialslett.4c00600

[40]   Dadwal A, Baldi A, Kumar Narang R. Nanoparticles as carriers for drug delivery in cancer. Artif Cells Nanomed Biotechnol 2018; 46(sup2): 295-305.
http://dx.doi.org/10.1080/21691401.2018.1457039 PMID: 30043651

[41]   Palazzolo S, Bayda S, Hadla M, *et al.* The clinical translation of organic nanomaterials for cancer therapy: a focus on polymeric nanoparticles, micelles, liposomes and exosomes. Curr Med Chem 2018; 25(34): 4224-68.
http://dx.doi.org/10.2174/0929867324666170830113755 PMID: 28875844

[42]   Chen Y, Gao DY, Huang L. *In vivo* delivery of miRNAs for cancer therapy: Challenges and strategies. Adv Drug Deliv Rev 2015; 81: 128-41.
http://dx.doi.org/10.1016/j.addr.2014.05.009 PMID: 24859533

[43]   Kalyane D, Raval N, Maheshwari R, Tambe V, Kalia K, Tekade RK. Employment of enhanced permeability and retention effect (EPR): Nanoparticle-based precision tools for targeting of therapeutic and diagnostic agent in cancer. Mater Sci Eng C 2019; 98: 1252-76.
http://dx.doi.org/10.1016/j.msec.2019.01.066 PMID: 30813007

[44]   Zhang RX, Ahmed T, Li LY, Li J, Abbasi AZ, Wu XY. Design of nanocarriers for nanoscale drug delivery to enhance cancer treatment using hybrid polymer and lipid building blocks. Nanoscale 2017; 9(4): 1334-55.
http://dx.doi.org/10.1039/C6NR08486A PMID: 27973629

[45]   Yoon HY, Selvan ST, Yang Y, *et al.* Engineering nanoparticle strategies for effective cancer immunotherapy. Biomaterials 2018; 178: 597-607.
http://dx.doi.org/10.1016/j.biomaterials.2018.03.036 PMID: 29576282

[46]   Bahrami B, Hojjat-Farsangi M, Mohammadi H, *et al.* Nanoparticles and targeted drug delivery in cancer therapy. Immunol Lett 2017; 190: 64-83.
http://dx.doi.org/10.1016/j.imlet.2017.07.015 PMID: 28760499

[47]   Decuzzi P, Pasqualini R, Arap W, Ferrari M. Intravascular delivery of particulate systems: does geometry really matter? Pharm Res 2009; 26(1): 235-43.
http://dx.doi.org/10.1007/s11095-008-9697-x PMID: 18712584

[48]   Yang Q, Jones SW, Parker CL, Zamboni WC, Bear JE, Lai SK. Evading immune cell uptake and clearance requires PEG grafting at densities substantially exceeding the minimum for brush conformation. Mol Pharm 2014; 11(4): 1250-8.
http://dx.doi.org/10.1021/mp400703d PMID: 24521246

[49]   Zylberberg C, Matosevic S. Pharmaceutical liposomal drug delivery: a review of new delivery systems and a look at the regulatory landscape. Drug Deliv 2016; 23(9): 3319-29.
http://dx.doi.org/10.1080/10717544.2016.1177136 PMID: 27145899

[50]   Lemière J, Carvalho K, Sykes C. Cell-sized liposomes that mimic cell motility and the cell cortex. Methods Cell Biol 2015; 128: 271-85.
http://dx.doi.org/10.1016/bs.mcb.2015.01.013 PMID: 25997352

[51]   Tang B, Peng Y, Yue Q, *et al.* Design, preparation and evaluation of different branched biotin modified liposomes for targeting breast cancer. Eur J Med Chem 2020; 193: 112204.
http://dx.doi.org/10.1016/j.ejmech.2020.112204 PMID: 32172035

[52]   Satsangi A, Roy SS, Satsangi RK, *et al.* Synthesis of a novel, sequentially active-targeted drug delivery nanoplatform for breast cancer therapy. Biomaterials 2015; 59: 88-101.
       http://dx.doi.org/10.1016/j.biomaterials.2015.03.039 PMID: 25956854

[53]   Moon HK, Lee SH, Choi HC. *In vivo* near-infrared mediated tumor destruction by photothermal effect of carbon nanotubes. ACS Nano 2009; 3(11): 3707-13.
       http://dx.doi.org/10.1021/nn900904h PMID: 19877694

[54]   Liu Y, Bhattarai P, Dai Z, Chen X. Photothermal therapy and photoacoustic imaging *via* nanotheranostics in fighting cancer. Chem Soc Rev 2019; 48(7): 2053-108.
       http://dx.doi.org/10.1039/C8CS00618K PMID: 30259015

[55]   Zheng R, Wang S, Tian Y, *et al.* Polydopamine-coated magnetic composite particles with an enhanced photothermal effect. ACS Appl Mater Interfaces 2015; 7(29): 15876-84.
       http://dx.doi.org/10.1021/acsami.5b03201 PMID: 26151502

[56]   Li J, Xiao H, Yoon SJ, *et al.* Functional photoacoustic imaging of gastric acid secretion using pH-responsive polyaniline nanoprobes. Small 2016; 12(34): 4690-6.
       http://dx.doi.org/10.1002/smll.201601359 PMID: 27357055

[57]   Nam J, La WG, Hwang S, *et al.* pH-responsive assembly of gold nanoparticles and "spatiotemporally concerted" drug release for synergistic cancer therapy. ACS Nano 2013; 7(4): 3388-402.
       http://dx.doi.org/10.1021/nn400223a PMID: 23530622

[58]   Dolmans DEJGJ, Fukumura D, Jain RK. Photodynamic therapy for cancer. Nat Rev Cancer 2003; 3(5): 380-7.
       http://dx.doi.org/10.1038/nrc1071 PMID: 12724736

[59]   Dos Santos AF, De Almeida DRQ, Terra LF, Baptista MS, Labriola L. Photodynamic therapy in cancer treatment - an update review. J Cancer Metastasis Treat 2019; 2019(25): 10-20517.
       http://dx.doi.org/10.20517/2394-4722.2018.83

[60]   Buytaert E, Dewaele M, Agostinis P. Molecular effectors of multiple cell death pathways initiated by photodynamic therapy. Biochim Biophys Acta 2007; 1776(1): 86-107.
       PMID: 17693025

[61]   Gunaydin G, Gedik ME, Ayan S. Photodynamic therapy for the treatment and diagnosis of cancer–a review of the current clinical status. Front Chem 2021; 9: 686303.
       http://dx.doi.org/10.3389/fchem.2021.686303 PMID: 34409014

[62]   Agostinis P, Berg K, Cengel KA, *et al.* Photodynamic therapy of cancer: An update. CA Cancer J Clin 2011; 61(4): 250-81.
       http://dx.doi.org/10.3322/caac.20114 PMID: 21617154

[63]   Sun X, Li Y, Liu T, Li Z, Zhang X, Chen X. Peptide-based imaging agents for cancer detection. Adv Drug Deliv Rev 2017; 110-111: 38-51.
       http://dx.doi.org/10.1016/j.addr.2016.06.007 PMID: 27327937

[64]   Mieog JSD, Achterberg FB, Zlitni A, *et al.* Fundamentals and developments in fluorescence-guided cancer surgery. Nat Rev Clin Oncol 2022; 19(1): 9-22.
       http://dx.doi.org/10.1038/s41571-021-00548-3 PMID: 34493858

[65]  Yap TA, Omlin A, de Bono JS. Development of therapeutic combinations targeting major cancer signaling pathways. J Clin Oncol 2013; 31(12): 1592-605.
http://dx.doi.org/10.1200/JCO.2011.37.6418 PMID: 23509311

[66]  Blagosklonny MV. Overcoming limitations of natural anticancer drugs by combining with artificial agents. Trends Pharmacol Sci 2005; 26(2): 77-81.
http://dx.doi.org/10.1016/j.tips.2004.12.002 PMID: 15681024

[67]  Hanahan D, Bergers G, Bergsland E. Less is more, regularly: metronomic dosing of cytotoxic drugs can target tumor angiogenesis in mice. J Clin Invest 2000; 105(8): 1045-7.
http://dx.doi.org/10.1172/JCI9872 PMID: 10772648

[68]  Nam JM, Thaxton CS, Mirkin CA. Nanoparticle-based bio-bar codes for the ultrasensitive detection of proteins. Science 2003; 301(5641): 1884-6.
http://dx.doi.org/10.1126/science.1088755 PMID: 14512622

[69]  Findeisen P, Neumaier M. Functional protease profiling for diagnosis of malignant disease. Proteomics Clin Appl 2012; 6(1-2): 60-78.
http://dx.doi.org/10.1002/prca.201100058 PMID: 22213637

[70]  Schwarzenbach H, Hoon DSB, Pantel K. Cell-free nucleic acids as biomarkers in cancer patients. Nat Rev Cancer 2011; 11(6): 426-37.
http://dx.doi.org/10.1038/nrc3066 PMID: 21562580

[71]  Daniels TR, Delgado T, Helguera G, Penichet ML. The transferrin receptor part II: Targeted delivery of therapeutic agents into cancer cells. Clin Immunol 2006; 121(2): 159-76.
http://dx.doi.org/10.1016/j.clim.2006.06.006 PMID: 16920030

[72]  Bhardwaj P, Arora B, Saxena S, *et al.* Paper-based point of care diagnostics for cancer biomarkers. Sens Diagn 2024; 3(4): 504-35.
http://dx.doi.org/10.1039/D3SD00340J

[73]  Zhang Q, Chen WW, Sun X, *et al.* The versatile emodin: A natural easily acquired anthraquinone possesses promising anticancer properties against a variety of cancers. Int J Biol Sci 2022; 18(8): 3498-527.
http://dx.doi.org/10.7150/ijbs.70447 PMID: 35637953

[74]  Chua HM, Moshawih S, Kifli N, Goh HP, Ming LC. Insights into the computer-aided drug design and discovery based on anthraquinone scaffold for cancer treatment: A systematic review. PLoS One 2024; 19(5): e0301396.
http://dx.doi.org/10.1371/journal.pone.0301396 PMID: 38776291

[75]  Arrousse N, Harras MF, El Kadiri S, *et al.* New anthraquinone drugs and their anticancer activities: Cytotoxicity, DFT, docking and ADMET properties. Results Chem 2023; 6: 100996.
http://dx.doi.org/10.1016/j.rechem.2023.100996

[76]  Zhao L, Zhang X, Wang X, Guan X, Zhang W, Ma J. Recent advances in selective photothermal therapy of tumor. J Nanobiotechnology 2021; 19(1): 335.
http://dx.doi.org/10.1186/s12951-021-01080-3 PMID: 34689765

[77]  Chen YC, Chiu WT, Chang C, *et al.* Chemo-photothermal effects of doxorubicin/silica–carbon hollow spheres on liver cancer. RSC Advances 2018; 8(64): 36775-84.

http://dx.doi.org/10.1039/C8RA08538B PMID: 35558959

[78]   Nowak-Perlak M, Ziółkowski P, Woźniak M. A promising natural anthraquinones mediated by photodynamic therapy for anti-cancer therapy. Phytomedicine 2023; 119: 155035.
http://dx.doi.org/10.1016/j.phymed.2023.155035 PMID: 37603973

[79]   Wang K, Meng X, Chen J, *et al.* Emodin-induced necroptosis and inhibited glycolysis in the renal cancer cells by enhancing ROS. Oxid Med Cell Longev 2021; 2021(1): 8840590.
http://dx.doi.org/10.1155/2021/8840590 PMID: 33532038

[80]   Liu Y, Meng P, Zhang H, *et al.* Inhibitory effect of aloe emodin mediated photodynamic therapy on human oral mucosa carcinoma *in vitro* and *in vivo*. Biomed Pharmacother 2018; 97: 697-707.
http://dx.doi.org/10.1016/j.biopha.2017.10.080 PMID: 29102913

[81]   Comini LR, Fernandez IM, Vittar NBR, Núñez Montoya SC, Cabrera JL, Rivarola VA. Photodynamic activity of anthraquinones isolated from Heterophyllaea pustulata Hook f. (Rubiaceae) on MCF-7c3 breast cancer cells. Phytomedicine 2011; 18(12): 1093-5.
http://dx.doi.org/10.1016/j.phymed.2011.05.008 PMID: 21665453

[82]   Staničová J, Verebová V, Beneš JJ. Interaction of a potential anticancer 990 agent hypericin and its model compound emodin with DNA and bovine 991 serum albumin. *In vivo* 2018; 32(5): 1063-70.

[83]   Malik EM, Müller CE. Anthraquinones as pharmacological tools and drugs. Med Res Rev 2016; 36(4): 705-48.
http://dx.doi.org/10.1002/med.21391 PMID: 27111664

[84]   Martins-Teixeira MB, Carvalho I. Antitumour anthracyclines: progress and perspectives. ChemMedChem 2020; 15(11): 933-48.
http://dx.doi.org/10.1002/cmdc.202000131 PMID: 32314528

[85]   Nisiewicz MK, Gajda A, Kowalczyk A, *et al.* Novel electrogravimetric biosensors for the ultrasensitive detection of plasma matrix metalloproteinase-2 considered a potential tumor biomarker. Anal Chim Acta 2022; 1191: 339290.
http://dx.doi.org/10.1016/j.aca.2021.339290 PMID: 35033237

[86]   Jiang H, Li H. Prognostic values of tumoral MMP2 and MMP9 overexpression in breast cancer: a systematic review and meta-analysis. BMC Cancer 2021; 21(1): 149.
http://dx.doi.org/10.1186/s12885-021-07860-2 PMID: 33568081

[87]   Roy R, Yang J, Moses MA. Matrix metalloproteinases as novel biomarkers and potential therapeutic targets in human cancer. J Clin Oncol 2009; 27(31): 5287-97.
http://dx.doi.org/10.1200/JCO.2009.23.5556 PMID: 19738110

[88]   Babichev Y, Kabaroff L, Datti A, *et al.* PI3K/AKT/mTOR inhibition in combination with doxorubicin is an effective therapy for leiomyosarcoma. J Transl Med 2016; 14(1): 67.
http://dx.doi.org/10.1186/s12967-016-0814-z PMID: 26952093

[89]   Rizvi SAA, Saleh AM. Applications of nanoparticle systems in drug delivery technology. Saudi Pharm J 2018; 26(1): 64-70.
http://dx.doi.org/10.1016/j.jsps.2017.10.012 PMID: 29379334

[90] Mirzaei S, Gholami MH, Hashemi F, *et al*. Advances in understanding the role of P-gp in doxorubicin resistance: Molecular pathways, therapeutic strategies, and prospects. Drug Discov Today 2022; 27(2): 436-55.
http://dx.doi.org/10.1016/j.drudis.2021.09.020 PMID: 34624510

[91] Tao X, Tao T, Wen Y, *et al*. Novel delivery of mitoxantrone with hydrophobically modified pullulan nanoparticles to inhibit bladder cancer cells and the effect of nano-drug size on inhibition efficiency. Nanoscale Res Lett 2018; 13(1): 345.
http://dx.doi.org/10.1186/s11671-018-2769-x

[92] Khuda F, Zahir I, Khalil AAK, *et al*. Preparation, characterization, and evaluation of physcion nanoparticles for enhanced oral bioavailability: An attempt to improve its antioxidant and anticancer potential. ACS Omega 2023; 8(37): 33955-65.
http://dx.doi.org/10.1021/acsomega.3c04821 PMID: 37744808

[93] Prateeksha MA. Yusuf, B.N. Singh, *et al*. Biomolecules 2019; 9: 68.
http://dx.doi.org/10.3390/biom9020068

[94] Thorn CF, Oshiro C, Marsh S, *et al*. Doxorubicin pathways. Pharmacogenet Genomics 2011; 21(7): 440-6.
http://dx.doi.org/10.1097/FPC.0b013e32833ffb56 PMID: 21048526

[95] Tao X, Tao T, Wen Y, *et al*. Novel delivery of mitoxantrone with hydrophobically modified pullulan nanoparticles to inhibit bladder cancer cell and the effect of nano-drug size on inhibition efficiency. Nanoscale Res Lett 2018; 13(1): 345.
http://dx.doi.org/10.1186/s11671-018-2769-x PMID: 30377872

[96] Wijesekara I, Zhang C, Van Ta Q, Vo TS, Li YX, Kim SK. Physcion from marine-derived fungus Microsporum sp. induces apoptosis in human cervical carcinoma HeLa cells. Microbiol Res 2014; 169(4): 255-61.
http://dx.doi.org/10.1016/j.micres.2013.09.001 PMID: 24071573

[97] Nualsanit T, Rojanapanthu P, Gritsanapan W, Lee SH, Lawson D, Baek SJ. Damnacanthal, a noni component, exhibits antitumorigenic activity in human colorectal cancer cells. J Nutr Biochem 2012; 23(8): 915-23.
http://dx.doi.org/10.1016/j.jnutbio.2011.04.017 PMID: 21852088

[98] Abu N, Zamberi NR, Yeap SK, *et al*. Subchronic toxicity, immunoregulation and anti-breast tumor effect of Nordamnacantal, an anthraquinone extracted from the stems of Morinda citrifolia L. BMC Complement Altern Med 2018; 18(1): 31.
http://dx.doi.org/10.1186/s12906-018-2102-3 PMID: 29374471

[99] Ren L, Li Z, Dai C, *et al*. Chrysophanol inhibits proliferation and induces apoptosis through NF-κB/cyclin D1 and NF-κB/Bcl-2 signaling cascade in breast cancer cell lines. Mol Med Rep 2018; 17(3): 4376-82.
http://dx.doi.org/10.3892/mmr.2018.8443 PMID: 29344652

[100] Chen YY, Chiang SY, Lin JG, *et al*. Emodin, aloe-emodin and rhein inhibit migration and invasion in human tongue cancer SCC-4 cells through the inhibition of gene expression of matrix metalloproteinase-9. Int J Oncol 2010; 36(5): 1113-20.
PMID: 20372784

[101] Paarakh PM, Sreeram DC, *et al. In vitro* antiproliferative and *in silico* activity of rubiadin isolated from roots of Rubia cordifolia. Mintage J Pharm Med Sci 2016(5): 1.

[102] Guo Y, Wang L, Lv P, Zhang P. Transferrin-conjugated doxorubicin-loaded lipid-coated nanoparticles for the targeting and therapy of lung cancer. Oncol Lett 2015; 9(3): 1065-72.
http://dx.doi.org/10.3892/ol.2014.2840 PMID: 25663858

[103] Bao W, Liu R, Wang Y, *et al.* PLGA-PLL-PEG-Tf-based targeted nanoparticles drug delivery system enhance antitumor efficacy *via* intrinsic apoptosis pathway. Int J Nanomedicine 2015; 10: 557-66.
PMID: 25609961

[104] Cozza G, Mazzorana M, Papinutto E, *et al.* Quinalizarin as a potent, selective and cell-permeable inhibitor of protein kinase CK2. Biochem J 2009; 421(3): 387-95.
http://dx.doi.org/10.1042/BJ20090069 PMID: 19432557

[105] Ettorre A, Cirilli M, Ughetto G. Degradation of the morpholino ring in the crystal structure of cyanomorpholinodoxorubicin complexed with d(CGATCG). Eur J Biochem 1998; 258(2): 350-4.
http://dx.doi.org/10.1046/j.1432-1327.1998.2580350.x PMID: 9874199

[106] De Moliner E, Moro S, Sarno S, *et al.* Inhibition of protein kinase CK2 by anthraquinone-related compounds. A structural insight. J Biol Chem 2003; 278(3): 1831-6.
http://dx.doi.org/10.1074/jbc.M209367200 PMID: 12419810

[107] Ding W, Guo L. Immobilized transferrin $Fe_3O_4@SiO_2$ nanoparticle with high doxorubicin loading for dual-targeted tumor drug delivery. Int J Nanomedicine 2013; 8: 4631-9.
PMID: 24348038

[108] Jordheim LP, Marton Z, Rhimi M, *et al.* Identification and characterization of inhibitors of cytoplasmic 5′-nucleotidase cN-II issued from virtual screening. Biochem Pharmacol 2013; 85(4): 497-506.
http://dx.doi.org/10.1016/j.bcp.2012.11.024 PMID: 23220537

[109] McDaniel JR, MacEwan SR, Dewhirst M, Chilkoti A. Doxorubicin-conjugated chimeric polypeptide nanoparticles that respond to mild hyperthermia. J Control Release 2012; 159(3): 362-7.
http://dx.doi.org/10.1016/j.jconrel.2012.02.030 PMID: 22421424

[110] Fazilati M. Folate decorated magnetite nanoparticles: synthesis and targeted therapy against ovarian cancer Cell boil Int 2014; 38(2): 154-63.
http://dx.doi.org/10.1002/cbin.10167

[111] Aftab S, Shah A, Nadhman A, *et al.* Nanomedicine: An effective tool in cancer therapy. Int J Pharm 2018; 540(1-2): 132-49.
http://dx.doi.org/10.1016/j.ijpharm.2018.02.007 PMID: 29427746

[112] Sohail M, Sun Z, Li Y, Gu X, Xu H. Research progress in strategies to improve the efficacy and safety of doxorubicin for cancer chemotherapy. Expert Rev Anticancer Ther 2021; 21(12): 1385-98.
http://dx.doi.org/10.1080/14737140.2021.1991316 PMID: 34636282

[113]    Wang X, Liu Y, Xu W, *et al.* Irinotecan and berberine co-delivery liposomes showed improved efficacy and reduced intestinal toxicity compared with Onivyde for pancreatic cancer. Drug Deliv Transl Res 2021; 11(5): 2186-97.
            http://dx.doi.org/10.1007/s13346-020-00884-4 PMID: 33452654

[114]    Crain ML. Daunorubicin & Cytarabine liposome (vyxeos™). Oncol Times 2018; 40(10): 30.
            http://dx.doi.org/10.1097/01.COT.0000534146.30839.ec

[115]    Sanadgol N, Wackerlig J. Developments of smart drug-delivery systems based on magnetic molecularly imprinted polymers for targeted cancer therapy: a short review. Pharmaceutics 2020; 12(9): 831.
            http://dx.doi.org/10.3390/pharmaceutics12090831 PMID: 32878127

[116]    Malek R, Wu ST, Serrano D, *et al.* ELIGANT: a Phase 4, interventional, safety study of leuprorelin acetate (ELIGARD®) in Asian men with prostate cancer. Transl Androl Urol 2022; 11(2): 179-89.
            http://dx.doi.org/10.21037/tau-21-723 PMID: 35280654

[117]    De Luca R, Blasi L, Alù M, Gristina V, Cicero G. Clinical efficacy of nab-paclitaxel in patients with metastatic pancreatic cancer. Drug Des Devel Ther 2018; 12: 1769-75.
            http://dx.doi.org/10.2147/DDDT.S165851 PMID: 29950811

[118]    Boix-Montesinos P, Soriano-Teruel PM, Armiñán A, Orzáez M, Vicent MJ. The past, present, and future of breast cancer models for nanomedicine development. Adv Drug Deliv Rev 2021; 173: 306-30.
            http://dx.doi.org/10.1016/j.addr.2021.03.018 PMID: 33798642

[119]    Norouzi M, Hardy P. Clinical applications of nanomedicines in lung cancer treatment. Acta Biomater 2021; 121: 134-42.
            http://dx.doi.org/10.1016/j.actbio.2020.12.009 PMID: 33301981

[120]    Desai N. Nanoparticle albumin-bound paclitaxel (Abraxane®). Albumin in medicine: Pathological and Clinical Applications 2016: 101-119.

[121]    Youn YS, Bae YH. Perspectives on the past, present, and future of cancer nanomedicine. Adv Drug Deliv Rev 2018; 130: 3-11.
            http://dx.doi.org/10.1016/j.addr.2018.05.008 PMID: 29778902

[122]    Lee KJ, Ko EJ, Park YY, *et al.* A novel nanoparticle-based theranostic agent targeting LRP-1 enhances the efficacy of neoadjuvant radiotherapy in colorectal cancer. Biomaterials 2020; 255: 120151.
            http://dx.doi.org/10.1016/j.biomaterials.2020.120151 PMID: 32505033

[123]    Wolf NK, Blaj C, Picton LK, *et al.* Synergy of a STING agonist and an IL-2 superkine in cancer immunotherapy against MHC I–deficient and MHC I $^{+}$ tumors. Proc Natl Acad Sci USA 2022; 119(22): e2200568119.
            http://dx.doi.org/10.1073/pnas.2200568119 PMID: 35588144

[124]    Yen J, Ying H, Wang H, Yin L, Uckun F, Cheng J. CD44 mediated nonviral gene delivery into human embryonic stem cells *via* hyaluronic-acid-coated nanoparticles. ACS Biomater Sci Eng 2016; 2(3): 326-35.
            http://dx.doi.org/10.1021/acsbiomaterials.5b00393 PMID: 33429536

[125] Guo S, Lin CM, Xu Z, Miao L, Wang Y, Huang L. Co-delivery of cisplatin and rapamycin for enhanced anticancer therapy through synergistic effects and microenvironment modulation. ACS Nano 2014; 8(5): 4996-5009.
http://dx.doi.org/10.1021/nn5010815 PMID: 24720540

[126] Yin H, Song CQ, Dorkin JR, *et al.* Therapeutic genome editing by combined viral and non-viral delivery of CRISPR system components *in vivo*. Nat Biotechnol 2016; 34(3): 328-33.
http://dx.doi.org/10.1038/nbt.3471 PMID: 26829318

[127] Chen G, Abdeen AA, Wang Y, *et al.* A biodegradable nanocapsule delivers a Cas9 ribonucleoprotein complex for *in vivo* genome editing. Nat Nanotechnol 2019; 14(10): 974-80.
http://dx.doi.org/10.1038/s41565-019-0539-2 PMID: 31501532

[128] Liu J, Chang J, Jiang Y, *et al.* Fast and efficient CRISPR/Cas9 genome editing *in vivo* enabled by bioreducible lipid and messenger RNA nanoparticles. Adv Mater 2019; 31(33): 1902575.
http://dx.doi.org/10.1002/adma.201902575 PMID: 31215123

[129] Xu CF, Iqbal S, Shen S, Luo YL, Yang X, Wang J. Development of 'CLAN' nanomedicine for nucleic acid therapeutics. Small 2019; 15(16): 1900055.
http://dx.doi.org/10.1002/smll.201900055 PMID: 30884095

[130] Malik MS, Alsantali RI, Jassas RS, *et al.* Journey of anthraquinones as anticancer agents – a systematic review of recent literature. RSC Advances 2021; 11(57): 35806-27.
http://dx.doi.org/10.1039/D1RA05686G PMID: 35492773

[131] Şeker Karatoprak G, Küpeli Akkol E, Yücel Ç, Bahadır Acıkara Ö, Sobarzo-Sánchez E. Advances in understanding the role of aloe-emodin and targeted drug delivery systems in cancer. Oxid Med Cell Longev 2022; 2022: 1-20.
http://dx.doi.org/10.1155/2022/7928200 PMID: 35087619

[132] Nakamura Y, Mochida A, Choyke PL, Kobayashi H. Nano drug delivery: is the enhanced permeability and retention effect sufficient for curing cancer? Bioconjug Chem 2016; 27(10): 2225-38.
http://dx.doi.org/10.1021/acs.bioconjchem.6b00437 PMID: 27547843

[133] Peng F, Liao M, Qin R, *et al.* Regulated cell death (RCD) in cancer: key pathways and targeted therapies. Signal Transduct Target Ther 2022; 7(1): 286.
http://dx.doi.org/10.1038/s41392-022-01110-y PMID: 35963853

[134] Castro DTH, Leite DF, da Silva Baldivia D *et al.* structural characterization and anticancer activity of a new anthraquinone from senna velutina (Fabaceae). Pharmaceuticals 2023; 16(7): 951.
https://doi.org/10.3390/ph16070951 PMID: 37513863

[135] Montecucco A, Zanetta F, Biamonti G. Molecular mechanisms of etoposide. EXCLI J 2015; 14: 95-108.
PMID: 26600742

[136] Yadav M, Dhagat S, Eswari JS. Structure-based drug design and molecular docking studies of anticancer molecules paclitaxel, etoposide and topotecan using novel ligands. Curr Drug Discov Technol 2020; 17(2): 183-90.
http://dx.doi.org/10.2174/1570163816666190307102033 PMID: 30848204

[137]   Diaz-Muñoz G, Miranda IL, Sartori SK, de Rezende DC, Diaz MAN. Anthraquinones: An overview. Studies in natural products chemistry 2018; 58: 313-38.
        http://dx.doi.org/10.1016/B978-0-444-64056-7.00011-8

[138]   Abdullah S, Mukherjee S, Shweta, Debnath B. The prevention of multi-drug resistance in cancers through the application of nanotechnology-based targeted delivery systems for combination therapies involving traditional Chinese medicine. Pharmacol Res - Mod Chin Med 2024; 10: 100386.
        https://doi.org/10.1016/j.prmcm.2024.100386

[139]   Hoesel B, Schmid JA. The complexity of NF-κB signaling in inflammation and cancer. Mol Cancer 2013; 12(1): 86.
        http://dx.doi.org/10.1186/1476-4598-12-86 PMID: 23915189

[140]   Guo YJ, Pan WW, Liu SB, Shen ZF, Xu Y, Hu LL. ERK/MAPK signalling pathway and tumorigenesis. Exp Ther Med 2020; 19(3): 1997-2007.
        PMID: 32104259

[141]   Downward J. Targeting RAS signalling pathways in cancer therapy. Nat Rev Cancer 2003; 3(1): 11-22.
        http://dx.doi.org/10.1038/nrc969 PMID: 12509763

[142]   Yamaoka T, Kusumoto S, Ando K, Ohba M, Ohmori T. Receptor tyrosine kinase-targeted cancer therapy. Int J Mol Sci 2018; 19(11): 3491.
        http://dx.doi.org/10.3390/ijms19113491 PMID: 30404198

[143]   Jha NK, Arfin S, Jha SK, *et al.* Re-establishing the comprehension of phytomedicine and nanomedicine in inflammation-mediated cancer signaling. Semin Cancer Biol 2022; 86(Pt 2): 1086-104.
        http://dx.doi.org/10.1016/j.semcancer.2022.02.022 PMID: 35218902

[144]   El-Kenawi AE, El-Remessy AB. Angiogenesis inhibitors in cancer therapy: mechanistic perspective on classification and treatment rationales. Br J Pharmacol 2013; 170(4): 712-29.
        http://dx.doi.org/10.1111/bph.12344 PMID: 23962094

[145]   Mroz RM, Schins RPF, Li H, *et al.* Nanoparticle-driven DNA damage mimics irradiation-related carcinogenesis pathways. Eur Respir J 2008; 31(2): 241-51.
        http://dx.doi.org/10.1183/09031936.00006707 PMID: 18057054

[146]   Paul S, Ghosh S, Maity T, *et al.* Photocleavable visible light-triggered anthraquinone-derived water-soluble block copolymer for peroxynitrite generation in cancer therapy. ACS Macro Lett 2024; 13(3): 288-95.
        https://doi.org/10.1021/acsmacrolett.3c00728

[147]   Campora M, Francesconi V, Schenone S, Tasso B, Tonelli M. Journey on naphthoquinone and anthraquinone derivatives: new insights in Alzheimer's disease. Pharmaceuticals (Basel) 2021; 14(1): 33.
        http://dx.doi.org/10.3390/ph14010033 PMID: 33466332

[148]   Adlakha K, Koul B, Kumar A. Value-added products of Aloe species: Panacea to several maladies. S Afr J Bot 2022; 147: 1124-35.
        http://dx.doi.org/10.1016/j.sajb.2020.12.025

[149] Francis AL, Namasivayam SK, Kavisri M, *et al.* Anti-microbial efficacy and notable biocompatibility of Rosa damascene and Citrus sinensis biomass-derived metabolites. Biomass Convers Biorefin 2023; 1-21.

[150] Yetisgin AA, Cetinel S, Zuvin M, Kosar A, Kutlu O. Therapeutic nanoparticles and their targeted delivery applications. Molecules 2020; 25(9): 2193.
http://dx.doi.org/10.3390/molecules25092193 PMID: 32397080

[151] Qiu, C., Zhang, J.Z., Wu, B. *et al.* Advanced application of nanotechnology in active constituents of Traditional Chinese Medicines. J Nanobiotechnol 2023; 21: 456.
https://doi.org/10.1186/s12951-023-02165-x PMID: 38017573

[152] Jia L, Zhang P, Sun H, *et al.* Optimization of nanoparticles for smart drug delivery: A review. Nanomaterials (Basel) 2021; 11(11): 2790.
http://dx.doi.org/10.3390/nano11112790 PMID: 34835553

[153] Holbein MEB. Understanding FDA regulatory requirements for investigational new drug applications for sponsor-investigators. J Investig Med 2009; 57(6): 688-94.
http://dx.doi.org/10.2310/JIM.0b013e3181afdb26 PMID: 19602987

[154] Umscheid CA, Margolis DJ, Grossman CE. Key concepts of clinical trials: a narrative review. Postgrad Med 2011; 123(5): 194-204.
http://dx.doi.org/10.3810/pgm.2011.09.2475 PMID: 21904102

[155] Chenthamara D, Subramaniam S, Ramakrishnan SG, *et al.* Therapeutic efficacy of nanoparticles and routes of administration. Biomater Res 2019; 23(1): 20.
http://dx.doi.org/10.1186/s40824-019-0166-x PMID: 31832232

[156] Patra JK, Das G, Fraceto LF, *et al.* Nano based drug delivery systems: recent developments and future prospects. J Nanobiotechnology 2018; 16(1): 71.
http://dx.doi.org/10.1186/s12951-018-0392-8 PMID: 30231877

[157] Chen CX, Yang SS, Pang JW, *et al.* Anthraquinones-based photocatalysis: A comprehensive review. Environ Sci Ecotechnol 2024; 22: 100449.
https://doi.org/10.1016/j.ese.2024.100449 PMID: 39104553

[158] Chavda VP, Patel AB, Mistry KJ, *et al.* Nano-drug delivery systems entrapping natural bioactive compounds for cancer: recent progress and future challenges. Front Oncol 2022; 12: 867655.
https://doi.org/10.3389/fonc.2022.867655 PMID: 35425710

[159] Xin D, Li H, Zhou S, Zhong H, Pu W. Effects of anthraquinones on immune responses and inflammatory diseases. Molecules 2022; 27(12): 3831.
http://dx.doi.org/10.3390/molecules27123831 PMID: 35744949

[160] Cui Y, Chen LJ, Huang T, Ying JQ, Li J. The pharmacology, toxicology and therapeutic potential of anthraquinone derivative emodin. Chin J Nat Med 2020; 18(6): 425-35.
http://dx.doi.org/10.1016/S1875-5364(20)30050-9 PMID: 32503734

[161] Svechkarev D, Mohs AM. Organic fluorescent dye-based nanomaterials: advances in the rational design for imaging and sensing applications. Curr Med Chem 2019; 26(21): 4042-64.
http://dx.doi.org/10.2174/0929867325666180226111716 PMID: 29484973

[162]  Mahato KD, Kumar U. A review of organic dye-based nanoparticles: preparation, properties, and engineering/technical applications. Mini Rev Org Chem 2023; 20(7): 655-74.

http://dx.doi.org/10.2174/1570193X19666220629103920

[163]  Chaudhary P, Ahamad L, Chaudhary A, Kumar G, Chen W-J, Chen S. Nanoparticle-mediated bioremediation as a powerful weapon in the removal of environmental pollutants. J Environ Chem Eng 2023; 11(2): 109591.

http://dx.doi.org/10.1016/j.jece.2023.109591

[164]  Ventura-Aguilar RI, Bautista-Baños S, Mendoza-Acevedo S, Bosquez-Molina E. Nanomaterials for designing biosensors to detect fungi and bacteria related to food safety of agricultural products. Postharvest Biol Technol 2023; 195: 112116.

http://dx.doi.org/10.1016/j.postharvbio.2022.112116

[165]  Lim JW, Ha D, Lee J, Lee SK, Kim T. Review of micro/nanotechnologies for microbial biosensors. Front Bioeng Biotechnol 2015; 3: 61.

http://dx.doi.org/10.3389/fbioe.2015.00061 PMID: 26029689

[166]  Sun H, Luo G, Chen D, Xiang Z. A comprehensive and system review for the pharmacological mechanism of action of rhein, an active anthraquinone ingredient. Front Pharmacol 2016; 7: 247.

http://dx.doi.org/10.3389/fphar.2016.00247 PMID: 27582705s

# SUBJECT INDEX

## A

Absorption 19, 20, 21, 22, 23, 217
  bands 19, 20, 21
Abstraction, hydrogen atom 86
Acetyl-CoA carboxylase 68
Acid 3, 4, 8, 16, 17, 42, 60, 61, 65, 83, 87, 97, 102, 218, 249, 252
  benzoic 83, 87
  boxylic 83
  carboxylic 83
  cassic 102
  chitosan polylactic 218
  choline chloride-lactic 17
  hyaluronic 252
  isochorismic 61
  kermesic 8
  lactic 16, 97
  lactic-co-glycolic 252
  methacrylic 42
  mevalonic 3, 4
  sulfonic 87
  sulfuric 97
  tricarboxylic 60, 65
Action 65, 67, 74, 100, 168
  anti-inflammatory 74
  antiproliferative 168
  catalytic 65
  enzymatic 67
  therapeutic 100
Activity 45, 74, 81, 83, 85, 86, 87, 88, 98, 118, 120, 125, 126, 127, 145, 163, 166, 177, 179, 182, 186, 190, 212, 213, 220, 232, 254
  anti-arthritic 98, 254
  anti-fibrotic 212
  anti-inflammatory 74
  antineoplastic 232
  anti-tumor 127
  antiviral 45
  inhibiting efflux pump 74
  medicinal 190

  metabolic 220
  photosensitizing 213
  phytoestrogen 81
  protein 186
  radical scavenging 85
  urease 83
Acute myeloid leukemia (AML) 99, 100
Acyltransferases 60
Adenocarcinoma 90, 91, 163
  pancreatic 163
Adenosine generation 187
Age macular diseases (AMD) 209, 222
Agents 45, 70, 73, 113, 219, 221, 241, 242, 243, 244
  activatable fluorescent 242
  anthraquinone-based chemotherapy 45
  cytotoxic 113
  photosensitizing 241
  photothermal 241, 243
Anthraquinone 4, 20, 24, 38, 40, 41, 42, 67, 213, 216, 222, 234, 235, 238, 239
  nanoparticles (AQNPs) 213, 216, 222, 234, 235, 238, 239
  nucleus 67
  pigments 4, 20
  pollutants 24, 38, 41, 42
  reduction 40
Anti-cancer 7, 112, 114, 127, 139, 189, 215, 231
  activity 139
  agents 112, 114, 127, 189, 215, 231
  treatments 7
Antibacterial 83, 86, 87, 88, 89, 219
  activity 83, 86, 87, 88, 89
  agent 219
Anticancer and antibacterial properties 92
Antimicrobial 1, 6, 7, 45, 81, 83, 84, 88, 103, 114, 167, 212, 213, 220
  activity 83, 84, 88, 212
  effects 83
  properties 1, 7